Total Socioenvironmental Systems

Yichun Xie

Total Socioenvironmental Systems

An Analytical Framework for Human-Natural Interactions and Sustainability

Springer

Yichun Xie
Eastern Michigan University
Ypsilanti, MI, USA

ISBN 978-3-031-39593-2 ISBN 978-3-031-39594-9 (eBook)
https://doi.org/10.1007/978-3-031-39594-9

This Springer imprint is published by the registered company Springer Nature Switzerland AG
The registered company address is: Gewerbestrasse 11, 6330 Cham, Switzerland

Paper in this product is recyclable.

I dedicate this work to my wife, who devoted all her time and energy unselfishly to support my research and teaching. I further dedicate this work to my father, who taught me that I should keep my dream alive for pursuing higher education, and to my mother and daughter for their unconditional love and support.

Foreword

The industrial revolution which began in England in the early eighteenth century propelled the invention of numerous technologies that have endowed humankind with the tools to generate ever-increasing wealth. This in turn has spurred newer and more powerful technologies as the world has turned from physical inventions based on harnessing increasing energies to digital tools, enabling us to develop new forms of communication which in turn have become essential to the way we organize production and consumption using ever more information technology. Information has begun to replace energy or at least complement it in numerous ways. The recent focus of this revolution has been the beginnings of weak forms of artificial intelligence (AI) which is succeeding in automating our interactions with one another in ways that were undreamed of three or four generations ago. To an extent, this might seem like science fiction, but it is based on powerful digital models that are being adopted and developed in many areas of modern society.

Yet despite such advances in science and technology which have been accompanied by accelerating globalization, major challenges have arisen with respect to the contemporary world order. Climate change is amongst the most severe, but aging in all societies is changing the demographic profile which in turn is and will have a major impact on the type of work we do, how we work, and who works. The environment too is being degraded as pollution increases and as human actions continue to destroy our complex ecosystems. The way we communicate is also becoming ever more complex, and the rise of the network society is making it more and more difficult to identify good data from bad data, real news from fake news. Health care of course has been revolutionized in the last two centuries by new technologies built around clean water, but it too is facing barriers as the relative costs of new medical interventions rise and as new treatments are invented.

The current hype about AI is the latest twist in this fairy tale story of increasing automation and prosperity, but it is now being recognized for the first time that these challenges are colliding with one another. Although each might be attacked on its own with the prospect of some possible resolution, together the overall challenge to maintain the momentum established in the last two centuries is ever more difficult to ensure. A much more integrated science is required to see all these challenges in perspective and to begin to make progress in facing them head-on. We need clever methods to generate new solutions which in turn are likely to begin to mitigate these issues, and to this end, new theories about how such widely different systems might interact with one another are required.

In this book, Yichun Xie begins to assemble such a framework for what he calls 'Total Socio-Environmental Systems': TSES. He defines this as an analytical methodology for coupling very different systems pertaining to these grand challenges together. TSES is a framework which is built around human and material systems and is developed with respect to interaction between the human and socio-economic and the physical and environmental worlds. His focus is on developing a series of methods built around geospatial models so that the bigger questions of sustainability for regional, city, and local systems can be explored and solutions to problems they pose identified. His approach defines 'sustainability' as the core concept relevant to meeting the grand challenges, and this orientation is essential to illuminate these questions using a range of socio-environmental models built around paradigms of spatial analysis and GIS which pertain to the examples that he develops in the book: grassland systems associated with their continuing deterioration, urban sustainability assessment, shrinking cities, and related questions of urbanization and population growth. He bases many of these simulations on new developments in linear modelling, particularly those that relax the assumptions of linearity through various kinds of machine learning, support vector machines, autoregressive structures, and related econometric methods. His focus in developing these models identifies five key missions, and he develops them from Chap. 1 onwards: these are "breaking silos between disciplines, integrating the human dimension, dealing with different spatial and temporal scales, combining quantitative and qualitative data, and pushing the adoption of socio-environmental system models and their use in policy".

The models that Xie develops throughout the book are based on linear structures that are generally embellished to simulate non-linear relationships by reducing their linearity. This is achieved by setting them up as structural equations, introducing multi-levels, building hierarchies into their structures, and elaborating their form in terms of building on different layers of simulation in the manner of machine learning. Spatial analysis of various kinds involving spatial autocorrelation, multicriteria analysis, ecological economics, and various types of coupling between these techniques and systems are prominently featured throughout the book, and first summarized in Chap. 2. There are some very clear model-based logics developed here, some in visually appealing diagrammatic form. Throughout the book, Xie uses YinYang philosophy to impress his point that although human and physical systems often imply contradictory goals, total socio-environmental systems must become the *modus operandi* of many of the models that need to be used to synthesize the key methods he applies to the different grand challenges he identifies.

The range of examples covered in the book is wide in that they span national entities such as the Inner Mongolia Autonomous Region (IMAR) to urban metropolitan systems such as southeast Michigan which includes Detroit all the way to China's 278 Prefecture Cities. The focus in each of these examples is somewhat different although each reflects an integration of features across the environmental-socio-economic spectrum. As is pointed out in Chap. 3, different temporal rhythms and spatial scales are involved and the focus in understanding how these systems work ranges from near data modelling focussed on data mining to much more theory-laden approaches such as mixed linear models of the econometric variety. Arguably, the range is wider still because the various models embrace microsimulation, aggregate spatial modelling, and agent-based models in part.

This range of examples that Xie has included in his book constitute applications that he and his team have worked on for nearly two decades in China and in Michigan. Although he deals with IMAR in Chap. 3, he extends this in Chap. 4 where he focuses on how one can examine the deterioration of the grassland economy and agronomy in defining key determinants of change such as urbanization, transport, as well as ecological balance over time. Readers need to study his original papers that are referenced here to take this analysis further. In Chap. 5, he switches to look at the sustainability of the different 278 prefecture cities in China between 2000 and 2015. These show all the characteristics of urban China during the last two generations or longer, where the focus is on urban growth in the coastal regions, the middle provinces, and of course, the far west which is now opening up extremely rapidly as new technologies and populations develop in these more remote provinces. China's Belt and Road initiative is important here.

He changes tack again in Chaps. 6 and 7, first illustrating the whole question of defining spatial units in space-time models such as those that permeate many of the examples in this book. He goes in search of consistent spatial units which are critical to developing robust models in all geospatial analysis. This chapter is dominated by developing models based on interpolating missing or poor data, articulated with respect to key environmental, transportation, and land use/cover data. In Chap. 7, he presents a three-level hierarchical linear mixed model which he uses to empirically examine spatial processes of urban decline in south-eastern Michigan centred around Detroit, introducing a focus on shrinking cities or rather areas of cities that face massive decline set within much larger hinterlands whose suburbs are beginning to face the same sort of decline. It is clear here that climate plays a big part in many of his examples, and this poses the challenge of how to link such physical and environmental factors to social and economic. This is a long-standing issue that represents a strong subtheme through the text.

In the penultimate chapter, the focus is back at IMAR but involving a town-based fine-scale panel regression model to examine grassland deterioration. In this sense, this represents a further unpacking of the way in which large regions change through urbanization and the way urban lifestyles and technologies transform the countryside. In the last chapter, Xie points to the limitations of his framework TSES: Total Socio-Environmental Systems. It is clear that there is now a groundswell of opinion that is being embodied in research, and indeed in practice that is beginning to tip the balance towards more comprehensive models integrating human and physical perspectives on systems and towards models that are able to address the grand challenges that need to be strongly integrated. This is a good book with some very well-balanced spatial analytic work, and readers should be aware that the text is full of interesting insights. Questions range from breaking up silos and then integrating the human physical dimension, handling change across different scales, focussing on qualitative and quantitative data, and last but not least figuring out the impact of all this on policies for greater sustainability which are addressed in the last chapter. In progressing further, let me alert the astute reader to the many messages about how to model, what to model, and where to model in gaining an understanding of problems of sustainability that need to be addressed in a comprehensive and integrated way. The urban and rural, national, regional, and city scales are critical in progressing our understanding of the problems that now dominate our quest to build better models and to use these to generate better policies that address the whole range of questions surrounding sustainability.

Centre for Advanced Spatial Analysis (CASA) Michael Batty
University College London
London, UK
14 August 2023

Introduction

> Today, some 55% of the world's population – 4.2 billion inhabitants – live in cities. This trend is expected to continue. By 2050, with the urban population more than doubling its current size, nearly 7 of 10 people in the world will live in cities.... (World Bank, 2020)

Fast and unplanned urbanization generates numerous environmental challenges, such as deterioration of natural resources, accelerated air and water pollution, climate change, excessive emission of greenhouse gases, etc. Subsequently, the urgency for advocating sustainable urban growth (SUG) and sustainable science has been recognized since the mid-1980s (Brundtland, 1987). However, SUG and its scientific research have confronted many challenges. SUG science concerns numerous environmental, socioeconomic, management, and policy disciplines. There is no common understanding and consistent conceptual framework to study sustainability in these fields. There are many analytical methods and assessment tools to evaluate sustainability, but they do not conform with each other and often contradict each other. Furthermore, many of these research methods focus on environmental and natural resource dimensions but neglect human factors and social science elements.

Second, SUG science involves increasingly growing volumes of data covering various topics, consisting of a true sense of big data. However, current big data analytics lack adequate methods or tools to assimilate or integrate various types of data together to support sustainability research. For instance, the data describing socioeconomic and environmental systems show distinct characteristics. Economic, demographic, and social statistics are collected over a nation, state, county, or administrative enumeration area such as Census tracts in the USA. These data are often enumerated yearly or during a fixed time duration of 5 years or 10 years.

Environmental data, such as air, plant, soil, and water samples, are usually obtained through field samplings following natural gradients such as precipitation and temperature at irregular intervals or for special projects (Brown et al., 2013). Remote sensing technologies have become new data sources in recent decades. Environmental and socioeconomic datasets do not share common geographic boundaries and synchronize over consistent time patterns. The unconformity in space and the inconsistency over time between socioeconomic and environmental datasets are the biggest challenges to big data analytics for examining sustainability. Moreover, geographic phenomena show varied spatial scales and temporal rhythms. There is no systematic exploration of what impacts of various spatial and temporal data scales on sustainable system studies.

Third, no commonly recognized conceptual frameworks and no comprehensive, systematic, and innovative analytical methods exist to examine SUG as an integrated or total socio-environmental system. Narrow-discipline-oriented or domain-specific analytical techniques cannot adequately or comprehensively describe, explain, explore, quantify, or predict complexity, causality, co-variance, dynamics, feedback, or interaction between various elements underlying sustainable socio-environmental systems. Non-comparable spatial and temporal characteristics and scales of environmental and socioeconomic datasets further drive their analytical methods apart from each other.

The book aims to explore solutions to the shortcomings of current sustainability research. The book intends to develop a total socio-environmental (TSES) analytical framework to examine the sustainability of an integrated socio-environmental system in its totality. The book adopts the socio-environmental coevolution theory as the conceptual foundation of the TSES analytical framework. Therefore, the TSES conceptual foundation and analytical framework are interdisciplinary and inclusive.

The book designs and implements several efficient, systematic, and constructive tools to address spatial nonconformity and temporal inconsistency of big geospatial data. This geo-intelligence (GI)-based data mining and processing toolset jointly construct unified socio-environmental big datasets to support the sustainability examination of TSES systems.

Moreover, the book synthesizes and experiments with several advanced statistical models to demonstrate how to analyze the unified TSES big data for assessing the sustainability characteristics of several exemplary socioeconomic systems (i.e., the grassland deterioration, the urban sustainability assessment, the shrinking cities, etc.).

In brief, the book attempts to address grand challenges in modeling socio-environmental systems for sustainability studies. The book mainly focuses on five of the eight grand challenges identified in socio-environmental systems modeling (Elsawah et al., 2020):

1. Breaking silos between disciplines
2. Integrating the human dimension
3. Dealing with (spatial and temporal) scales
4. Combing quantitative and qualitative data
5. Elevating the adoption of socio-environmental system models and impacts on policy

The book contains nine chapters. Chapters 1 and 2 jointly provide the foundation for the analytical framework of the new total socio-environmental system (TSES). Chapter 1 focuses on the conceptual foundation, while Chapter 2 emphasizes the methodological or technical foundation.

Chapter 3 discusses matching socioeconomic and environmental data with distinct temporal rhythms and scales. This chapter introduces and synthesizes time-series data mining techniques to temporarily match the socioeconomic and environmental data.

Chapter 4 constructs a unified (temporally matched) socio-grassland system dataset based on the time-series data mining techniques developed in Chap. 3. In the original dataset, the climate and grassland productivity variables contain seasonal changes as well as long-term change trends. The socioeconomic and land use and land cover (LULC) data only show yearly changes. Five time-series data mining techniques are applied to generate five temporally filtered datasets from the original data. Chapter 4 develops a formal socio-grassland system model to examine grassland deterioration under global climate change and socioeconomic transformation. This chapter fits six datasets (including the five filtered and the original data) into three panel regression models (the conventional, the dependent-lagged, and the dynamic-lagged). The chapter uses the 18 combinations to assess which variety of the temporally filtered data and panel regression models can reveal and augment the grassland response to climate changes and socioeconomic drivers.

Moreover, the temporally filtered data are further spatially interpolated in Chap. 8 to interpolate the enhanced vegetation index and socioeconomic data from the counties to the towns. As a result, we can construct a two-scale mixed panel data model that integrates the town-based interpolated data and the county-based published statistics to detect local and regional variations of grassland deterioration processes.

Chapter 5 describes a cross-lagged and integrated structural equation model (SEM) and panel regression model to measure sustainable urban system levels among 278 prefecture cities in China. The cross-lagged SEM/panel model takes qualitative considerations to formalize a causal construct underlying urban sustainability. In other words, Chap. 5 translates the recommendation of measuring urban sustainability from the literature review into a cross-lagged structural equation model fitted with panel data. The model integrates the cross-section and time-series panel data by adding lagged effects of urban sustainability and its drivers to fit the SEM model to evaluate urban sustainability levels. The integrated model combines the advantages of SEM, panel data analysis, and cross-lagged autoregressive models. The integrated model can generate a set of important quantitative and qualitative descriptions of urban sustainability, including urban sustainability evolutionary path diagrams, sustainability scores, city sustainability rankings, causal relationships between urban sustainability rankings and underlying socio-environmental structures, legacy effects of the past urban sustainability levels and their driving factors on current or future urban sustainability, regional and spatial disparities of urban sustainability, and a multi-objective assessment of the city sustainability rankings.

Chapter 6 illustrates how to match varied geographic scales of socioeconomic and environmental data into a consistent spatial unit. This chapter develops several new and improved areal interpolation techniques for achieving this goal, including land use and land cover (LULC) natural-patches-based areal interpolation

(NP-interpolation), natural communities (the groups of natural patches classified by their topographic, environmental, and transportation characteristics) based interpolation (NC-interpolation), natural districts (the groups of Census Tracts classified by their socioeconomic attributions and topographic, environmental, and transportation features) based interpolation (ND-interpolation), and road network-based interpolation (RN-interpolation).

Chapter 7 presents a three-level hierarchical linear mixed model (3L-HLM) to empirically examine spatial processes of urban decline in southeastern Michigan centered around Detroit. The Census Tracts (CT), the road network analysis zones (RNAZ), and the counties (CS) consist of the three spatial hierarchies. RNAZ is derived from the RN-interpolation technique developed in Chap. 6. The RN-interpolation computes gravity scores of the geographic centers of the Census Tracts (CT) using the socioeconomic statistics collected from the ACS (American Community Survey) dataset. Then it classifies the CT centers into adjacent clusters using the GIS spatially constrained classification function. These spatial clusters are called the road network analysis zones (RNAZ), where the areal interpolation is carried out. The 3L-HLM in Chap. 7 is using RNAZ as the most critical spatial unit to explore how the urban decline in Detroit spreads to its suburban areas, the spreading processes, and their spatial disparities. In other words, the 3L-HLM model treats RNAZ as the fundamental random effect and the socioeconomic and environmental variables as the fixed effects on city shrinking (i.e., the urban vacancy rate). The model confirms that the fixed effect counts 72.6% vacancy variation between tracts within RNAZ. The random effects explain 16.4% of the vacancy variation between counties and 11.0% between RNAZs within counties.

Chapter 8 presents a town-based fine-scale panel regression model to examine grassland deterioration in China's Inner Mongolia Autonomous Regions (IMAR). In addition, Chap. 8 illustrates how fitting temporally filtered data in the panel regression model affects causal relationships between grassland deterioration and climate, LULC, and socioeconomic driving factors. The socioeconomic statistics do not exist at the town level in IMAR, and there is a lack of fine-scale data in most countries worldwide. Chapter 8 applies the temporal filtering techniques developed in Chap. 3 and the areal interpolation techniques developed in Chap. 6 to interpolate the enhanced vegetation index and socioeconomic data from the counties to the towns. Eventually, the panel regression model is a two-scale (or two-leveled) mixed panel data model that integrates the town-based interpolated data and the county-based published statistics. The mixed scale panel model results reveal that the town-based variables have different regression coefficients from the county-based similar variables. The two sets of regression coefficients indicate distinct socioeconomic and physical meanings concerning how LULC and socioeconomic variables affect grassland deterioration at different scales. These differences disclosed at the different scales clarify many confusions about the impacts of climate, LULC, and socioeconomic drivers on grasslands in the current literature.

Chapter 9 summarizes the innovations of TSES and points out the TSES's limitations. It also discusses future solutions to these limitations from the perspective of the grand challenges that socio-environmental system (SES) modeling faces.

References

Brown, D. G., Agrawal, A., Sass, D. A., Wang, J., Hua, J., & Xie, Y. (2013). Responses to climate and economic risks and opportunities across national and ecological boundaries: Changing household strategies on the Mongolian Plateau. *Environmental Research Letters*, *8*(045011), 9. https://doi.org/10.1088/1748-9326/8/4/045011.

Brundtland, G. H. (1987). *Report of the world commission on environment and development: Our common future*. http://www.un-documents.net/our-common-future.pdf. Accessed on 12-5-2022.

Elsawah, S., Filatova, T., Jakeman, A. J., Kettner, A. J., Zellner, M. L., Athanasiadis, I. N., Hamilton, S. H., Axtell, R. L., Brown, D. G., Gilligan, J. M., Janssen, M. A., Robinson, D. T., Rozenberg, J., Ullah, I. I. T., & Lade, S. J. (2020). Eight grand challenges in socio-environmental systems modeling. *Socio-Environmental Systems Modelling*, *2*, 16226. https://doi.org/10.18174/sesmo.2020a16226.

World Bank Urban Development (2020). Available online: https://www.worldbank.org/en/topic/urbandevelopment/overview#1. Accessed on 29 Dec 2021.

Acknowledgments

I want to express my sincere gratitude and appreciation to my colleagues who have collaborated with me on the research projects and publications that have built the foundations for this monograph.

I immensely thank academician Chenghu Zhou at the Institute of Geographic Sciences and Natural Resources of the *Chinese Academy of Sciences* (CAS) for encouraging me to pursue this monography project. I want to express my gratitude and appreciation to Dr. Xueliang Zhang and Dr. Yang Yang, Shanghai University of Finance and Economics, for sharing the urban statistics of the Chinese prefecture cities in 2000, 2005, 2010, and 2015. I appreciate the assistance from Dr. Hua Jin and Dr. Shujuan Chang, Inner Mongolia Forestry and Grassland Monitoring and Planning Institute, for collecting the socioeconomic statistics in the Inner Mongolia Autonomous Regions of China. I also appreciate Dr. Bin Jian, University of Gävle, Sweden, who provides insightful comments about Chaps. 1 and 2.

In addition, I express my sincere appreciation to Dr. Chao Liu, Mr. Binbin Tang, and Ms. Fenglian Xia at Anhui University of Science and Technology. They have helped code the multiple-objective optimization problems in Chap. 5, debugged the Python codes in Chaps. 3 and 8, and checked the references in Chaps. 1, 2, 3, 4, 5 and 6.

Furthermore, my heartfelt thanks go to my colleagues at Eastern Michigan University (EMU) for their generous support. The sabbatical leave and research activity release I got from EMU has given me time to complete the book.

Finally, I would like to thank Mr. Jianyuan Liang, an MS GIS graduate at EMU and a doctoral GIS student at Wuhan University, and Ms. Trupti Lokhande, an MS GIS graduate at EMU and a GIS Analyst at Detroit Metro Consulting Associates, for their contributions to the development of Python codes in time-series data processing and aerial interpolation.

Contents

Chapter 1
Theoretical Foundations of Total Socioenvironmental System (TSES)

1.1 Socioenvironmental Coevolution Theory

The socioenvironmental coevolution theory is the theoretical foundation of the total socioenvironmental system (TSES). Socioenvironmental coevolution accounts for coevolutionary processes between biological species, the abiotic environment, and culture. The theoretical concepts of socioenvironmental coevolution originated from the Darwin's coevolution theory. Naturalist Darwin developed a broad spectrum of ideas concerning species transmutation or evolution in his seminal book, *On the Origin of Species* (Darwin & Kebler, 1859). However, biologists did not seriously accept the concept of species coevolution until almost a century later (Ehrlich & Raven, 1964). Ehrlich and Raven define coevolution as an interaction between two major groups of organisms that have palpable biological relations. They recommend that a considerable amount of data be available to convince this type of relationship between butterflies and their food plants. Through a systematic analysis rather than long-term experimentation, they can convince that the coevolution approach is a reasonable starting point for understanding community evolution (Ehrlich & Raven, 1964, p. 586). Their most significant contribution is to formalize a generalized form of Darwinian evolution and expand the concept of coevolution from its original intention to relations between species to dynamic interactions between coevolving complex systems (Weisz & Clark, 2011). The methodology of systematic analysis for studying coevolution has opened the door for scholars to adopt the evolution theory in many other disciplines in the years followed. The socioeconomic (human society) and environmental (including biological and ecological) coevolution (abbreviated as socioenvironmental coevolution) has been at the core of coevolution studies.

Economists have introduced the biological concept of coevolution to study sustainable economic development potential. They see analogies between ecological and socioeconomic systems at the early adoption stage. For instance, genetic mutation resembles social and technological innovation and the survival of the fittest

Y. Xie, *Total Socioenvironmental Systems*,
https://doi.org/10.1007/978-3-031-39594-9_1

looks like economic competition. Later, it became a dominant view that ecosystems and socio-systems share one common characteristic. Their sustainable development relies on numerous and complex feedback mechanisms. Norgaard (1984) invents the term coevolutionary development to assess the potential of sustainable socio-economic development. He argues that coevolutionary development can be envisioned as an interaction between ecological and socioeconomic systems involving a surplus of energy that is beyond maintaining the ecological and socioeconomic systems in their present states (Norgaard, 1984, p. 161). Coevolutionary development will happen only when this surplus is invested in future beneficial change. In this process, the human society alters the biophysical system and adapts to its transformation, called coevolution progress (Norgaard & Kallis, 2011).

The notion that sustainable development involves both environmental and socio-economic systems has led to a rapid increase of interest in examining interactions and feedback between socioeconomic development and its environmental impact. Turner and his colleagues have proposed a framework for a coupled human–environment system (Turner et al., 2003). They have examined the interactions within the coupled systems through three case studies to illustrate how coupling interactions affect the systems' responses to hazards and their vulnerability. They conclude that external forces have been essential in reshaping the systems' responses to hazards. The varied access capacities of stakeholders to social and biophysical capitals have determined the systems' vulnerability to hazards. They vigorously promote focusing attention on the interactions of the coupled system and seeking solutions to reduce the vulnerability of the entire system.

A different approach has been taken to look at the vulnerability of a coupled human–environment system. Walker and his colleagues suggest that future social–ecological systems' (SESs) trajectories are determined by three attributes: resilience, adaptability, and transformability (Walker et al., 2004). They strongly recommend that the concept of SESs advocates sustainable development through resilience analysis, adaptive resource management, and adaptive governance. The idea of adaptive here is analogous to the coevolution theory.

A variant term for the coupled human–environmental system, the coupled human and natural system, was developed with the support of six case studies on five continents (Liu et al., 2007). This seminal work has carried out an in-depth examination of the coupled human and natural systems. It points out that coupling human and natural systems can unveil complex patterns and processes that cannot be revealed by either a human or natural system alone. People and nature in a coupled system interact reciprocally and form complex feedback loops (Liu et al., 2007, p. 1513). These complex feedbacks exhibit several common features such as reciprocal loops, nonlinear dynamics, multi-scaled (organizationally and/or spatially), and time-lagged. The time-lagged or legacy effects affect the coupled human and natural system's present conditions and future directions (Hull et al., 2015).

The concept of coevolution remains the fundamental trait of these complex systems (Norgaard, 1984). An economic society evolves by tasking resources from a non-evolving biophysical system for social and economic welfare. Resource consumption or exploitation can be measured economically, leading to ecological footprint, environmental accounting, and ecological economics concepts. For example, the impact of a society on its environment can be expressed as the amount of land required to sustain the livelihood of the society. This quantification unveils the human's imprint on the land, which is initially meant by the term ecological footprint (Rees & Wackernagel, 1996). Later, the concept of ecological footprint is expanded as a general indicator of human pressure on the environment (Hoekstra & Wiedmann, 2014).

Another similar term is the system for integrated environmental and economic accounting (SEEA). SEEA accounts for the damages to environments and ecosystem assets and services caused by excessive human exploitation (Bartelmus et al., 1991). The introduction of accounting systems into ecosystem services can help different stakeholders understand the financial implications of environmental impacts. La Notte and Rhodes examined the theoretical frameworks behind integrated ecosystem, environmental response, and economic accounting systems (La Notte & Rhodes, 2020). They classified the underpinning theoretical frameworks on accounting into three categories: the economic accounting framework (EAF), the SEEA central framework (SEEA-CF), and the SEEA experimental ecosystem accounts (SEEA-EEA) framework. They also classified these theoretical frameworks on the ecosystem services into three categories: the Cascade framework, the EPA (US Environmental Protection Agency) Final Ecosystem Goods and Services (FEGS) framework, and the EPA National Ecosystem Services Classification System (NESCS) framework. They compared these six frameworks in a graphic form to connect them based on the System of National Accounts (SNA) (Fig. 1.1).

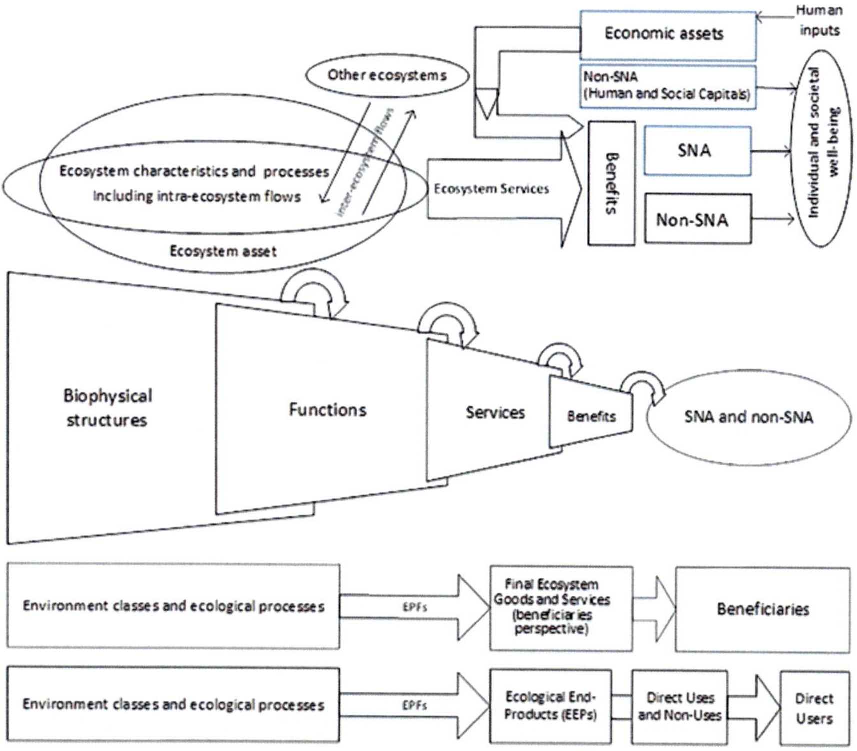

Fig. 1.1 Comparison between the SEEA-EEA theoretical framework and the three ecosystem services classification frameworks (Taken from Fig. 8 in La Notte & Rhodes, 2020)

It has been a long tradition to combine good economics and sound ecology (called ecological economics) to seek sustainable and efficient exploitation of biological resources (Armsworth & Roughgarden, 2001). This tradition invites the participation of economists in studies of the ecological and environmental implications of natural resource exploitation. It provides a critical interface that supports the integration of ecological science and understanding into policymaking. Ecological economics has made substantial contributions to modeling the transformations of nature and valuing their alternative trade-offs from the perspective of energy and material flows (Rosa-Schleich et al., 2019). In recent years, there has been a call to return to the root of ecological economics to understand and quantify the biophysical consequences of economic activities to address the socioecological challenges of the twenty-first century (Melgar-Melgar & Hall, 2020).

1.2 Coevolution in Geography

The development and application of coevolution theory have been very vibrant in geographical studies. The true tradition of geographical research is to keep culture and nature under one conceptual umbrella (Livingstone, 1992, p. 177). *Hägerstrand* and Harvey are two pioneering geographers who are painting a big picture of the society–nature coevolutionary development on a large intellectual canvas (Castree, 2009; Weisz & Clark, 2011). *Hägerstrand* published about 300 articles and essays (Lenntorp, 2004b). The essence of *Hägerstrand*'s publications advocates a need for a mutual language across natural and social science and humanities to understand fundamental ecological and social conditions and processes (Lenntorp, 2004a, p. 223). Harvey takes a different approach. He aims to identify possibilities for collective human activities that could generate significantly different outcomes through navigating or reconciling various elements of these ecological and social processes (Harvey, 2000). Harvey promotes the exchange of thoughts from both social and life scientists. He suggests that collective activities should go beyond the human social sphere and acknowledge that human beings can affect subsequent evolution through their different behavior (Harvey, 1996). More significantly, Harvey is one of the early geographers who advocates the concept of constituting a socioecological totality from interwoven elements, ensembles, moments, and processes in an open coevolutionary socioecological system (Harvey, 2009).

Recent geographic information science and systems (GIS) advancements have enabled geographical tools and technologies to play vital roles in monitoring and analyzing unprecedented challenges to our societies and biophysical systems. The geographic research communities have identified four dimensions and 11 strategic research questions for the next decade of geographers (USNRC, 2010, pp. 2–3):

D1. How to understand and respond to environmental change:

1. How are we changing the physical environment of the Earth's surface?
2. How can we best preserve biological diversity and protect endangered ecosystems?
3. How are climate and other environmental changes affecting the vulnerabilities of coupled human–environment systems?

D2. How to promote sustainability:

4. Where and how will 10 billion people live?
5. How will we sustainably feed everyone in the coming decade and beyond?
6. How does where we live affect our health?

D.3 How to recognize and cope with the rapid spatial reorganization of economy and society:

7. How is the movement of people, goods, and ideas changing the world?
8. How is economic globalization affecting inequality?
9. How are geopolitical shifts influencing peace and stability?

D4. How to leverage technological change for the benefit of society and environment:

10. How might we better observe, analyze, and visualize a changing world?
11. What are the societal implications of citizen mapping and mapping citizens?

Most of these 11 strategic questions are related to concerns and processes of the coevolutionary socioecological or socioenvironmental systems.

It is worth pointing out that geography or space has an essential role in socioenvironmental coevolution. The injection of space into the socioenvironmental coevolution process breaks traditional approaches' boundaries. The increasing geopolitical and economic globalization generates unprecedented volumes of commodities, communication, information, monetary, and people flows. In this globalization process, the land and natural resources that produce goods and provide ecosystem services in one place are no longer simply interacting with its local socioeconomic conditions but are affecting and also are affected by socioenvironmental processes in distant places through tele-coupling (Liu et al., 2018) or teleconnection (Seto et al., 2012).

A tele-coupling coevolutionary system comprises three subsystems (Fig. 1.2). Commodities are produced at an origin subsystem for transporting to or exporting to a target subsystem. For example, Brazil exports a large volume of soybean to China (Liu et al., 2018). The soybean production involves land, soil, irrigation, climate condition, fertilization, labor, storage, transportation, and so on. The soybean exportation affects or is affected by individuals, households, corporations, financial systems, and organizations that have an interest or a role in the production, transportation, or exporting process. In other words, the environmental or biophysical conditions interact with socioeconomic factors through complex causal chains, feedback, and interaction loops to comprise a coupled socioenvironmental system at the origin.

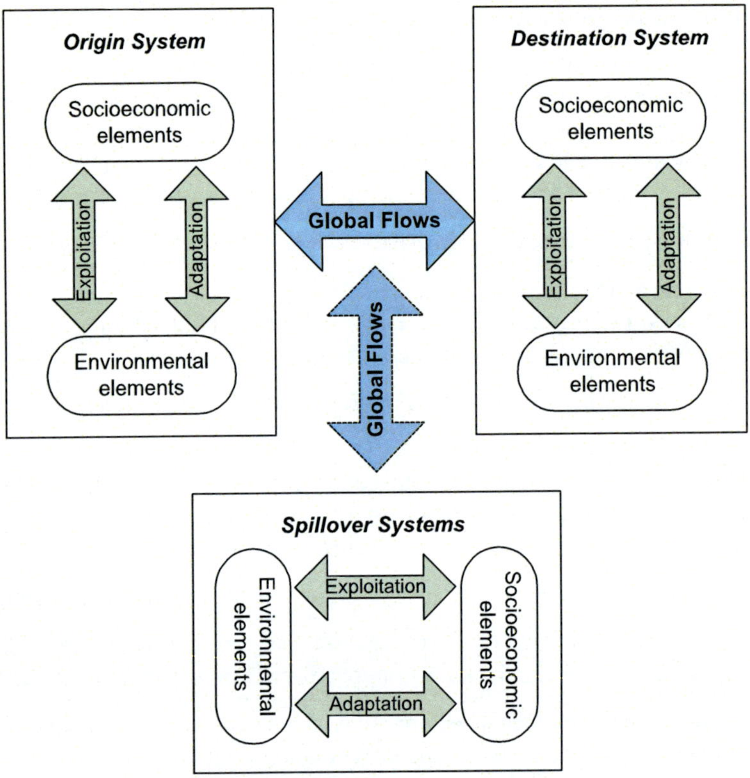

Fig. 1.2 A sketch diagram of tele-coupling socioenvironmental system

On the other hand, at the destination or the importing location (country), the land and natural resources spared from the soybean production will be used for planting other economic crops or reserved for some type of socioeconomic or environmental welfare. Consequently, a different coupled socioeconomic system will occur at the destination location. The soybean trade between Brazil and China plays a vital role in global trade markets and affects both the origin and destination subsystems.

The third component in the tele-coupled system is the spillover subsystem, consisting of multiple instances or spillover subsystems. A spillover subsystem appears in one of three primary forms: an intermediate stopover or transfer site between the origin and the destination locations; an adjacent site in the vicinity of the original or the destination that exhibits interactions; and places that play roles of competition or collaboration with the origin or the destination, or both.

Geography matters to the tele-coupling socioeconomic system. In addition to the global view of the spillover, external and internal spillovers affect the geography of innovation and production (Gersbach & Schmutzler, 1999). Productivity in a region is affected by human capital, research and development capacity, and public infrastructure of neighboring areas (Bronzini & Piselli, 2009). Geographic spillover initially indicates that firms' production processes in one location impact the productivity of other firms in adjacent locations. Geographic spillover effects are also observed in banking (Pino et al., 2019), economic crisis (Oakley & Tsao, 2007), housing market (Zhang et al., 2021), pandemic threat (Walsh et al., 2020), and urban shrinking (Lokhande and Xie, 2023). There are many methods developed to investigate geographic spillover effects, such as the Granger causality test (Profillidis & Botzoris, 2019), spatial autoregressive model (Anselin & Smirnov, 1996), spatial error model (De Blander, 2020), spatial lag of X model (Shumway & Stoffer, 2017), and spatial Durbin model (Mur & Angulo, 2006).

1.3 Coevolution in Sustainable Development and Urban Design

Another evident improvement of socioenvironmental coevolution is treating human social behavior as an artifact of human biology and culture (Durham, 1976). Human is specifically dominant in cultural evolution (Sieferle, 2011). In the process of evolution, culture has developed into a system that includes several subsystems: the economic subsystem, the symbolic-cognitive subsystem, and the political subsystem (Sieferle, 2011, p. 315). Cultural evolution does not replace organic evolution but merges with it to constitute one coevolutionary complex total system. This notion provides the theoretical foundation for integrating socioeconomic and policy factors into sustainable growth management. In the early stages, regional and urban sustainability research has borrowed many ideas concerning ecosystem sustainability. For example, the environment, economy, and social system consist of three pillars of the first ecosystem sustainability model (Costanza, 1991; Munasinghe, 1993). The environment term here encompasses a much broader meaning, including ecology, natural resource, climate, and other physical conditions. The four-pillar model of sustainability adds a new dimension of culture (Hawkes, 2001), governance (Lozano, 2008), or institutional function (Higgins, 2015). The contents of public health and community safety are newly added contents in regional and urban sustainability evaluation. Public concerns about human well-being, safety, and environmental health are emergent issues facing municipalities seeking sustainable development (Mapar et al., 2017). In recent years, more focus has been placed on governance and policy implications for sustainability. It is strongly recommended that urban sustainability is facing four intertwined challenges: (1) smart management of environmental hazards and climate change impacts; (2) science-driven analysis of impacts of urbanization on land transformation and biodiversity deterioration; (3) equal access to public services and urban resources; and (4) joint administrative and public response to decarbonization imperatives (UN-Habitat, 2016). This vision of urban sustainability well integrates the perceptions of sustainability from the points of review of ecosystems, urban systems, and governance objectives.

It is worth pointing out that urban sustainability research has incrementally incorporated many urban administration, design, development, and planning ideas.

UN-Habitat launched the City Prosperity Initiative in 2012 and started to develop a conceptual matrix to measure the overall achievement of a city called the City Prosperity Index (CPI) (UN-Habitat, 2015). The CPI is a compound index that embrace six dimensions of city prosperity: environmental sustainability, equity and social inclusion, infrastructure, governance and legislation, productivity, and quality of life. Each dimension includes two to four measurable indicators. Moreover, the CPI can be formularized or measured at four scales from a global ranking purpose to a local monitoring focus. The first scale is the global city ranking for global and regional monitoring. The second one is the basic CPI as an internationally comparable initial diagnosis. The third scale is the extended CPI for in-depth diagnosis that is comparable within a specific country. The fourth one is the contextual CPI, which is used as an urban monitoring tool for policy implementation. In short, the CPI establishes an adaptable and inclusive indicator framework. The CPI provides conceptual and methodological solutions to formulate a comprehensive index that connects healthy cities' indicators with the needs of governance and policy (Wong, 2015).

In parallel to sustainable urban development, sustainable watershed management is also very enthusiastic under the increasing agony of global climate change and water scarcity. Water scarcity has induced many ecological and environmental concerns, which, in turn, have been hastening the stresses of water scarcity and quality (de Paul Obade & Moore, 2018). Irrigation is often the first economic sector threatened by water scarcity and the primary cause of groundwater depletion and land degradation (Hanjra & Qureshi, 2010). Land use and land cover (LULC) changes near a river and lakefronts often associate with excessive tourism, luxury real estate development, soil erosion, and sedimentation. As a result, waterfront development dramatically decreases ecological service values and leads to ecological destruction and water pollution (Kertésza et al., 2019; Liu et al., 2019). Human adaptation strategies and intervention policies affect the direction and strength of ecological and environmental responses to water scarcity. Unfortunately, residents and communities are challenged by inadequate coping capacity, no access to complete information on the driving factors of water scarcity, and little awareness of the impacts on human well-being (Srinivasan et al., 2012). In some places, gaps and inconsistencies often exist between the coping strategies adopted by the central government, local governments, and residents (Yuan et al., 2019). Even in places where rational adaptation policies exist, there are no credible enforcement mechanisms or too many abstractors that prevent effective implementation of these policies (Srinivasan et al., 2012).

As a result, researchers and policymakers advocate and develop fully integrated and coupled theoretical and analytical frameworks to address sustainable watershed management. For instance, an integrated impact modeling framework (IIMF) was created to model how climate change affected water resources in the context of water protection policies (Schonhart et al., 2018). The IIMF integrated four models dealing with the biophysical process, land-use optimization, quantitative precipitation/runoff, and nutrient emission. These models were examined under three climate scenarios (from dry to wet) and two water protection policies (region-oriented and national protection policies). The former considered regional characteristics and variations in water environments, agri-environmental programs, and management measures, and the latter only considered several overall national goals. The model outcomes from the combined regional policy and decreasing precipitation revealed a great benefit of regional policy intervention.

Another analytical framework, called DPSIR (Driving forces, Pressures, States, Impacts, and Responses), was developed to analyze the interactions and causal feedback between these elements to provide advice for sustainable oasis management in Tunisia (Mekki et al., 2013). Zhou and his colleagues (2022) adopted DPSIR to examine regional water sustainability management (Fig. 1.3). Zhou's integrated socioenvironmental and spatiotemporal (ISES) framework consists of four interconnected regional water scarcity causal chains. Each chain or block represents one dimension or domain of driving factors and exhibits regional disparity (space) and temporal variability (time). Each chain involves a group of variables, which can be entered into a panel regression analysis as a block of variables. The influences of these blocks of variables on the water scarcity can be differentiated by interpreting the changes of the adjusted R squares and the t-test values with additions of these blocks of variables.

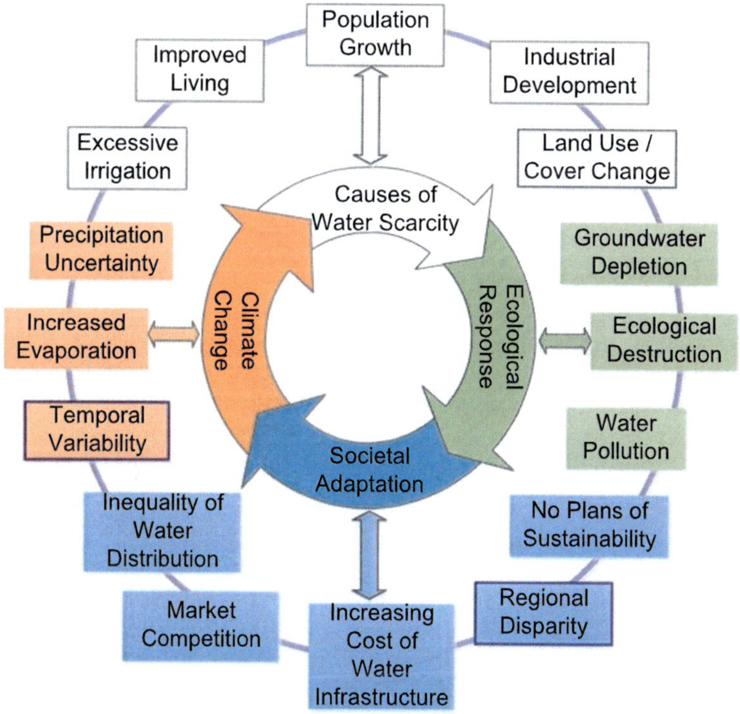

Fig. 1.3 A watershed coevolutionary system

Another big school assesses and measures urban sustainability from inner city-built structures to a broad urban environment. Urban planners, architects, and civil and environmental engineers have invented several celebrated ranking systems to certify or rate sustainability levels of buildings and constructions. A couple of comprehensive rating systems are introduced below as illustrations. The Building Research Establishment Environmental Assessment Method (BREEAM) was originally developed by the Building Research Establishment (BRE) in 1990 in the United Kingdom for assessing and certifying buildings sustainability (Cole & Valdebenito, 2013). The newest version of BREEAM is called BREEAM Communities, which covers economic, social, and environmental sustainability assessment at neighborhood scale and beyond (BRE, 2020). The second system is LEED-ND; LEED is an abbreviation for Leadership in Energy and Environmental Design, while ND stands for Neighborhood Development. LEED-ND is a US-based building rating system that applies the concepts of green building, smart growth, and urbanism into neighborhood design and development (Benson & Bereitschaft, 2020). The third system is DGNB-NSQ, a neighborhood sustainability assessment tool; DGNB stands for the German Sustainable Building Council, in German, while NSQ is the German abbreviation for New City District (Komeily & Srinivasan, 2015). CAMSUD (Comprehensive Assessment Method for Sustainable Urban Development) is another new multi-criteria rating system for sustainable urban planning. Good reviews of the neighborhood or inner city-built-based urban sustainability assessment concepts and implementation tools can be found in the papers by Sharifi and Murayama (Sharifi & Murayama, 2013) and Ali-Toudert and his colleagues (Ali-Toudert et al., 2020).

1.4 A Total Socioenvironmental System (TSES)

The previous sections reviewed the origin and development of coupled environmental/ecological and socioeconomic studies from several classic research fields, environmental and socioeconomic interactions, ecological economics, geographic perspectives of natural, cultural, and human connections, regional and urban sustainable growth, and sustainable building construction and city design. It is important to point out that this chapter conducted the literature review about coupled socioeconomic and environmental studies from a historical perspective of the references that are most related to scientific explorations of total socioenvironmental systems. The review was not exhaustive but focused on the mainstreams of coupled human and nature interaction studies.

It is worth noting that the studies of coupled human–nature systems are increasingly expanding. There are numerous case studies of coupled human and natural interactions. Several new trends are emerging. First, growing interest is in supporting meta-coupling framework (MCF) for exploring human–natural interactions. An excellent review favors the broader applicability of MCF than Tobler's first law of geography for examining global sustainability (Manning et al., 2023). Second, there is a call for advocating positive actions for reversing or reducing biodiversity loss and environmental deterioration. A review paper assesses the necessity of taking six simultaneous sets of actions to address these challenges (Obura et al., 2022). Several of them emphasize human behavior, including eliminating overconsumption and luxury living styles caused by increasing wealthy, and a respect of equalities of all communities. The third one is to allow suboptimal systems. In other words, realistic solutions to achieve sustainable ecosystems are the strategies that favor the overall benefits of the population by restraining individuals' desires (Grumbach & Hamant, 2020).

In the arena of ecological economics studies for sustainable human–nature systems, the concept of payments for ecosystem services (PES) becomes a foremost strategy for promoting sustainable social and economic development (Yan et al., 2022; Li et al., 2023). The underlying scientific concepts are essential ecosystem service values and variables (EESVs) (Liang et al., 2020; Balvanera et al., 2022).

Urban sustainability and related research fronts are the most diversified. Environmental scientists continue their tradition of identifying minimum acceptable thresholds for both ecological stability and standard of living (Goodwin et al., 2021). Some examine how urban and peri-urban agriculture affects urban sustainability (Rao et al., 2022). Civil engineers and architects maintain their focus on building designs, their functions, and urban-scale sustainability assessments (Purvis et al., 2022). Planners and policymakers examine urban sustainability from rural–urban interactions and transitions (Gutierrez et al., 2022), community capitals (Panzarella et al., 2023), garden city (Sharifi, 2016), urban green technologies (Nassary et al., 2022), neighborhood equality and disparity (Moroke et al., 2019), and human health (Tonne et al., 2021). Information technologists argue that state-of-the-art information technologies can empower stakeholders to pursue sustainable objectives and production methods through increased information access and thus can make informed decisions. For instance, the Digital Twin (DT) is a powerful tool to help urban planners and managers to achieve the urban sustainability goals (Purcell et al., 2023).

These new references add numerous case studies, expand exploration topics, and explode the literature on coupled human–nature system research. However, they roughly fall into similar structures and face the analogous challenges as this chapter's classic literature analyzed previously.

In summary, the current literature frequently adopts the following terms – coupling, integrating, tele-connecting, tele-coupling, causal loops, feedbacks, interactions, and trade-off – to examine relationships between nature and society. However, these terminologies have varied definitions, explanations, and interpretations in different disciplines or application fields. There is no common understanding or agreement about conceptualizing, formularizing, and quantitatively analyzing the relationships between environments and human beings. There exist disciplinary boundaries or silos that are critical for keeping their disciplinary signatures but fail to promote interdisciplinary approaches.

Coevolution is the common characteristic of the interrelationships between nature and society. Though many researchers do not use the term coevolution explicitly in their studies, the applications, approaches, and scopes of these studies can be explained by coevolution, coexistence, or cointeractions. The term coevolution between ecological and socioeconomic systems, invented by Norgaard (1984), correctly summarizes what these studies intend to describe, examine, and interpret. Moreover, the ecological–socioeconomic coevolution theory calls for a robust quantitative approach supporting a unified analytical framework (Kallis, 2007; Nick, 2005; Waring & Richerson, 2011). The term unified (unification) is analogous to the *Hägerstrand*'s notion of mutual language across natural and social science and humanities to understand fundamental ecological and social conditions and processes (Lenntorp, 2004a, p. 223). The term unification is also equivalent to the concept of totaling in studying the interactions between nature and society. These concepts call to study the interactions of nature and society in their totality rather than partially. For instance, socioecological totality takes into account all interconnected elements, ensembles, moments, and processes in an open socioecological system (Harvey, 2009). Cultural evolution merges with organic evolution to form a complex coevolutionary system (i.e., total) (Sieferle, 2011). The totality promoted by the coevolution theory advocates examining the relationships between socioeconomic and environmental systems as a whole (i.e., total) system. Therefore, the coevolution theory unites the studies of socioeconomic and environmental interactions from all disciplines and can serve as the theoretical foundation for all of them. For this argument, we propose the term the total socioenvironmental system (TSES) as the key phrase of this book title. TSES focuses on the totality when studying feedback, interactions, and trade-offs between hierarchically interconnected causal chains of varied nature and societal elements (Xie et al., 2021).

A sketch diagram of TSES is depicted in Fig. 1.4. A TSES includes two coevolutionary indispensable counterparts: the society (human) and the environment (nature). A TSES consists of n number of subsystems such as land use and land cover (LULC), water, ecosystems, environmental status, demographics, economy, infrastructure, and institution. Some are explicitly identified and measured, called a named subsystem. In contrast, others implicitly exist and may not be known to a research group or not measured directly at present, called an unnamed subsystem. For example, a spillover subsystem in tele-coupling socioenvironmental systems and a latent variable in structural equation models belong to this subsystem. They may exist, but their roles are not fully understood or measured at present. They could also exist at remote distances and are challenging to identify, observe, measure, or quantify.

Each subsystem consists of a set of m variables. Some variables are observed and measured, while others are latent causal structures. TSES contains interactions, feedback, and trade-offs within the subsystems, between the subsystems, and across the subsystems.

TSES requires a unified quantitative analytical framework for its implementation. A TSES analytical framework must be built with a fully assimilated spatial and temporal socioenvironmental dataset. Socioeconomic and environmental datasets display different spatial scales and disclose distinct temporal rhythms. The first task for supporting a TSES analysis is to make the two divergent socioeconomic and

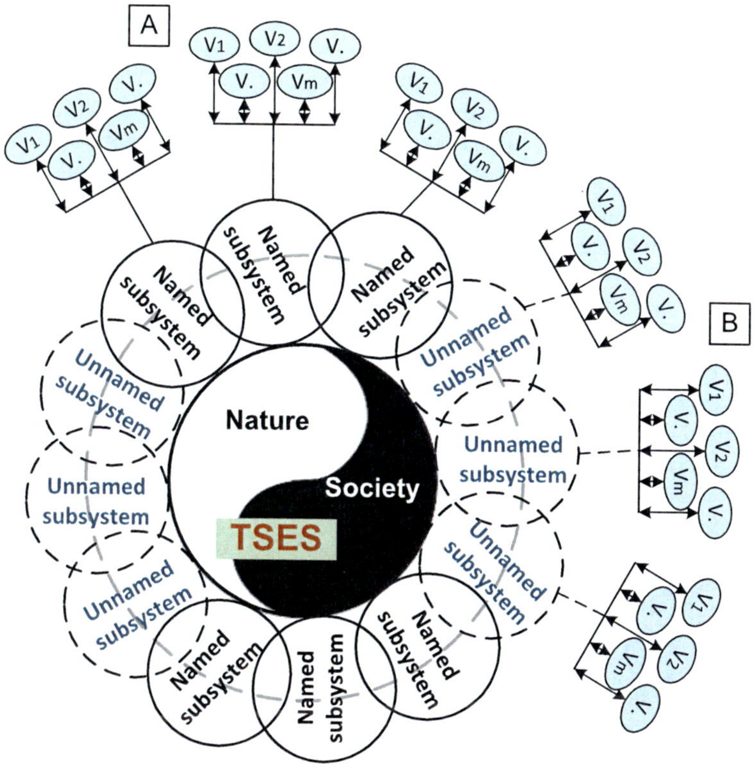

Fig. 1.4 The structural diagram of a TSES system: *A* indicates a named subsystem that is known and identified to a research project; *B* is an unnamed subsystem that is not fully understood or measured to a research project

environmental datasets compatible in space and time. Chapter 3 will illustrate how to build a temporally compatible TSES dataset, while Chap. 6 will construct a spatially compatible TSES dataset. Chapter 8 will demonstrate how to build a spatially and temporally compatible dataset.

The book also develops several unified TSES analytical frameworks. Chapter 2 will introduce and explain how these TSES analytical frameworks are evolved. Conventional, time-lagged, and dynamic-lagged panel regression models as the first TSES analytical framework are examined in Chap. 4. The cross-lagged structural equation models (SEM) with panel data as the second TSES analytical framework will be experimented with and analyzed in Chap. 5. The hierarchical mixed (multi-level) linear models as the third TSES analytical framework will be developed in Chap. 7. The two-level panel regression models as the fourth TSES analytical method will be analyzed and discussed in Chap. 8.

The book makes several important contributions. The book makes the first effort to establish that ecological–socioeconomic coevolution is the theoretical foundation for studying total (coupled) socioeconomic and environmental systems. The coevolution theory breaks silos constructed by individual disciplines and can provide a common ground for these disciplines to communicate and collaborate. Second, this book offers a comprehensive literature review, develops a set of analytical methods, and conducts several case studies to establish and validate TSES as a systematic and viable solution to examine socioeconomic and environmental interactions in the context of sustainability science. The author and his colleagues proposed the concept of TSES a couple of years ago (Xie et al., 2021). However, its idea, method, and application had no comprehensive development. In other words, this book systematically develops and implements the TSES analytical framework. Third, the book conducts the first extensive experiment for constructing a unified socioeconomic and environmental database by spatially and temporally matching and merging two divergent socioeconomic and environmental datasets in space and time. Lastly, the book is the first effort to design and develop a suite of tools to conduct natural patch-based areal interpolation to transform socioeconomic and environmental datasets across nonconformable geographic boundaries.

References

Ali-Toudert, F., Ji, L. M., Fahrmann, L., & Czempik, S. (2020). Comprehensive assessment method for sustainable urban development (CAMSUD) – A new multi-criteria system for planning, evaluation and decision-making. *Progress in Planning, 140*, 100430. https://doi.org/10.1016/j.progress.2019.03.001

Anselin, L., & Smirnov, O. (1996). Efficient algorithms for constructing proper higher order spatial lag operators. *Journal of Regional Science, 36*(1), 67–89. https://doi.org/10.1111/j.1467-9787.1996.tb01101.x

Armsworth, P. R., & Roughgarden, J. E. (2001). An invitation to ecological economics. *Trends in Ecology & Evolution, 16*(5), 229–234.

Balvanera, P., Brauman, K., Cord, A., Drakou, E., Geijzendorffer, I. R., Karp, D., Martín-López, B., Mwampamba, T., & Schröter, M. (2022). Essential ecosystem service variables for monitoring progress towards sustainability. *Current Opinion in Environmental Sustainability., 54*, 101152. https://doi.org/10.1016/j.cosust.2022.101152

Bartelmus, P., Stahmer, C., & Vantongeren, J. (1991). Integrated environmental and economic accounting – Framework for a SNA satellite system. *Review of Income and Wealth, 2*, 111–148. Retrieved from <Go to ISI>://WOS:A1991FX78700001.

Benson, E. M., & Bereitschaft, B. (2020). Are LEED-ND developments catalysts of neighborhood gentrification? *International Journal of Urban Sustainable Development, 12*(1), 73–88. https://doi.org/10.1080/19463138.2019.1658588

BRE. (2020). *BREEAM Communities international technical standard: Integrating sustainable design into the masterplanning of new communities or regeneration projects*. Retrieved from https://www.breeam.com/discover/technical-standards/communities/

Bronzini, R., & Piselli, P. (2009). Determinants of long-run regional productivity with geographical spillovers: The role of R&D, human capital and public infrastructure. *Regional Science and Urban Economics, 39*(2), 187–199. https://doi.org/10.1016/j.regsciurbeco.2008.07.002

Castree, N. (2009). Charles Darwin and the geographers. *Environment and Planning A, 41*(10), 2293–2298. https://doi.org/10.1068/a42224

Cole, R. J., & Valdebenito, M. J. (2013). The importation of building environmental certification systems: International usages of BREEAM and LEED. *Building Research and Information, 41*(6), 662–676. https://doi.org/10.1080/09613218.2013.802115

Costanza, R. (1991). *Ecological economics: The science and management of sustainability*. Columbia University Press.

Darwin, C., & Kebler, L. (1859). *On the origin of species by means of natural selection, or, the preservation of favoured races in the struggle for life*. J. Murray.

De Blander, R. (2020). Iterative estimation correcting for error auto-correlation in short panels, applied to lagged dependent variable models. *Econometrics and Statistics, 15*, 3–29. https://doi.org/10.1016/j.ecosta.2020.02.001

de Paul Obade, V., & Moore, R. (2018). Synthesizing water quality indicators from standardized geospatial information to remedy water security challenges: A review. *Environment International, 119*, 220–231. https://doi.org/10.1016/j.envint.2018.06.026

Durham, W. H. (1976). The adaptive significance of cultural behavior. *Human Ecology, 4*(2), 89–121. https://doi.org/10.1007/bf01531215

Ehrlich, P. R., & Raven, P. H. (1964). Butterflies and plants: A study in coevolution. *Evolution, 18*, 586–698. Retrieved from http://www.jstor.org/stable/2406212

Gersbach, H., & Schmutzler, A. (1999). External spillovers, internal spillovers and the geography of production and innovation. *Regional Science and Urban Economics, 29*, 679–696. Retrieved from www.elsevier.nl/locate/econbase

Goodwin, K., Wiedmann, T., Chen, G., & Teh, S. H. (2021). Benchmarking urban performance against absolute measures of sustainability – A review. *Journal of Cleaner Production, 314*, 128020. https://doi.org/10.1016/j.jclepro.2021.128020

Grumbach, S., & Hamant, O. (2020). How humans may co-exist with earth? The case for suboptimal systems. *Anthropocene, 30*, 100245. https://doi.org/10.1016/j.ancene.2020.100245

Gutierrez, V., Gilbert, M., Kinsey, D., & Behm, J. (2022). Beyond the 'urban' and the 'rural': Conceptualizing a new generation of infrastructure systems to enable rural–urban sustainability. *Current Opinion in Environmental Sustainability, 56*, 101177. https://doi.org/10.1016/j.cosust.2022.101177

Hanjra, M., & Qureshi, M. (2010). Global water crisis and future food security in an era of climate change. *Food Policy, 35*, 365–377. https://doi.org/10.1016/j.foodpol.2010.05.006

Harvey, D. (1996). *Justice, nature, and the geography of difference*. Blackwell Publishers.

Harvey, D. (2000). *Spaces of hope*. University of California Press.

Harvey, D. (2009). *Cosmopolitanism and the geographies of freedom*. Columbia University Press.

Hawkes, J. (2001). *The fourth pillar of sustainability: Culture's essential role in public planning*. Common Ground Publishing.

Higgins, K. L. (2015). From bud to blossom: Nurturing sustainable stewardship. In K. L. Higgins (Ed.), *Economic growth and sustainability* (pp. 167–180). Academic Press.

Hoekstra, A. Y., & Wiedmann, T. O. (2014). Humanity's unsustainable environmental footprint. *Science, 344*(6188), 1114–1117. https://doi.org/10.1126/science.1248365

Hull, V., Tuanniu, M. N., & Liu, J. G. (2015). Synthesis of human-nature feedbacks. *Ecology and Society, 20*(3), 17. https://doi.org/10.5751/Es-07404-200317

Kallis, G. (2007). Socio-environmental coevolution: Some ideas for an analytical approach. *International Journal of Sustainable Development and World Ecology, 14*(1), 4–13.

Kertésza, A., Nagyb, L., & Balázsb, B. (2019). Effect of land use change on ecosystem services in Lake Balaton catchment. *Land Use Policy, 80*, 430–438. https://doi.org/10.1016/j.landusepol.2018.04.005

Komeily, A., & Srinivasan, R. S. (2015). A need for balanced approach to neighborhood sustainability assessments: A critical review and analysis. *Sustainable Cities and Society, 18*, 32–43. https://doi.org/10.1016/j.scs.2015.05.004

La Notte, A., & Rhodes, C. (2020). The theoretical frameworks behind integrated environmental, ecosystem, and economic accounting systems and their classifications. *Environmental Impact Assessment Review, 80*, 106317. https://doi.org/10.1016/j.eiar.2019.106317

Lenntorp, B. (2004a). Path, prism, project, pocket and population: An introduction. *Geografiska Annaler: Series B, Human Geography, 86*(4), 223–226.

Lenntorp, B. (2004b). Publications by Torsten Hagerstrand 1938–2004. *Geografiska Annaler: Series B, Human Geography, 86*(4), 327–334.

Li, F., Liu, H., Wu, S., Wang, Y., Xu, Z., Yu, P., & Yan, D. (2023). A PES framework coupling socioeconomic and ecosystem dynamics from a sustainable development perspective. *Journal of Environmental Management, 329*, 117043. https://doi.org/10.1016/j.jenvman.2022.117043

Liang, J., Xie, Y., Sha, Z. and Zhou, A., (2020). Modeling urban growth sustainability in the cloud by augmenting Google Earth Engine (GEE). *Computer, Environment, and Urban Systems 84*(2020) 101542.

Liu, J. G., Dietz, T., Carpenter, S. R., Alberti, M., Folke, C., Moran, E., et al. (2007). Complexity of coupled human and natural systems. *Science, 317*(5844), 1513–1516. https://doi.org/10.1126/science.1144004

Liu, J. G., Dou, Y., Batistella, M., Challies, E., Connor, T., Friis, C., et al. (2018). Spillover systems in a telecoupled Anthropocene: Typology, methods, and governance for global sustainability. *Current Opinion in Environmental Sustainability, 33*, 58–69. https://doi.org/10.1016/j.cosust.2018.04.009

Liu, W., Zhan, J. Y., Zhao, F., Yan, H. M., Zhang, F., & Wei, X. Q. (2019). Impacts of urbanization-induced land-use changes on ecosystem services: A case study of the Pearl River Delta metropolitan region, China. *Ecological Indicators, 98*, 228–238. https://doi.org/10.1016/j.ecolind.2018.10.054

Livingstone, D. N. (1992). *The geographical tradition: Episodes in the history of a contested Enterprise*. Blackwell.

Lokhande, T. and Xie, Y. (2023). Spatial spillover effects of urban decline in Southeast Michigan. *Applied Geography 158*(2023) 103031. https://doi.org/10.1016/j.apgeog.2023.103031.

Lozano, R. (2008). Envisioning sustainability three-dimensionally. *Journal of Cleaner Production, 16*(17), 1838–1846. https://doi.org/10.1016/j.jclepro.2008.02.008

Manning, N., Li, Y., & Liu, J. (2023). Broader applicability of the metacoupling framework than Tobler's first law of geography for global sustainability: A systematic review. *Geography and Sustainability, 4*(1), 6–18. https://doi.org/10.1016/j.geosus.2022.11.003

Mapar, M., Jafari, M. J., Mansouri, N., Arjmandi, R., Azizinejad, R., & Ramos, T. B. (2017). Sustainability indicators for municipalities of megacities: Integrating health, safety and environmental performance. *Ecological Indicators, 83*, 271–291. https://doi.org/10.1016/j.ecolind.2017.08.012

Mekki, I., Jacob, F., Marlet, S., & Ghazouani, W. (2013). Management of groundwater resources in relation to oasis sustainability: The case of the Nefzawa region in Tunisia. *Journal of Environmental Management, 121*, 142–151. https://doi.org/10.1016/j.jenvman.2013.02.041

Melgar-Melgar, R. E., & Hall, C. A. S. (2020). Why ecological economics needs to return to its roots: The biophysical foundation of socio-economic systems. *Ecological Economics, 169*. https://doi.org/10.1016/j.ecolecon.2019.106567

Moroke, T., Schoeman, C., & Schoeman, I. (2019). Developing a neighbourhood sustainability assessment model: An approach to sustainable urban development. *Sustainable Cities and Society, 48*. https://doi.org/10.1016/j.scs.2019.101433

Munasinghe, M. (1993). *Environmental economics and sustainable development*. World Bank.

Mur, J., & Angulo, A. (2006). The spatial Durbin model and the common factor tests. *Spatial Economic Analysis, 1*(2), 207–226. https://doi.org/10.1080/17421770601009841

Nassary, E. K., Msomba, B. H., Masele, W. E., Ndaki, P. M., & Kahangwa, C. A. (2022). Exploring urban green packages as part of nature-based solutions for climate change adaptation measures in rapidly growing cities of the Global South. *Journal of Environmental Management, 310*, 114786. https://doi.org/10.1016/j.jenvman.2022.114786

Nick, W. (2005). Modernism, evolution and vaporous visions of future unity: Clarification in response to Norgaard. *Ecological Economics, 54*(4), 366–369.

Norgaard, R. B. (1984). Coevolutionary development potential. *Land Economics, 60*(2), 160–173.

Norgaard, B., & Kallis, G. (2011). Coevolutionary contradictions: Prospects for a research Programme on Social and Environmental Change. *Geografiska Annaler Series B-Human Geography, 93b*(4), 289–300. https://doi.org/10.1111/j.1468-0467.2011.00383.x

Oakley, D., & Tsao, H.-S. (2007). Socioeconomic gains and spillover effects of geographically targeted initiatives to combat economic distress: An examination of Chicago's Empowerment Zone. *Cities, 24*(1), 43–59. https://doi.org/10.1016/j.cities.2006.10.003

Obura, D. O., DeClerck, F., Verburg, P. H., Gupta, J., Abrams, J. F., Bai, X., Bunn, S., Ebi, K. L., et al. (2022). Achieving a nature- and people-positive future. *One Earth, 6*(2), 105–117. https://doi.org/10.1016/j.oneear.2022.11.013

Panzarella, F., Turcanu, C., Abelshausen, B., & Cappuyns, V. (2023). Community capitals and (social) sustainability: Use and misuse of asset-based approaches in environmental management. *Journal of Environmental Management, 329*, 117122. https://doi.org/10.1016/j.jenvman.2022.117122

Pino, G., Herrera, R., & Rodríguez, A. (2019). Geographical spillovers on the relation between risk-taking and market power in the US banking sector. *The North American Journal of Economics and Finance, 47*, 351–364. https://doi.org/10.1016/j.najef.2018.05.005

Profillidis, V. A., & Botzoris, G. N. (2019). Chapter 7 – Econometric, gravity, and the 4-step methods. In V. A. Profillidis & G. N. Botzoris (Eds.), *Modeling of transport demand* (pp. 271–351). Elsevier.

Purcell, W., Neubauer, T., & Mallinger, K. (2023). Digital twins in agriculture: Challenges and opportunities for environmental sustainability. *Current Opinion in Environmental Sustainability, 61*, 101252. https://doi.org/10.1016/j.cosust.2022.101252

Purvis, B., Mao, Y., & Robinson, D. (2022). A multi-scale integrated assessment model to support urban sustainability. *Sustainability Science*. https://doi.org/10.1007/s11625-021-01080-0

Rao, N., Patil, S., Singh, C., Roy, P., Pryor, C., Poonacha, P., & Genes, M. (2022). Cultivating sustainable and healthy cities: A systematic literature review of the outcomes of urban and peri-urban agriculture. *Sustainable Cities and Society, 85*, 104063. https://doi.org/10.1016/j.scs.2022.104063

Rees, W., & Wackernagel, M. (1996). Urban ecological footprints: Why cities cannot be sustainable – And why they are a key to sustainability. *Environmental Impact Assessment Review, 16*(4–6), 223–248. https://doi.org/10.1016/S0195-9255(96)00022-4

Rosa-Schleich, J., Loos, J., Mußhoff, O., & Tscharntke, T. (2019). Ecological-economic trade-offs of diversified farming systems – A review. *Ecological Economics, 160*, 251–263. https://doi.org/10.1016/j.ecolecon.2019.03.002

Schonhart, M., Trautvetter, H., Parajka, J., Blaschke, A. P., Hepp, G., Kirchner, M., et al. (2018). Modelled impacts of policies and climate change on land use and water quality in Austria. *Land Use Policy, 76*, 500–514. https://doi.org/10.1016/j.landusepol.2018.02.031

Seto, K. C., Reenberg, A., Boone, C. G., Fragkias, M., Haase, D., Langanke, T., et al. (2012). Urban land teleconnections and sustainability. *Proceedings of the National Academy of Sciences of The United States of America, 109*(20), 7687–7692. Retrieved from <Go to ISI>://WOS:000304369800033.

Sharifi, A. (2016). From garden city to eco-urbanism: The quest for sustainable neighborhood development. *Sustainable Cities and Society, 20*, 1–16. https://doi.org/10.1016/j.scs.2015.09.002

Sharifi, A., & Murayama, A. (2013). A critical review of seven selected neighborhood sustainability assessment tools. *Environmental Impact Assessment Review, 38*, 73–87. https://doi.org/10.1016/j.eiar.2012.06.006

Shumway, R. H., & Stoffer, D. S. (2017). *Time series analysis and its applications: With R examples* (Springer texts in statistics) (4th ed., p. 1 online resource (XIII, 562 pages 148 illustrations, 570 illustrations in color). https://doi.org/10.1007/978-3-319-52452-8

Sieferle, R. P. (2011). Cultural evolution and social metabolism. *Geografiska Annaler Series B-Human Geography, 93*(4), 315–324. https://doi.org/10.1111/j.1468-0467.2011.00383.x

Srinivasan, V., Lambin, E. F., Gorelick, S. M., Thompson, B. H., & Rozelle, S. (2012). The nature and causes of the global water crisis: Syndromes from a meta-analysis of coupled human-water studies. *Water Resources Research, 48*. https://doi.org/10.1029/2011wr011087

Tonne, C., Adair, L., Adlakha, D., Anguelovski, I., Belesova, K., Berger, M., Brelsford, C., Dadvand, P., Dimitrova, A., Giles-Corti, B., Heinz, A., Mehran, N., Nieuwenhuijsen, M., Pelletier, F., Ranzani, O., Rodenstein, M., Rybski, D., Samavati, S., Satterthwaite, D., Schöndorf, J., et al. (2021). Defining pathways to healthy sustainable urban development. *Environment International, 146*, 106236. https://doi.org/10.1016/j.envint.2020.106236

Turner, B. L., Matson, P. A., McCarthy, J. J., Corell, R. W., Christensen, L., Eckley, N., et al. (2003). Illustrating the coupled human-environment system for vulnerability analysis: Three case studies. *Proceedings of the National Academy of Sciences of the United States of America, 100*(14), 8080–8085. https://doi.org/10.1073/pnas.1231334100

UN-Habitat. (2015). *The city prosperity initiative (CPI) global city report – 2015*. Retrieved from https://smartnet.niua.org/content/ba3a1dcb-3012-44d6-87b5-fbaa28318de7

UN-Habitat. (2016). *World cities report 2016: urbanization and development – Emerging futures*. Retrieved from https://unhabitat.org/world-cities-report

USNRC (U.S. National Research Council), (2010). *Understanding the changing planet: strategic directions for the geographical sciences*. Washington D.C., The National Academies Press, pp.172.

Walker, B., Hollin, C. S., Carpenter, S. R., & Kinzig, A. (2004). Resilience, adaptability and transformability in social-ecological systems. *Ecology and Society, 9*(2) Retrieved from <Go to ISI>://WOS:000228062200010.

Walsh, M. G., Sawleshwarkar, S., Hossain, S., & Mor, S. M. (2020). Whence the next pandemic? The intersecting global geography of the animal-human interface, poor health systems and air transit centrality reveals conduits for high-impact spillover. *One Health, 11*, 100177. https://doi.org/10.1016/j.onehlt.2020.100177

Waring, T. M., & Richerson, P. J. (2011). Towards unification of the socio-ecological sciences: The value of coupled models. *Geografiska Annaler Series B-Human Geography, 93b*(4), 301–314. https://doi.org/10.1111/j.1468-0467.2011.00384.x

Weisz, H., & Clark, E. (2011). Society-nature coevolution: Interdisciplinary concept for sustainability. *Geografiska Annaler: Series B, Human Geography, 93*(4), 281–287.

Wong, C. (2015). A framework for 'City Prosperity Index': Linking indicators, analysis and policy. *Habitat International, 45*, 3–9. https://doi.org/10.1016/j.habitatint.2014.06.018

Xie, Y. C., Fan, S. Y., & Zhou, C. H. (2021). Examining ecosystem deterioration using a total socioenvironmental system approach. *Science of the Total Environment, 784*. https://doi.org/10.1016/j.scitotenv.2021.147171

Yan, H., Yang, H., Guo, X., Zhao, S., & Jiang, Q. (2022). Payments for ecosystem services as an essential approach to improving ecosystem services: A review. *Ecological Economics, 201*, 107591. https://doi.org/10.1016/j.ecolecon.2022.107591

Yuan, F., Wei, Y. D., Gao, J. L., & Chen, W. (2019). Water crisis, environmental regulations and location dynamics of pollution-intensive industries in China: A study of the Taihu Lake watershed. *Journal of Cleaner Production, 216*, 311–322. https://doi.org/10.1016/j.jclepro.2019.01.177

Zhang, H., Chen, J., & Wang, Z. (2021). Spatial heterogeneity in spillover effect of air pollution on housing prices: Evidence from China. *Cities, 113*. https://doi.org/10.1016/j.cities.2021.103145

Zhou, C., Xie, Y., Zhang, A., Liu, C., & Yang, J. (2022). Spatiotemporal analysis of interactions between seasonal water, climate, land use, policy, and socioeconomic changes: Hulun-Buir steppe as a case study. *Water Research*, 117937. https://authors.elsevier.com/a/1eFRN9pi-SbLn

Chapter 2
Analytical Frameworks of TSESs

2.1 Current Analytical Frameworks

There is a rich collection of analytical approaches to studying the interactions, feedback, and trade-offs of total socioenvironmental systems (TSESs). This chapter introduces several typical ones as a summative review of popular analytical frameworks. Based on this brief review, the chapter then points out the limitations of the existing approaches. The chapter further presents possible solutions to these limitations from newly developed analytical methods and research explorations in which the author has experienced or participated for several decades.

2.1.1 Coupling Several Models into a Modeling Framework

It is conventional wisdom to couple several models concerning crucial subsystems of a TSES or link-related research fields that center around the primary themes of the TSES to quantitatively examine interactions embedded in the TSES of interest (Schönhart et al., 2018). For instance, this approach is adopted to develop a Water-Vision Agriculture Model Framework to achieve sustainable dairy farming under severe water constraints. A hydrological simulation model, SWAP (Soil, Water, Atmosphere, and Plant), a whole-farm dairy model, DairyWise, and a crop growth model, WOFOST (WOrld FOod STudies), were linked to model water vision agriculture (Hack-ten Broeke et al., 2019). SWAP was used to simulate water flow for estimating water transport, dissolved substances, and soil temperature. WOFOST was adopted to take the outputs of SWAP and to predict potential crop production, including dry weights of the various plant organs. SWAP and WOFOST jointly simulated atmosphere–soil–water–plant interactions. A meta-model based on Random Forest regression trees was applied to accurately predict crop production

Y. Xie, *Total Socioenvironmental Systems*,
https://doi.org/10.1007/978-3-031-39594-9_2

through many SWAP and WOFOST model combinational iterations. DairyWise simulated the impacts of technical, environmental, and financial processes on dairy farming.

2.1.2 *Multi-Criteria Evaluation Framework*

Multi-criteria-based evaluation systems are popular and conventional for modeling and measuring urban sustainability. An excellent systematic review of 12 national sustainability models is conducted (Ali-Toudert & Ji, 2017). All models were built on the theoretical foundation of either the three sustainability pillars model (environment, economy, and social system) (Costanza, 1991) or the four sustainability pillars model (environment, economy, social system, and culture) (Hawkes, 2001). Most models are constructed by the following rigorous procedures: (1) formulating principles of selecting criteria list; (2) classifying the selected criteria; (3) identifying interactions between the selected criteria; (4) weighting and scaling the criteria and the criteria interactions; and (5) scoring by point allocations to the criteria and the criteria interactions to determine a final score. For instance, the maximum sustainability score of the Building Research Establishment Environmental Assessment Method (BREEAM) system and the DGNB (German abbreviation for German Sustainable Building Council) system is 100. The Leadership in Energy and Environmental Design (LEED) system can come up to 110 points because it offers bonus points up to 10 (Ali-Toudert & Ji, 2017).

2.1.3 *Remote Sensing, Spatial Analysis, and Indicator-Based Modeling Framework*

Remote sensing (RS) makes it economical and feasible to study ecological and environmental phenomena at regional and global scales (Xie et al., 2008; Sun et al., 2021). Geographic information systems (GIS) and spatial analytical tools based on GIS are increasingly playing a fundamental role in ecological and environmental studies, emergency services, transportation and urban planning, and many other economic sectors (NRC, 2010). In addition, it is prevalent to use RS and GIS data and tools to calculate some indicators to describe the states of various subsystems of a TSES and then analyze the relationships between these indicators to examine the feedback or trade-off of the TSES or its subsystems. Public policies, collective decisions, and individual preferences interact with environmental constraints and socioeconomic conditions affecting TSES management and sustainable development. Their interactions and feedback can be described through synthetic indicators (Singh et al., 2009; Al Sayah et al., 2019).

For example, Landsat images downloaded from the National Land Cover Database (NLCD) were classified into nine land-use classes in Kentucky, USA. Average annual precipitation and temperature were obtained from the PRISM Climate Group. This large consortium gathers climate observations from a wide range of monitoring networks to support public and academic research (Bai et al., 2019). Various spatial analyses and environmental models were applied to assess ecosystem services, including soil retention and water purification. Land-use change and climate change were analyzed through four different scenarios. Two indicators, the relative importance index (RII) and the combined effect index (CEI), were developed to differentiate the effects of land use and climate change on ecosystem services under these four scenarios (Bai et al., 2019). The analysis concluded that it was an efficient and powerful way to evaluate the combined effects of land use and climate changes on ecosystem services by using indicators that integrate scenario analysis with ecosystem service evaluation.

2.1.4 Machine Learning and Computation-Based Modeling Frameworks

These frameworks belong to an increasingly growing field of TSES models. They include many newly developed modeling paradigms and machine learning techniques such as artificial intelligence, artificial neural network, genetic algorithms, evolutionary computing, cellular automata, and multi-agent-based geo-simulation. A good review can be found in the seminal book *Geospatial Analysis – A Comprehensive Guide to Principles, Techniques, and Software Tools* (De Smith et al., 2021). For example, a spatially explicit agent-based model was designed to simulate city decline in Michigan's Detroit Tri-County area (Jiang et al., 2021). The model contained two types of agents: households buying and selling houses and banks taking over the unsold homes and continuing to sell these houses. The model created an environment consisting of artificial housing markets and their geographical boundaries. The environment determined how household and bank agents moved around and where the homes were located. Each different submarket also determined average and median house prices. Households were making their decisions based on their income status, that is, to either stay or leave their current locations (Fig. 2.1). When they decided to move, they became buyers and started choosing sellers within the filtered list and offering a bid (i.e., a bid-price) to either sellers or banks. Their buying abilities were also relying on whether they could get bank mortgages. However, for unemployed sellers, if they could not sell the house in four consecutive years, they would be removed from the system (akin to foreclosure). The bank agents took over their houses and kept selling them at that time. As a result, the model demonstrated how the buying and selling of homes could lead to urban shrinkage through a bottom-up approach. This agent-based simulation potentially offers a means to test policies to achieve urban sustainability.

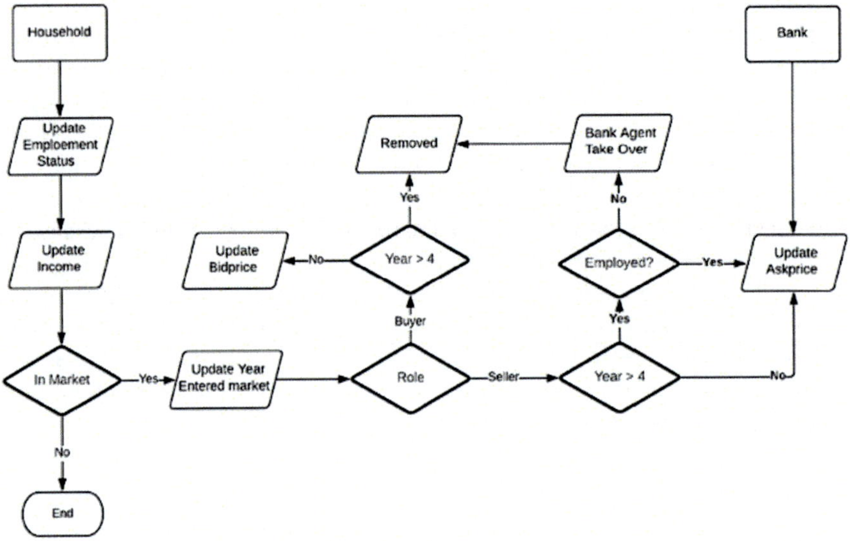

Fig. 2.1 Household and bank agents in urban shrinkage simulation. (Taken from Fig. 6. Household Dynamics, Jiang, N., Crooks, A., Wang, W., & Xie, Y. (2021). Simulating urban shrinkage in Detroit via agent-based modeling. *Sustainability*, *13*(4), 2283)

2.1.5 Ecological-Economics Modeling Framework

Ecological economics has played an essential role in understanding and quantifying the biophysical consequences of economic activities (Melgar-Melgar & Hall, 2020). Among numerous ecological-economic analysis approaches, the combined ecological network analysis with the economic input–output model is a powerful and effective method to examine the sustainable characteristics of a TSES. Ecological network analysis (ENA) is a systems-oriented methodology to analyze within-system interactions and identify an ecological system's holistic properties (Fath et al., 2007). An ecological system model includes all compartments and interactions of the system, analyzes their overall relationships, and determines their significance levels. The interactions and relationships resemble an ecological food web quantified through energy-matter flow (Ulanowicz, 2004). An ecological system is confined within a boundary, and energy-matter transfers within the system boundary comprise the network. Crossing the boundary is either input or output to the network (Fath et al., 2007, p. 50). The integration with an economic input–output model adopts the economic input–output analysis to compile a physical input–output table between various compartments of the ecological systems (Zhai et al., 2019). Then this economic input–output table is used to substitute the energy-matter flows to analyze the holistic properties of this ecological system (Zhang et al., 2019).

2.1.6 Spatial Autocorrelation Modeling Framework

The Toler's First Law of Geography states that everything is related to everything else, but near things are more related than distant things (Tobler, 1970, p. 236). Analytically, the statistic of spatial autocorrelation is developed to measure the strength of the first law of geography (Goodchild, 1986; Griffith, 1987; Getis & Ord, 1992). Ordinary least-square (OLS) regression models are commonly used to identify linear relationships between a dependent variable and a set of independent variables, for instance, how socioeconomic factors driving agricultural land conversion into urban land use (Xie et al., 2005). The concept of spatial autocorrelation is introduced into OLS to model geographic effect, spatial disparity, or spatial process, leading to the development of spatial autoregressive models (Leichenko, 2003).

Spatial autoregression models are further expanded into spatial econometric models, which address both spatial autocorrelation (dependency) and spatial heterogeneity (structure) (Anselin, 2003). Spatial autocorrelation contains several specifications (Anselin, 2010). The first one is that the value of a dependent variable (the response variable) at a location is impacted by the values of the independent variables (the regressors) at the same location, called the direct effect. Second, the dependent variable value at a location is affected by the values of the dependent variables in its neighboring locations, which is named the spatial autoregressive effect or spatial lag effect. The third one is that the value of the response variable is influenced by the values of the regressors in its neighboring locations, called the spillover effect. Finally, the spatial heterogeneity implies that the prediction errors of the response variable are spatially correlated, called the spatial error interaction effect.

The traditional ordinary least-square (OLS) regression model only accounts for the direct effect of the regressors on the response variable. Applying a spatial weight matrix to OLS can support examining four effects: (1) spatial autoregressive effect (spatial lag effect), (2) spatial spillover effect, (3) spatial error interaction effect, and (4) their mixed effects (Lokhande & Yichun, 2023). As a result, there are five spatial econometric models. The first three effects correspond to spatial lag model (SLM), spatial lag X model (SLX), and spatial error model (SEM). The spatial mixed effects lead to the spatial Durbin model (SDM) and the spatial Durbin error model (SDEM). SDM integrates SLM and SLX, thus incorporating the spatial lags of the response variable and regressors in a single model (Elhorst & Vega, 2017). SDEM combines SLX and SEM models, accounting for spatial autocorrelation in regressors and spatial heterogeneity in error term (Lokhande & Yichun, 2023).

Geographically weighted regression (GWR) models also deal with spatial autocorrelation, and GWR models significantly enhance OLS for examining spatial varying relationships in great detail (Li et al., 2021). The GWR can generate a set of local-specific coefficients, including individual parameter estimates, R^2, corresponding t-test values, and model residuals. These local-specific coefficients can be mapped in a GIS software package to visualize the varying spatial relationships between response and regressor variables (Gao & Li, 2011).

2.2 Limitations of Current Frameworks and Unique Requirements of TSESs

2.2.1 Limitations of Current Frameworks

Current analytical frameworks are generally confronted with the following limitations. TSESs are usually examined as a collection of several human and environmental subsystems based on traditional disciplinary boundaries (Gibson et al., 2005). These subsystems are evaluated using their state variables (Ness et al., 2010) or connected with feedback-based indicators (Dawson et al., 2010). Each of these subsystems could have a domain-specific model. These subsystems or corresponding quantitative models are sequentially linked into a TSES system. The output of an early model could serve as the input to a next model. The feedback or trade-offs between these subsystems could be modeled with an integration software module, which searches for the best parameters by iterating these sub-models (Hack-ten Broeke et al., 2019). However, these subsystem models are loosely coupled to describe the entire system's state. The loosely coupled approaches fail to explore a connected and integrated socioenvironmental system that includes explicit analysis of the processes and dynamics between the environmental and human components of the system (NSF, 2020).

The multi-criteria-based frameworks are sophisticated in selecting the list of factors included in a model, comprehensively classifying them into trade-off loops or causal feedbacks, as well as rigorous scoring points for various variables. However, most of these models end up with a total score, which is no longer meaningful to reflect the dynamics embedded within complex TSES systems. Furthermore, the selection, classification, and scoring criteria primarily rely on qualitative analysis and domain knowledge. As a result, multi-criteria-based models are labor-intensive (Elo & Kyngas, 2008). Their interpretations are prone to personal experience and expertise (Bowen, 2006). It is difficult to verify these models' findings (Kirkwood, 2018). Thus, the final scores from different multi-criteria evaluation systems are not easily comparable (Ali-Toudert & Ji, 2017).

These limitations also apply to the indicator-based analytical frameworks. In addition, the indicator selections are often challenged by the limitation of available data. Many indicators are selected based on specific project goals, local contexts, or researcher disciplinary training and experience, and do not apply to other regions or research topics (Xie et al., 2022). As a result, the indicator formulations are often mystified by the ambiguity of sustainability targets and thresholds, and the lack of a conceptual framework for indicator selection (Verma & Raghubanshi, 2018).

The artificial intelligence and computational learning-based frameworks are new and innovative. However, they are facing several challenges. Usually, a typical separation exists between these new modeling paradigms and the real-world decisions they intend to solve (Smajgl et al., 2011). In many case studies, these algorithms or toolboxes, just like many other computer-based models, belong to black or gray boxes. It is tough to interpret or understand what they intend to tell in the context of the disciplinary theory or knowledge (Vuculescu & Bergenholtz, 2014). Next, many of these models are constructed to assume that less data is needed than other traditional models because they are capable of self-learning, training, and adaptation. The empirical data supporting these models are not available or easy to collect. There exist serious data gaps (Bell et al., 2017). Third, many of these models are developed or designed to seek metaphor rather than searching for practical solutions (Batty, 2013). Fourth, for the same reason, these models are retained mainly in academic circles and are seldom receptive to practitioners and policymakers. Finally, validating these models' results is difficult due to the previously stated limitations (Fagiolo et al., 2007).

The ecological economics-based approaches are data intensive. For instance, the data needed to feed cost–benefit analysis, ecological network analysis, and economic input–output analysis are either available at national or provincial scales or must be collected through intensive field sampling or experimental activities. Although access to large or even enormous datasets is dramatically improving these years, most of these datasets vary widely across time and space (Martin et al., 2014). These datasets' incompleteness, inconsistency, and incompatibility constitute severe limitations to wide applications of ecological-economic models (Bretschger & Pittel, 2020).

Spatial econometric and geographically weighted regression models are powerful for revealing spatial autocorrelation, spatial clustering, and spatial processes. Particularly, spatial econometric models are useful to empirically assess the magnitude and significance of spatial spillover effects, which significantly improves the standard econometric models (Elhorst & Vega, 2013). However, these spatial models are often applied in geographical studies but less incorporated in integrated socioeconomic and environmental system explorations. On the other hand, there are some critics about the applicability of spatial econometric models for count data (Glaser, 2017).

2.2.2 *Unique Requirements of TSESs*

TSESs consist of coevolutionary socioeconomic and environmental subsystems. This statement has several important implications:

1. A TSES system includes at least two specific subsystems: one for economic, cultural, or social and the other for biophysical, ecological, or environmental (Fig. 1.4 in Chap. 1).
2. Two specific subsystems interact in a geographical space, a territory, or a habitat.
3. Two distinctive subsystems are coevolving and showing longitudinal (temporal) dynamics.

These two distinctive subsystems function differently. Water, temperature, light, and nutrients primarily affect biophysical growth. Precipitation and temperature display prominent yearly, seasonal, abrupt, and long-term changes and determine the formation of biomes and their geographical distributions. On the other hand, human resources, physical capitals, political systems, and technologies are vital elements shaping economic development and growth beyond natural resources. The temporal changes of these factors reflect long-term or short-term financial plans, market stabilities or fluctuations, administrative terms or shifts, and particular political or socioeconomic events.

Let us take grassland degradation and people's adaptation strategies in the Mongolian Plateau under global climate change and socioeconomic transformation as an illustration (Xie et al., 2012). The Mongolian Plateau comprises substantial portions of the independent country of Mongolia and the Inner Mongolia Autonomous Region (IMAR) of China. Mongolia has a long history of sustainable pastoralism. Yet, recent political-economic changes and a dramatic rise in poverty combined with accelerating degradation, climate, and land-use change threaten the balance of Mongolia's rangelands (Brown et al., 2013). IMAR, with a very similar grassland environment, has experienced very different socioeconomic dynamics. IMAR's rapid population growth during the 1950s and 1960s resulted in a much larger population and led to widespread conversion of livelihoods from nomadic grazing to settled animal husbandry and intensive agriculture. IMAR's grassland has been overgrazed, and productivity declined by more than 50% from the 1960s to 2010 (Xing et al., 2010).

Mongolia and IMAR have continental climates and disturbances related to grazing, farming, and mining, compounded with climate change, create ecosystem resilience and sustainability challenges, and interact in ways that are not entirely understood. IMAR and Mongolia seek policy strategies to help reverse negative trends, but their institutional, economic, and social contexts are quite different, as are the policy tools they employ. IMAR has taken a technological and top-down approach and implemented two nationwide ecological engineering projects (Wang et al., 2013). In contrast, Mongolia has seen minimal government involvement in rangeland management since the transition to democracy (Fernandez-Gimenez & Batbuyan, 2004; Fernandez-Gimenez et al., 2011).

In the terminology of TSES, there are two questions from the perspectives of environment and society, respectively (Fig. 2.2). What are the distinct pathways leading to grassland degradation in Mongolia and IMAR? What are strategies of grassland restoration possible and practical given the contrasting ecological, political, and socioeconomic conditions? Analytically, two distinct sets of spatial and temporal scales and feedback must be examined across the environmental and societal subsystems. Spatial heterogeneity is an important characteristic influencing the diversity and resilience of ecosystems (Parrott & Meyer, 2012). The ecological processes of grassland degradation operate at local, regional, and global scales. A multi-scale sampling and analysis are needed to examine the dynamics and feedback of ecological processes of grassland degradation (Seipel et al., 2012; Münkemüller et al., 2014; Soranno et al., 2014). For example, the Mongolian grassland degradation studies consider at least four scales and their feedback. The plant and soil samples were taken at the sampling plots. The degradation processes and states were examined at the patches. These analyses were downscaled by the

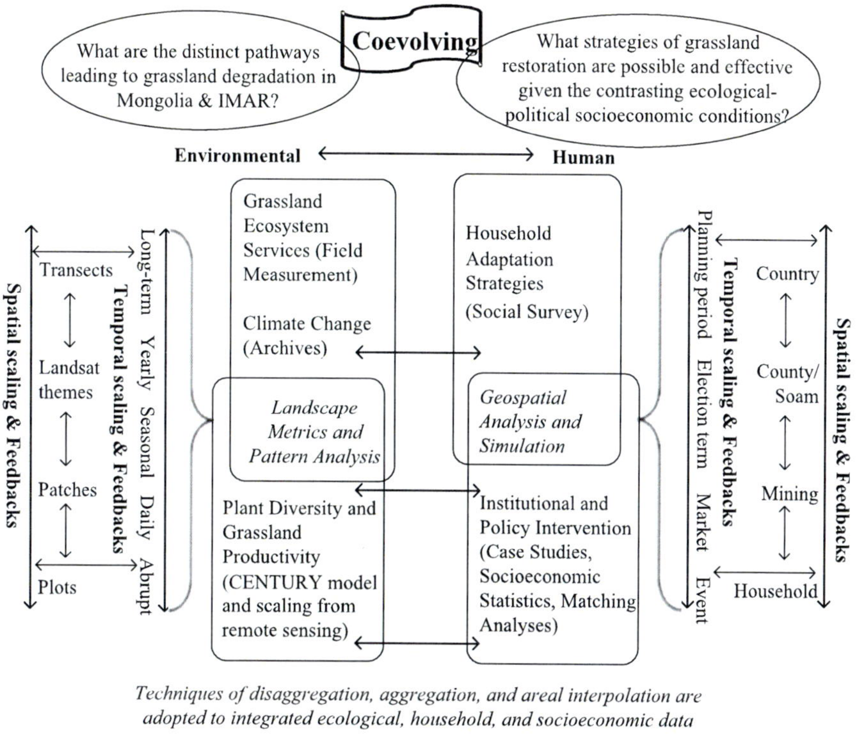

Fig. 2.2 The coevolving grasslands – socioeconomic system in Inner Mongolia, China

Landsat themes, and the spatial patterns of grassland degradation were identified across the water-heat or biomes transects (Fig. 2.2).

The spatial heterogeneity of the human or social dimension differed from the environmental one. Households were located at the bottom scale. Mining exploitations were significant disturbances to the environment and grassland ecosystems, although they exhibited more extraordinary attractions for households to abandon their traditional grazing activities. The choices of households and the destructions of mining varied across counties in IMAR and soums (an administrative unit in Mongolia, equivalent to a county). Mongolia and IMAR contrasted with each other at the country scale.

There exists a spatial mismatch between environmental and social systems. Critical boundaries of environmental systems or ecosystems such as vegetation zones or plant communities are influenced by precipitation and temperature gradients at macroscales and by topographical compositions at local scales (Xie et al., 2019; Sha et al., 2022). However, social systems' boundaries are administrative, military, or political. These two systems of boundaries are incompatible with each other. Subsequently, the attributes of a vegetation zone, such as temperature, precipitation, and plant species, cannot easily be assembled with the attributes of a county, including population, demographics, GDP, and other socioeconomic indices. Even in an administrative area, the distributions of socioeconomic activities are heterogeneous. Residents and their economic activities may concentrate in a single city or a few large urban centers. Given this spatial mismatch between environmental and socioeconomic data sources, it is imperative to assimilate datasets of different spatial areas and inconsistent scales into a unified database.

Furthermore, time dimensions or temporal rhythms between environmental and socioeconomic systems are also distinct. Environmental phenomena such as climatic events, ecosystems, and geophysical features display prominent daily, seasonal, yearly, and long-term trends, as well as abrupt short-term changes (Fig. 2.2). Social and political events primarily drive economic growth, pandemic eruption, policy change, social unrest, market fluctuations, leadership changes, and short- or long-term planned policy or marketing interventions.

Datasets pertaining to these systems are also collected with different methods. Economic, demographic, and social statistics are collected over a nation, state, county, or administrative enumeration area such as census tracts in the United States. These data are often enumerated yearly or during a fixed time of 5 or 10 years. Surveys on specific events or policies are common for collecting individual-level socioeconomic or policy information (Brown et al., 2013). Environmental data are usually obtained through field samplings, such as air, plant, soil, and water samples. For a long time, airborne and spaceborne remote-sensed data have been necessary to obtain ecological and environmental information (Xie et al., 2008). In recent years, drone-based remote sensing has become an emergent and effective technique for extracting biophysical data (Sarigai et al., 2021; Sun et al., 2021).

In brief, environmental and socioeconomic datasets mismatch spatially (no common boundaries), are asynchronous temporally, and are collected through distinct methods and processes. Researchers cannot simply combine them together to investigate their interactions and trade-offs without preprocessing them into a unified database with consistent spatial scale and temporal rhythm.

2.3 Technical Approaches for Implementing the TSES Analytical Framework

Current analytical frameworks face diverse limitations for effectively addressing the coevolutionary characteristics of TSESs. Current approaches of treating environmental and socioeconomic data in the same ways cannot bridge distinct data processing techniques and different spatiotemporal features between environmental and socioeconomic datasets. It is imperative to assimilate environmental and socioeconomic datasets into spatially and temporarily compatible (or spatiotemporally unified) datasets. Spatial data assimilation in geographical science is called areal interpolation, which assimilates environmental and socioeconomic data onto compatible spatial or areal units. Temporal data assimilation is to extract trends of change or filter out noises from time-series data through data mining techniques.

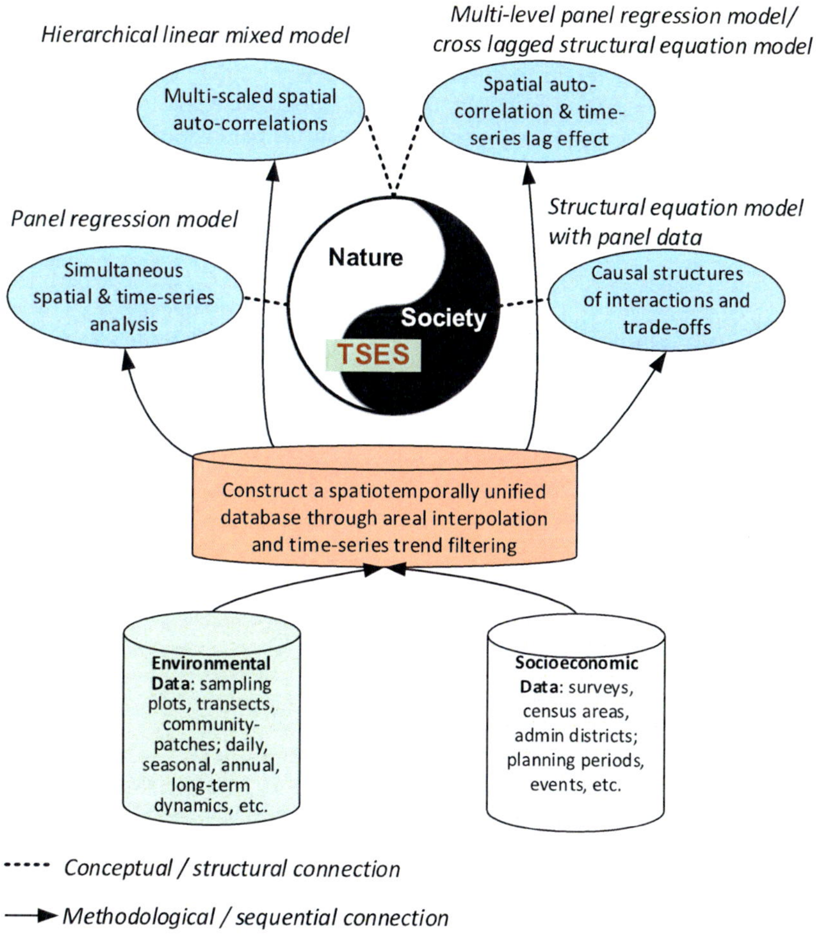

Fig. 2.3 Technical and methodological approaches for implementing the TSES analytical framework

Together the spatial and temporal assimilations can generate a spatiotemporally unified dataset (Fig. 2.3).

Afterward, there is an urgent need to develop or propose analytical methods that can use the spatiotemporally unified dataset to address the following challenges (Fig. 2.3):

1. Support a unified and quantitative examination of spatial disparities and temporal dynamics of coevolutionary TSESs. TSES deploys panel data analysis approaches (i.e., the panel regression models) to support this analytical function.
2. Support a unified and quantitative exploration of TSESs' interactions and trade-offs by revealing the underlying causal structures of the TSESs. TSES adopts the structural equation modeling approach to enable this analysis.

3. Support a unified and quantitative investigation of multi-scaled spatial autocorrelations of TSESs to reveal TSESs' multi-scaled interactions and trade-offs. TSES develops hierarchical linear mixed models to support this function.
4. Support a unified and simultaneous investigation of time-series lagged effects and spatial autocorrelations of TSESs to reveal interactions and trade-offs embedded in TSES systems. TSES develops cross-lagged structural equation models (CLSEM) (which is also called cross-lagged panel models (CLPM)) and multilevel panel models (MLPM) to implement these explorations.

Time-series trends extraction and filtering of TSES database will be experimented in Chap. 3. A case study of temporal (time-series) causality embedded in a grassland TSES system will be illustrated in Chap. 4. A special case study of examining coevolving urban sustainability based on its underlying causal structures will be demonstrated in Chap. 5. Spatial assimilation (geographical areal interpolation) of a shrinking city system will be examined in Chap. 6. An empirical examination of spatially lagged causal relationships between urban decline and socioenvironmental driving factors will be presented in Chap. 7. A comprehensive examination of time-series and spatial-lagged causal relationships will be illustrated in Chap. 8. The limitations of the illustrated TSESs analytical frameworks and methods and possible solutions will be discussed in Chap. 9.

References

Al Sayah, M. J., Nedjai, R., Kaffas, K., Abdallah, C., & Khouri, M. (2019). Assessing the impact of Man–Made ponds on soil erosion and sediment transport in limnological basins. *Water, 11*, 2526. https://doi.org/10.3390/w11122526

Ali-Toudert, F., & Ji, L. (2017). Modeling and measuring urban sustainability in multi-criteria based systems—A challenging issue. *Ecological Indicators, 73*, 597–611. https://doi.org/10.1016/j.ecolind.2016.09.046

Anselin, L. (2003). Chapter 14: Spatial econometrics. In B. H. Baltagi (Ed.), *A companion to theoretical econometrics* (pp. 310–330). Blackwell Publishing Ltd. http://web.pdx.edu/~crkl/WISE/SEAUG/papers/anselin01_CTE14.pdf

Anselin, L. (2010). Thirty years of spatial econometrics. *Papers in Regional Science, 89*(1), 3–25.

Bai, L., Jiang, L., Yang, D. Y., & Liu, Y. B. (2019). Quantifying the spatial heterogeneity influences of natural and socioeconomic factors and their interactions on air pollution using the geographical detector method: A case study of the Yangtze River Economic Belt, China. *Journal of Cleaner Production, 232*, 692–704. https://doi.org/10.1016/j.jclepro.2019.05.342

Batty, M. (2013). *The new science of cities*. MIT Press.

Bell, K., Hajireza, P., Shi, W., & Zemp, R. (2017). Temporal evolution of low-coherence reflectrometry signals in photoacoustic remote sensing microscopy. *Applied Optics, 56*, 5172–5181.

Bowen, G. (2006). Document analysis as a qualitative research method. *Qualitative Research Journal, 9*, 27–40. https://doi.org/10.3316/QRJ0902027

Bretschger, L., & Pittel, K. (2020). Twenty key challenges in environmental and resource economics. *Environmental and Resource Economics*, 1–26. https://doi.org/10.1007/s10640-020-00516-y

Brown, D. G., Agrawal, A., Sass, D. A., Wang, J., Hua, J., & Xie, Y. (2013). Responses to Climate and Economic Risks and Opportunities across National and Ecological Boundaries: Changing Household Strategies on the Mongolian Plateau. *Environmental Research Letters, 8*(045011). https://doi.org/10.1088/1748-9326/8/4/045011, 9pp.

Costanza, R. (1991). *Ecological economics: The science and management of sustainability*. Columbia University Press.
Dawson, T. P., Rounsevell, M. D. A., Kluvankova-Oravska, T., Chobotova, V., & Stirling, A. (2010). Dynamic properties of complex adaptive ecosystems: Implications for the sustainability of service provision. *Biodiversity and Conservation, 19*(10), 2843–2853. https://doi.org/10.1007/s10531-010-9892-z
De Smith, M. J., Goodchild, M. F., & Longley, P. A. (2021). *Geospatial analysis – A comprehensive guide to principles, techniques, and software tools* (6th ed.). Winchelsea Press.
Elhorst, P., & Vega, S. H. (2013). *On spatial econometric models, spillover effects, and W, 53rd congress of the European regional science association: "regional integration: Europe, the Mediterranean and the world economy", 27–31 August 2013, Palermo, Italy*. European Regional Science Association (ERSA).
Elhorst, J. P., & Vega, S. H. (2017). The SLX model: Extensions and the sensitivity of spatial spillovers to W. *Papeles de Economía Española, 152*, 34–50. https://spatial-panels.com/wp-content/uploads/2017/07/PEE_2017-Elhorst-Halleck-Vega-English-Translation.pdf
Elo, S., & Kyngas, H. (2008). The qualitative content analysis process. *Journal of Advanced Nursing, 62*(1), 107–115. https://doi.org/10.1111/j.1365-2648.2007.04569.x
Fagiolo, G., Moneta, A., & Windrum, P. (2007). A critical guide to empirical validation of agent-based models in economics: Methodologies, procedures, and open problems. *Computational Economics, 30*, 195–226. Retrieved from https://doi.org/10.1007/s10614-007-9104-4.
Fath, B. D., Scharler, U. M., Ulanowicz, R. E., & Hannon, B. (2007). Ecological network analysis: Network construction. *Ecological Modelling, 208*(1), 49–55. https://doi.org/10.1016/j.ecolmodel.2007.04.029
Fernandez-Gimenez, M. E., & Batbuyan, B. (2004). Law and disorder: Local Implementation of Mongolia's Land Law. *Development and Change, 35*(1), 141–166.
Fernandez-Gimenez, M. E., Wang, X., Batkhishig, B., Klein, J. A., & Reid, R. S. (Eds.). (2011). *Restoring community connections to the land: Building resilience through community-based rangeland management in China and Mongolia*. CABI.
Gao, J., & Li, S. (2011). Detecting spatially non-stationary and scale-dependent relationships between urban landscape fragmentation and related factors using geographically weighted regression. *Applied Geography, 31*(1), 292–302.
Getis, A., & Ord, J. K. (1992). The analysis of spatial association by use of distance statistics. *Geographical Analysis, 24*(3), 189–206.
Gibson, R. B., Hassan, S., Holtz, S., Tansey, J., & Whitelaw, G. (2005). *Sustainability assessment criteria and processes* (pp. xii, 254p.). Retrieved from http://ebookcentral.proquest.com/lib/sinciatw/detail.action?docID=429905
Glaser, S. (2017). *A review of spatial econometric models for count data* (Hohenheim discussion papers in business, economics and social sciences, No. 19-2017). Universität Hohenheim, Fakultät Wirtschafts- und Sozialwissenschaften. https://nbn-resolving.de/urn:nbn:de:bsz:100-opus-13975
Goodchild, M. F. (1986). *Spatial autocorrelation* (Catmog 47). Geo Books.
Griffith, D. (1987). *Spatial autocorrelation: A primer* (Resource publications in geography). Association of American Geographers.
Hack-ten Broeke, M. J. D., Mulder, H. M., Bartholomeus, R. P., van Dam, J. C., Holshof, G., Hoving, I. E., et al. (2019). Quantitative land evaluation implemented in Dutch water management. *Geoderma, 338*, 536–545. https://doi.org/10.1016/j.geoderma.2018.11.002
Hawkes, J. (2001). *The fourth pillar of sustainability: Culture's essential role in public planning*. Common Ground Publishing.
Jiang, N., Crooks, A., Wang, W., & Xie, Y. C. (2021). Simulating urban shrinkage in detroit via agent-based modeling. *Sustainability, 13*, 1–22. https://doi.org/10.3390/su13042283
Kirkwood, K. (2018). Review: Directed qualitative content analysis: The description and elaboration of its underpinning methods and data analysis process. *Journal of Research in Nursing, 23*(1), 56–57. https://doi.org/10.1177/1744987117742209

Leichenko, R. M. (2003). Does place still matter? Accounting for income variation across American Indian tribal areas. *Economic Geography, 79*(4), 365–386.

Li, S., Xie, Y., Yang, Z., Lu, Y., & Li, H. (2021). Examining winter fallow farmland from space and geography: A case study in Guizhou, China. *Journal of Spatial Science, 66*(1), 163–178.

Lokhande, T., & Xie, Y. (2023). Spatial spillover effects of urban decline in Southeast Michigan. *Applied Geography, 158*, 103031. https://doi.org/10.1016/j.apgeog.2023.103031

Martin, R., de Preux, L. B., & Wagner, U. J. (2014). The impact of a carbon tax on manufacturing: Evidence from microdata. *Journal of Public Economics, 117*, 1–14. https://doi.org/10.1016/j.jpubeco.2014.04.016

Melgar-Melgar, R. E., & Hall, C. A. S. (2020). Why ecological economics needs to return to its roots: The biophysical foundation of socioeconomic systems. *Ecological Economics, 169*. https://doi.org/10.1016/j.ecolecon.2019.106567

Münkemüller, T., Gallien, L., Lavergne, S., Renaud, J., Roquet, C., Abdulhak, S., Dullinger, S., Garraud, L., Guisan, A., Lenoir, J., Svenning, J. C., Van Es, J., Vittoz, P., Willner, W., Wohlgemuth, T., Zimmermann, N. E., & Thuiller, W. (2014). Scale decisions can reverse conclusions on community assembly processes. *Global Ecology and Biogeography, 23*(6), 620–632. https://doi.org/10.1111/geb.12137

Ness, B., Anderberg, S., & Olsson, L. (2010). Structuring problems in sustainability science: The multi-level DPSIR framework. *Geoforum, 41*(3), 479–488.

NRC (National Research Council). (2010). *Understanding the changing planet: Strategic directions for the geographic sciences, the committee on strategic directions for the geographic sciences in the next decade* (p. 172). The National Academies Press.

NSF. (2020). *Dynamics of integrated socio-environmental systems (DISES)* (NSF20579). https://www.nsf.gov/pubs/2020/nsf20579/nsf20579.htm. Accessed on 10 Mar 2023.

Parrott, L., & Meyer, W. S. (2012). Future landscapes: Managing within complexity. *Frontiers in Ecology and the Environment, 10*(7), 382–389. https://doi.org/10.1890/110082

Sarigai, E., Yang, J., Zhou, A., Han, L., Li, Y., & Xie, Y. (2021). Monitoring urban black-odorous water by using hyperspectral data and machine learning. *Environmental Pollution, 269*, 116166. https://doi.org/10.1016/j.envpol.2020.116166

Seipel, T., Kueffer, C., Rew, L. J., Daehler, C. C., Pauchard, A., Naylor, B. J., Alexander, J. M., Edwards, P. J., Parks, C. G., Arevalo, J. R., Cavieres, L. A., Dietz, H., Jakobs, G., McDougall, K., Otto, R., & Walsh, N. (2012). Processes at multiple scales affect richness and similarity of non-native plant species in mountains around the world. *Global Ecology and Biogeography, 21*(2), 236–246. https://doi.org/10.1111/j.14668238.2011.00664.x

Schönhart, M., Trautvetter, H., Parajka, J., Blaschke, A., Hepp, G., Kirchner, M., Mitter, H., Schmid, E., Strenn, B., & Zessner-Spitzenberg, M. (2018). Modelled impacts of policies and climate change on land use and water quality in Austria. *Land Use Policy, 76*, 500–514. https://doi.org/10.1016/j.landusepol.2018.02.031

Sha, Z., Bai, Y., Li, R., Lan, H., Zhang, X., Li, J., Liu, X., Chang, S., & Xie, Y. (2022). The global carbon sink potential of terrestrial vegetation can be increased substantially by optimal land management. *Nature Communication Earth and Environment, 3*, 8. https://doi.org/10.1038/s43247-021-00333-1

Singh, R. K., Murty, H. R., Gupta, S. K., & Dikshit, A. K. (2009). An overview of sustainability assessment methodologies. *Ecological Indicators, 9*(2), 189–212. https://doi.org/10.1016/j.ecolind.2008.05.011

Smajgl, A., Brown, D. G., Valbuena, D., & Huigen, M. G. A. (2011). Empirical characterisation of agent behaviours in socio-ecological systems. *Environmental Modelling & Software, 26*(7), 837–844. https://doi.org/10.1016/j.envsoft.2011.02.011

Soranno, P. A., Cheruvelil, K. S., Bissell, E. G., Bremigan, M. T., Downing, J. A., Fergus, C. E., Filstrup, C. T., Henry, E. N., Lottig, N. R., Stanley, E. H., Stow, C. A., Tan, P. N., Wagner, T., & Webster, K. E. (2014). Cross scale interactions: Quantifying multiscaled cause effect relationships in macrosystems. *Frontiers in Ecology and the Environment, 12*(1), 65–73. https://doi.org/10.1890/120366

Sun, Z., Wang, X., Wang, Z., Yang, L., Xie, Y., & Huang, Y. (2021). UAVs as remote sensing platforms in plant ecology: Review of applications and challenges. *Journal of Plant Ecology, 14*(6), 1003–1023.

Tobler, W. R. (1970). A computer movie simulating urban growth in the Detroit region. *Economic Geography, 46*, 234–240. https://doi.org/10.2307/143141

Ulanowicz, R. E. (2004). Quantitative methods for ecological network analysis. *Computational Biology and Chemistry, 28*(5–6), 321–339. https://doi.org/10.1016/j.compbiolchem.2004.09.001

Verma, P., & Raghubanshi, A. S. (2018). Urban sustainability indicators: Challenges and opportunities. *Ecological Indicators, 2018*(93), 282–291. https://doi.org/10.1016/j.ecolind.2018.05.007

Vuculescu, O., & Bergenholtz, C. (2014). How to solve problems with crowds: A computer-based simulation model. *Creativity and Innovation Management, 23*(2), 121–136. https://doi.org/10.1111/caim.12059

Wang, J., Brown, D. G., & Agrawal, A. (2013). Climate adaptation, local institutions, and rural livelihoods: A comparative study of herder communities in Mongolia and Inner Mongolia, China. *Global Environmental Change, 23*(6), 1673–1683. https://doi.org/10.1016/j.gloenvcha.2013.08.014

Xie, Y., Yu, M., Tian, G., & Xing, X. (2005). Socioeconomic driving forces of arable land conversion: A case study of Wuxian City, China. *Global Environmental Change, 15*, 238–252.

Xie, Y., Sha, Z., & Yu, M. (2008). Remote sensing imagery in vegetation mapping: A review. *Journal of Plant Ecology, 1*, 9–23.

Xing, X., Xu, X., Zhang, X., Zhou, C., Song, M., Shao, B., & Ouyang, H. (2010). Simulating net primary production of grasslands in northern Asia using MODIS data from 2000 to 2005. *Journal of Geographical Sciences, 20*(2), 193–204.

Xie, Y., & Sha, Z. (2012). Quantitative analysis of driving factors of grassland degradation: A case study in Xilin River Basin, Inner Mongolia. *The Scientific World Journal, 2012*, Article ID 169724, 14 pages. https://doi.org/10.1100/2012/169724

Xie, Y., Crary, D., Bai, Y., Cui, X., & Zhang, A. (2019). Modeling grassland ecosystem responses to coupled climate and socioeconomic influences in multi-spatial-and-temporal scales *Journal of Environmental Informatics*. https://doi.org/10.3808/jei.201600337

Xie, Y., Liu, C., Chang, S., & Jiang, B. (2022). Urban sustainability: Integrating socioeconomic and environmental data for multi-objective assessment. *Sustainability, 2022*(14), 9142. https://doi.org/10.3390/su14159142

Zhai, M., Huang, G., Liu, L., & Zhang, X. (2019). Ecological network analysis of an energy metabolism system based on input-output tables: Model development and case study for Guangdong. *Journal of Cleaner Production, 227*, 434–446. https://doi.org/10.1016/j.jclepro.2019.04.039

Zhang, G., Huang, G., Liu, L., Niu, G., Li, J., & McBean, E. (2019). Ecological network analysis of an urban water metabolic system based on input-output model: A case study of Guangdong, China. *Science of the Total Environment, 670*, 369–378. https://doi.org/10.1016/j.scitotenv.2019.03.132

Chapter 3
Building Temporally Compatible TSES Dataset

Total socioenvironmental systems (TSES) involve many socioeconomic and environmental subsystems and their interactions and feedback. Socioeconomic and environmental subsystems have varied temporal rhythms. For instance, many environmental phenomena and processes show notable variations when observed at a wide range of time scales, including hourly, daily, monthly, seasonally, yearly, decadal, and abrupt changes (Propastin et al., 2008; Watson et al., 2014). Likely, human behaviors and economic activities exhibit different time rhythms, such as statistical enumeration and policy-implementation phases and periods, as well as short-term, intermediate-term, and long-term targets and goals (Juchem Neto et al., 2018). Different temporal rhythms between environmental and socioeconomic systems make it inconsistent to combine environmental and socioeconomic datasets into one dataset. Unfortunately, current literature about synchronizing socioeconomic and environmental datasets into a temporally compatible one remains scattered. This chapter will illustrate how to generate temporally consistent socioeconomic and environmental datasets.

The process of building a temporally compatible TSES dataset will be illustrated using a coupled natural–human dataset in the Inner Mongolia Autonomous Region (IMAR) of China. The dataset includes three environmental variables with complex temporal rhythms. The variables of enhanced vegetation index (EVI; as a proxy of grassland productivity), precipitation, and temperature include 153 readings (9 periods in the growing season each year × 17 years from 2001 to 2017) (Table 3.1). The remaining 28 socioeconomic and land-use and land-cover (LULC) variables are yearly statistics (i.e., 17 readings each). Different approaches have to be developed to handle these two different time-series datasets. Several trend analysis methods will be explored to transform the complex time-series data into trend-based data to support examining temporal causal relationships between these two data types.

Supplementary Information The online version contains supplementary material available at https://doi.org/10.1007/978-3-031-39594-9_3.

Y. Xie, *Total Socioenvironmental Systems*,
https://doi.org/10.1007/978-3-031-39594-9_3

Table 3.1 List of variables in the case study datasets

Short name	Long name	Unit
EVI	Enhanced vegetation index (9 periods in growing seasons, 2001–2017)	Spectral Index (band ratio)
PRE	Precipitation (summered precipitation of each period in growing season)	mm
TEM	Temperature (averaged temperature of each period in growing season)	°C
DPop	Population density	100/sq.km
gdppc	Per capita gross domestic product (GDP)	1000 Yuan pc
gova	Gross output value of farming, forestry, animal husbandry, and fishery – Share of GDP	%
Paa	Percent of arable area to total land area	%
gpsk	Grain production per square kilometers	Ton/sq.km
lssk	The amount of the livestock per square kilometers	100 head/sq.km
fai	Fixed assets investment – share of GDP	%
lgr	Local government revenue – share of GDP	%
iofp	Net income of farmers and pastoralists - share of GDP	%
lohsk	Total length of highways per square kilometers	km/sq.km
Prpop	Percent of rural population to total population	%
HMP	Health and medical professionals per 10,000 people	Person/10,000 people
HPB	Hospital beds per 10,000 people	Bed/10,000 people
MHT	Middle and high school teachers per 10,000 people	Person/10,000 people
PTE	Professional and technical employment per 10,000 people	Person/10,000 people
PcUDI	Per capita disposable income of the permanent urban resident	1000 Yuan pc
PIO	Primary industry output value – share of GDP	%
SIO	Secondary industry output value – share of GDP	%
TIO	Tertiary industry output value – share of GDP	%
TCRV	Total consumer retail value – share of GDP	%
CropPG	Crop area + planted grassland for harvest (percentage to grassland)	%
Forest[a]	Forest area (percentage to grassland)	%
Grass	Grassland area (percentage to total land area)	%
Sand	Sand area (percentage to total land area)	%
Shrub	Shrub area (percentage to grassland)	%
Urban	Urban area (percentage to total land area)	%
Water	Water area (percentage to grassland)	%
Wetland	Wetland area (percentage to grassland)	%
Grazing	1 = agricultural, 2 = mixed agriculture and grazing, 3 = grazing	

[a]Land use data come from the NASA MCD12Q1 LULC data and China's impervious land surface data

On the other hand, stationarity tests and non-stationarity removal methods are needed to preprocess the simple time-series data. As a result, this chapter consists of five sections: (1) introduction of case study datasets; (2) trend extraction methods; (3) stationarity tests and non-stationarity removals; (4) discussions of data time-series transformation validity; (5) results of trend extraction, stationarity test, and non-stationarity removal.

3.1 The Case Study Datasets

The Inner Mongolia Autonomous Region (IMAR) (37°24′–53°23′N, 97°12′–126°04′E) is located in the northern border of China, belonging to the Mongolia Plateau, with a total area of about 1.18 million km^2. The average elevation of the study area was 1000–1200 m. The terrain is flat, with the Greater Khingan Range in the east and Yinshan and Henan Mountains in the south. The climate in the steppe area is a typical temperate continental climate, with an annual precipitation of 50–450 mm and an annual average temperature of −1.01 °C to 9.15 °C. The climate gradually transitions from east to west from humid and semi-humid to semi-arid and arid. The average precipitation decreases from the northeast to the southwest, but the temperature increases. The whole area extends diagonally from the northeast to the southwest in a long and narrow shape (Fig. 3.1). It is the primary

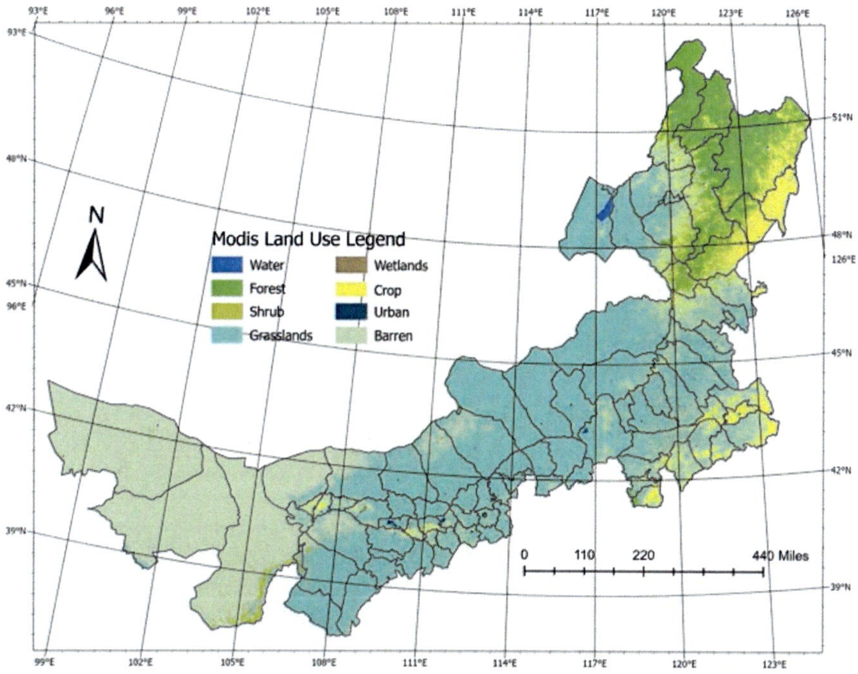

Fig. 3.1 The study area of IMAR

distribution area of temperate grasslands in China, which accounts for about 67% of the region's total area and 22% of the grassland area in China.

The case study datasets contain 32 variables reflecting grassland productivity, climate, socioeconomics, and land-use and land-cover (LULC) changes. ***EVI***, a better proxy of grassland productivity (Li et al., 2013), was downloaded from the U.S. Geological Survey (USGS) website (https://earthexplorer.usgs.gov/). The *EVI* data came from MODIS MOD13Q1 product and included nine consecutive 16-day growing season images between April and September from 2001 to 2017. The cell values of *EVI* at 250 m resolution were resampled within the boundary of a banner (the Mongolian term of county) for compiling a consistent and comprehensive dataset. In addition, the averaged value was assigned to this banner for each growing period of the study period. As a result, 13,617 observations of EVI (89 counties × 17 years × 9 periods per year × 1 imagery per period) were collected from the complete *EVI* data covering the entire IMAR (Table 3.1). *EVI* data processing and sampling were performed using ArcGIS Desktop 10.6.

The climatic variables were processed from the meteorological data at 45 ground stations across IMAR (http://cdc.cma.gov.cn). Precipitation (*PRE*) and temperature (*TEM*) are the two most critical variables affecting the productivity of grassland plants (Hao et al., 2014), and these two factors jointly influence vegetation growth. The daily temperature measurements were averaged over every 16-day period to be temporally consistent with the *EVI* data, while precipitation was summarized for each period so that the cumulative effects of precipitation could be assessed. There were 13,770 (45 stations × 17 years × 9 periods per year × 2 meteorological variables) meteorological observations.

The climate variables were collected from discrete weather stations in IMAR. In order to generate banner-based averages such as the *EVI* values, a spatial interpolation method, the inverse distance weighted (IDW) interpolation, was used to create a continuous climate grid surface for each climatic variable per period. The IDW assumes that the influence of events decreases with the distance from the sampled location, which makes it a more preferred interpolation method than Kriging or Spline as it creates a geographically weighted surface without gaps or values lower than the minimum observed values (Wilcox & Xie, 2007; Philip & Watson, 1982). We used the Cross-Validation tool from the ArcGIS Desktop Geostatistical Analyst extension to calibrate the IDW interpolation parameters (Chen et al., 2018). We removed one data location, predicted the associated data using the data at the rest of the locations, and repeated this remove-and-predict process for all data locations. Then, similar to the calculation of *EVI* value for an individual banner, an averaged value of each banner was computed using the cells within the banner's boundary for every climatic variable during each period (Table 3.1).

Twenty socioeconomic variables were collected from the statistical yearbooks of IMAR (China Statistical Press, 2001–2018). Total area, total population, rural population, and arable area disclose primary conditions of land and people of the study area (Yan et al., 2020). Gross domestic product (GDP), grain production, livestock volume, and farming and grazing income are conventional indicators of the grassland economy (Liu et al., 2012). Local government revenue, state investment, and length of highway reflect the state of regional economic growth (Xie et al., 2019). Health and medical professionals, hospital beds, middle and high school teachers, engineers, and technicians were important indicators reflecting overall social and educational development progress. In addition, total consumer retail value, disposable income, and primary, secondary, and tertiary industrial output values revealed wealth status and economic strengths (Roman & Thiry, 2016). The 20 socioeconomic indicators were standardized by the following rules (Xie et al., 2019): (1) total population was standardized as a density indicator by total land area, and rural population as a percentage by total population; (2) GDP was normalized as per capita GDP by total population, while other economic variables with values as the shares of GPD; (3) the economic variables measured by areas, lengths, and weights were standardized by total land area; and (4) the social development variables were standardized as per 10,000 people (Table 3.1).

Eight LULC variables reflected the main types of LULC in IMAR. The total land area was used to standardize the grassland and sandy land areas over each banner (the Mongolian term for the county). The grassland was the primary LULC type in the majority of IMAR banners. However, in a few banners, the desertification worsened in recent years, and the areas of sandy land surpassed grassland. Therefore, these types of LULC were normalized by the total land area. Other six LULC types were normalized by the area of grassland.

The data of LULC primarily came from the NASA MCD12Q1 Data Product (https://lpdaac.usgs.gov/products/mcd12q1v006/) at 500 m resolution. Sixteen out of the 17 International Geosphere–Biosphere Program, LULC types were found in the study area, except "Evergreen Broadleaf Forest" (http://www.igbp.net/). However, the urban land areas in the NASA MCD12Q1 Data Product were the values in 2000, and no further updates were provided. Therefore, we replaced the urban land data in the NASA MCD12Q1 Data Product with the impervious land surface dataset of China (Gong et al., 2019). The NASA MCD12Q1 Data Product contains 17 LULC types (Table 3.2) and we reclassified them into nine categories.

Grazing was a categorical variable indicating what the primary rural economy a banner was engaged with: 1 = Agricultural, 2 = Mixed agriculture and grazing, 3 = Grazing.

The original data are stored in the folder, \Chapter_3\DataProcess\IMAR_Data\Original. The names or explanations of the datasets are provided in Table 3.3. The socioeconomic data are combined with the LULC data as the csv file, IMAR_Socio_Land_Data_CSV. The socioeconomic, LULC, EVI, PRE (precipitation), and TEM (temperature) are also organized into a panel data format, IMAR_NewPanel_Data_Origin_NoFilter.

Table 3.2 Reclassification of the NASA MCD12Q1 LULC data

Name	Value	Recode	Description
Evergreen needleleaf forests	1	1	Dominated by evergreen conifer trees (canopy >2 m). Tree cover >60%
Evergreen broadleaf forests	2	1	Dominated by evergreen broadleaf and palmate trees (canopy >2 m). Tree cover >60%
Deciduous needleleaf forests	3	1	Dominated by deciduous needleleaf (larch) trees (canopy >2 m). Tree cover >60%
Deciduous broadleaf forests	4	1	Dominated by deciduous broadleaf trees (canopy >2 m). Tree cover >60%
Mixed forests	5	1	Dominated by neither deciduous nor evergreen (40–60% of each) tree type (canopy >2 m). Tree cover >60%
Closed shrublands	6	2	Dominated by woody perennials (1–2 m height). Tree cover >60%
Open shrublands	7	2	Dominated by woody perennials (1–2 m height). Tree cover 10–60%.
Woody savannas	8	2	Tree cover 30–60% (canopy >2 m)
Savannas	9	2	Tree cover 10–30% (canopy >2 m)
Grasslands	10	3	Dominated by herbaceous annuals (<2 m)
Permanent wetlands	11	4	Permanently inundated lands with 30–60% water cover and > 10% vegetated cover
Croplands	12	5	At least 60% of area is cultivated cropland
Urban and built-up lands[a]	13	6	At least 30% impervious surface area including building materials, asphalt, and vehicles
Cropland/natural vegetation mosaics	14	5	Mosaics of small-scale cultivation 40–60% with natural tree, shrub, or herbaceous vegetation
Permanent snow and ice	15	7	At least 60% of area is covered by snow and ice for at least 10 months of the year
Barren (sand)	16	8	At least 60% of area is non-vegetated barren (sand, rock, soil) areas with less than 10% vegetation
Water bodies	17	9	At least 60% of area is covered by permanent water bodies

[a] Urban and build-up lands were replaced with China's impervious land surface dataset (Gong et al., 2019)

3.2 Trend Extraction Methods

Ecological, environmental, and climate datasets usually contain noticeable seasonal rhythms and yearly fluctuations. Their time-series observations are much more complex than those data that only exhibit yearly changes. The mixed seasonal and yearly time-series fluctuations distort the change trends and complicate causal relationship analysis. There is a need to detect the trends of change or transform or filter the original mixed time-series observations into new time-series data that can reveal yearly changes. In other words, the time-series data transformation or filtering aims to generate new standardized data for reducing seasonal fluctuations and disclosing

Table 3.3 The data and folder structures

Folder	File	Explanation
IMAR_Data/ Original	IMAR_EVI_Orig	Enhanced vegetation index – original data
	IMAR_PRE_Orig	Precipitation – original data
	IMAR_Socio_Land_Data_CSV	Socioeconomic – original data
	IMAR_TEMP_Orig	Temperature – original data
	IMAR_NewPanel_Data_Origin_ NoFilter	The panel wide format of original data without time-series filtering
IMAR_Data/ Filtered_Step1	The filtered data and the python codes used to filter the original data	
IMAR_Data/ Filtered_Step2	The filtered data are reorganized for creating panel datasets for further analysis	
IMAR_Data/ Filtered_Final	The final panel dataset and the python codes for appending the filtered data into the panel format	
EMD	Python codes for EMD filtering and intermediate files	
HCL	Python codes for HCL filtering and intermediate files	
Stationarity	Python and Stata codes for stationarity Tests and intermediate files	
TSE	Python codes for TSE filtering and intermediate files	

yearly changes. Consequently, the new data are compatible with other datasets that are only collected yearly. As a result, all data items or variables are integrated as a unified dataset with compatible temporal changes and can be analyzed quantitatively to examine their interactions and feedbacks.

Many techniques can extract, filter, or transform trends of change from the original time-series data. Several representative techniques are depicted in Fig. 3.2 and their descriptions and discussions are provided in the following sections.

This chapter has several methodological improvements. First, the chapter first time synthesizes a comprehensive review of time-series trends extraction methods. Although the book author and his colleagues applied several of these methods to filter the trends of change (Zhou et al., 2022), this chapter provides detailed algorithmic foundations and illustrations of these methods, filling a gap in the current literature. Second, the chapter analyzes the advancement of the support vector regression (SVR) model for filtering trends of change compared with the traditional ordinary least squares (OLS) regression method. Third, the chapter integrates a thorough analysis of stationarity tests, including both uni-section and panel time-series data, bringing fragmented literature together. Fourth, the chapter provides a complete set of well-organized datasets, including the original, intermediate, and final processed data along with the accompanying Python codes. The readers can easily repeat the data processing procedures and adapt them in their own temporary data mining applications.

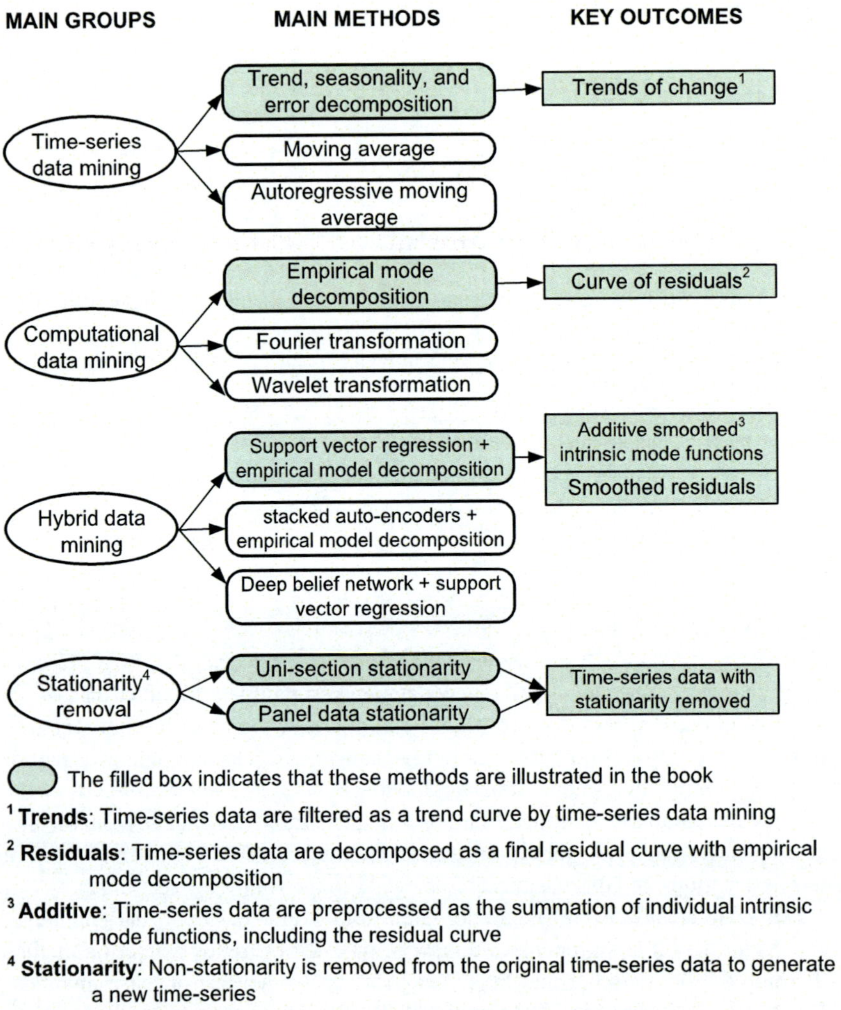

Fig. 3.2 Primary methods of extracting trends of change from time-series data

3.2.1 *Time-Series Data Mining for Extracting Seasonal Trends*

There are three groups of time-series data processing techniques for filtering change trends: classic time-series data mining (TDM), computational data mining (CDM), and hybrid computational learning (HCL)-based data mining. TDM provides a suite of tools for examining time-series data (Fu, 2011). From the perspective of TDM, time-series data can be extracted as noise, trend, seasonality, or cycle. TDM data extraction algorithms are standard ones in many statistical packages. They have been proven to be practical for measuring the strength of trend and seasonality in a time-series dataset (Wang et al., 2006). TDM has been applied in long-term trend analysis and forecasting, seasonal variation analysis, cyclic variation, and irregular variation analysis (Valenzuela et al., 2008; Verbesselt et al., 2010). TDM algorithms are uni-section and easy to use. Therefore, they are regaining popularity in current big data analytics (Bendre & Manthalkar, 2019) and have become one of the rapidly growing methods in data mining and machine learning (Bendre & Manthalkar, 2019; Wang et al., 2020).

TDM contains many different algorithms (Mills, 2019; Montgomery et al., 2015; Shumway & Stoffer, 2019). The essential method for time-series data transformation is the data decomposition into trend, seasonality, and error (noise), or called the TSE decomposition (Hayes, 2021). In our case studies, we adopted the TSE decomposition method with Python (Hayes, 2021). We used Alxa Right Banner as illustrations with three variables, *EVI*, *PRE*, and *TEM*.

All original observations of *EVI*, *PRE*, and *TEM* displayed relatively wide notched curves (Figs. 3.3, 3.4, and 3.5); the errors showed narrowly jagged curves; the seasonality curves were tall, narrow, and regular waves, revealing nine growing seasons each year. However, the trend curves showed different patterns. The EVI trend curve displayed a declining trend from the 10th to 72nd period (i.e., from 2002 to 2009), a rising curve from 2010 to 2013, a falling trend from 2014 to 2015, and a slightly increasing trend afterward. The PRE trend displayed a 3–4-year cyclical phenomenon. TEM decreased from 2001 to 2003, increased from 2004, and

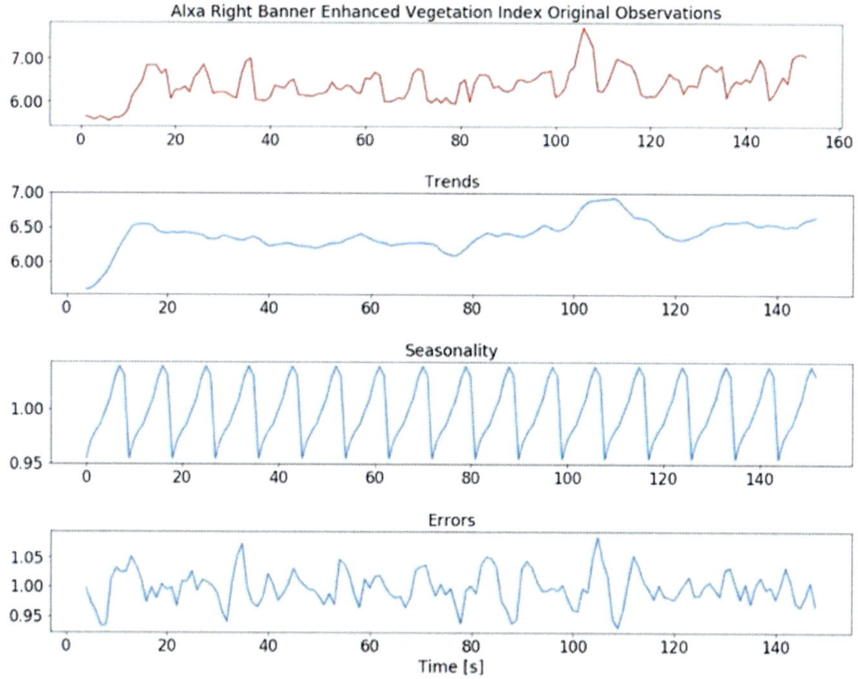

Fig. 3.3 TSE decomposition of enhanced vegetation index

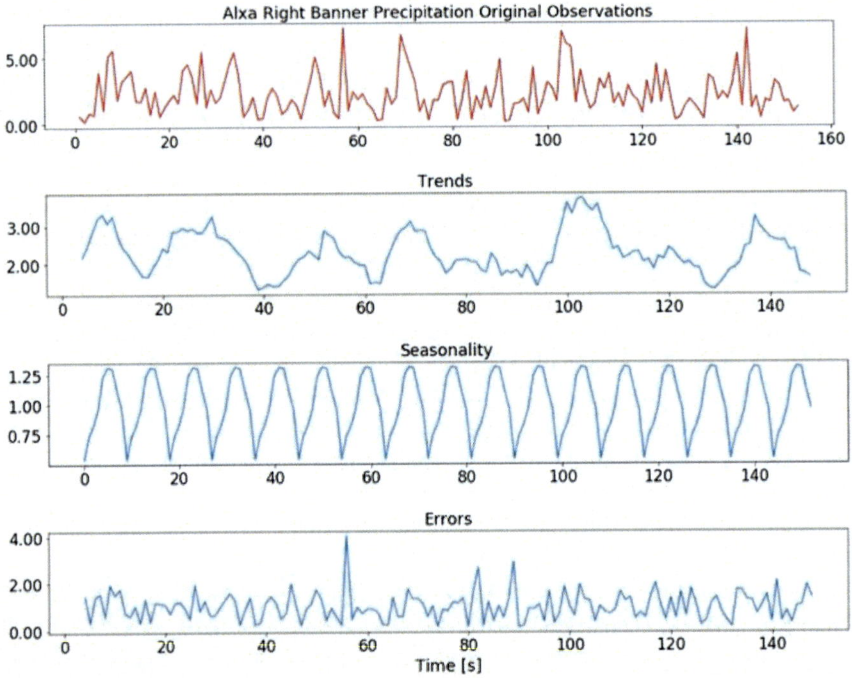

Fig. 3.4 TSE decomposition of precipitation

remained relatively high until 2011. Afterward, TEM dropped for several years and has increased since 2015. The testing dataset, the python codes, and the results of the TSE decomposition for *EVI*, *PRE*, and *TEM* over 89 banners were included in the Chapter Data Folder.

The input data, the Python codes for TSE filtering, and the TSE filtered data are stored in the folder, \Chapter_3\DataProcess\TSE. There is a readme text file that provides a brief explanation.

3.2.2 Computational Data Mining and Residuals through Empirical Mode Decomposition

Computational data mining (CDM) originates from a suite of techniques for reducing signal noises. Several computational noise reduction algorithms (Fourier, Wavelet, and Empirical Mode Decomposition – EMD) have been developed to decompose time-series data (Chen et al., 2018). Fourier analysis and decomposition techniques have long been widely used for stationary time-series decomposition (Bruns, 2004; Duan et al., 2016). Fourier transformations are also applied in environmental studies and remote sensing applications. They are used to describe

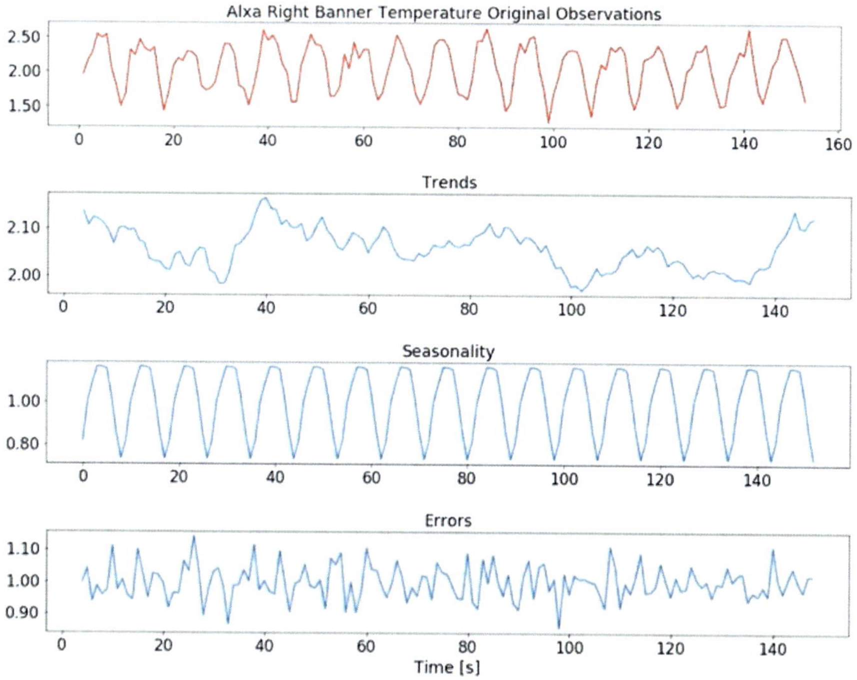

Fig. 3.5 TSE decomposition of temperature

land-surface phenologies (Moody & Johnson, 2001), classify rangeland vegetation types and coverages (Evans & Geerken, 2006), and predict and map mangrove biomass (Proisy et al., 2007). Basic concepts, algorithms, and applications of Fourier transform are well introduced and illustrated in Bracewell's (1978) and Körner's (1989) works.

Wavelet transformation is another technique initially used to extract information from audio signals and images and now expanded into many other types of time-series data. Martínez and Gilabert (2009) successfully detected the dynamics of land cover changes by applying wavelet transformation through normalized deviation vegetation index (NDVI) on the interannual scale. Demirel and Anbarjafari (2011) developed a new satellite image resolution enhancement technique based on interpolating the high-frequency sub-bands obtained by discrete wavelet transform. Hafez and colleagues (2012) applied a wavelet spectral analysis technique to detect sudden geomagnetic commencements automatically. Gaci (2014) used wavelet-based denoising techniques to enhance the first-arrival picking on seismic traces. Burrus and colleagues (1998) published a well-written book introducing wavelets and wavelet transforms. Compared to Fourier transformation, the wavelet transformation can simultaneously extract local spectral and time-series information (Mallat, 2008). Moreover, researchers can choose from a good library of wavelets to match the shapes of signals in studies (Martínez & Gilabert, 2009).

Empirical mode decomposition (EMD) has been frequently used for time-series decomposition, especially in the past two decades. EMD is a robust and adaptive technique for analyzing nonlinear and non-stationary time-series data. Compared with wavelet analysis and other methods, EMD has a significant advantage because it can decompose a complex signal into a series of relatively simple components referred to as intrinsic mode functions (IMFs) and a residual curve without the requirement for any predetermined basis functions.

EMD decomposes the non-stationary signals into IMFs based on the peculiarity of the signals (Huang et al., 1998). Kong et al. (Kong et al., 2015) presented a rigorous decomposition approach for MODIS NDVI with ensemble EMD. In this application, the trend is generated by the combination of lower frequency components decomposed from the original data and the residual, which is an overall trend of time series (Wu et al., 2007). Wu and Huang (Wu & Huang, 2009) claimed that the overall trend is derived as an intrinsically fitted monotonic function where there can be at most one extremum within a given span. The residual curve represents a long-term tendency of change embedded in the time-series data (Chen et al., 2018).

The algorithm of EMD was developed initially to decompose non-stationary and nonlinear signals with several zero mean components (Huang et al., 1998). This algorithm was improved by adding new stopping criteria to make EMD implementation more effective (Rilling et al., 2003). Rilling's EMD algorithm code was adopted in this book. The following is a brief description of the EMD approach. Signal $x(t)$ is the given time series of EVI, or precipitation or temperature (for details, see Rilling et al. (2003) and Chen et al. (2018)):

1. Create the top and bottom envelope with the cubic splines fitted by maximums and minimums extras of $x(t)$.
2. Calculate the mean of envelopes: $\text{mean}(t) = 0.5\times(e\text{max}(t) + e\text{min}(t))$ (where e stands for envelop).
3. Extract the mean value out: $x'(t) = x(t) - \text{mean}(t)$.
4. Repeat previous steps until stopping criteria are met and then extract out the intrinsic mode function (IMF): $\text{IMF} = x(t) - \sum(\text{mean}(t))$.
5. Extract IMF out of $x(t)$: $x' = x(t) - \text{IMF}$.
6. Repeat Steps 1–5 until no more IMF can be extracted.
7. The residual is what is left: $\text{Residual} = x(t) - \sum(\text{IMF})$.

The EMD IMFs decompositions were illustrated with three time-series variables, *EVI*, *PRE*, and *TEM*, in Alxa Right Banner (Figs. 3.6, 3.7, and 3.8). Compared with the TSE decomposition, the EMD IMFs gradually became less jagged and eventually turned into a smooth line. Especially, the last IMF, also called the EMU residual curve, was mono-shaped, either descending or ascending. For example, the EVI and TEM residual curves were increasing (Figs. 3.6 and 3.8), but the PRE residual line was descending (Fig. 3.7).

The input data, the Python codes for EMD decomposition, and the EMD filtered data are stored in the folder, \Chapter_3\DataProcess\EMD. There is a readme text file that provides a brief explanation.

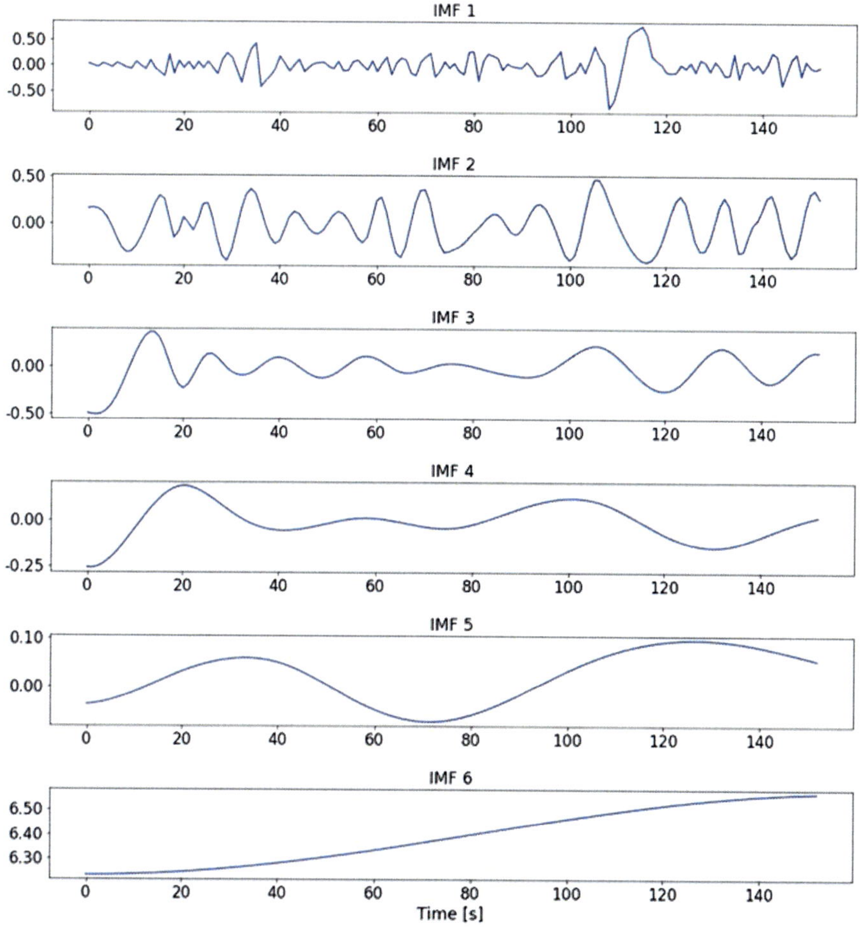

Fig. 3.6 EMD IMFs decompositions of enhanced vegetation index

3.2.3 *Hybrid Computational Learning (HCL)*

The TDM and CDM techniques extract or decompose linear curves from complex time-series data. Nevertheless, it is tough to substantiate those natural systems that comply with the linear model assumption. Therefore, more robust hybrid computational learning (HCL) approaches have been developed to process time-series data without prior knowledge about linear relationships in recent years (Duan et al., 2016). Many HCL algorithms analyze complex time-series phenomena (Yang & Chen, 2019). EMD-based hybrid computational learning models (that combine EMD with some computational learning technique) comprise a favored group. For example, an artificial neural network (ANN) was integrated with EMD to forecast wind speed (H. Liu et al., 2012). The support vector machine (SVM) was combined

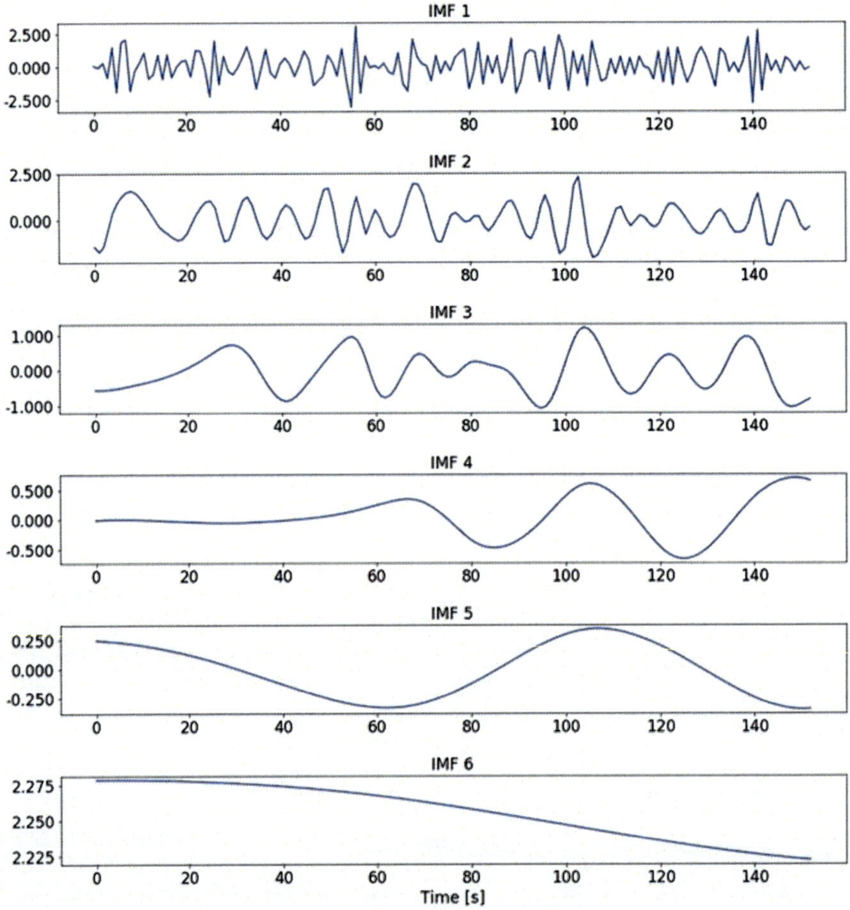

Fig. 3.7 EMD IMFs decompositions of precipitation

with EMD to predict monthly streamflow (Huang et al., 2014) and to estimate short-term wind power (Zhang et al., 2015a). Several ensemble machine learning methods were applied with EMD to forecast wind speed (Ren et al., 2015). Stacked auto-encoder (SAE) was combined with EMD to predict traffic flow (Yang & Chen, 2019). In these hybrid models, EMD was adopted to decompose the time-series data into a set of subseries, while the computational learning algorithms were applied to train each subseries to obtain optimal parameters for forecasting (Xie et al., 2019; Zhang et al., 2015b).

This chapter adopted a combined EMD and support vector regression (SVR) model to illustrate hybrid computational learning (HCL)-based data mining applications. SVR is a prediction model based on the support vector machine (SVM) theory developed by Cortes and Vapnik (1995) under the learning theory. Compared with the traditional ordinary least squares (OLS) regression, an SVR model tries to

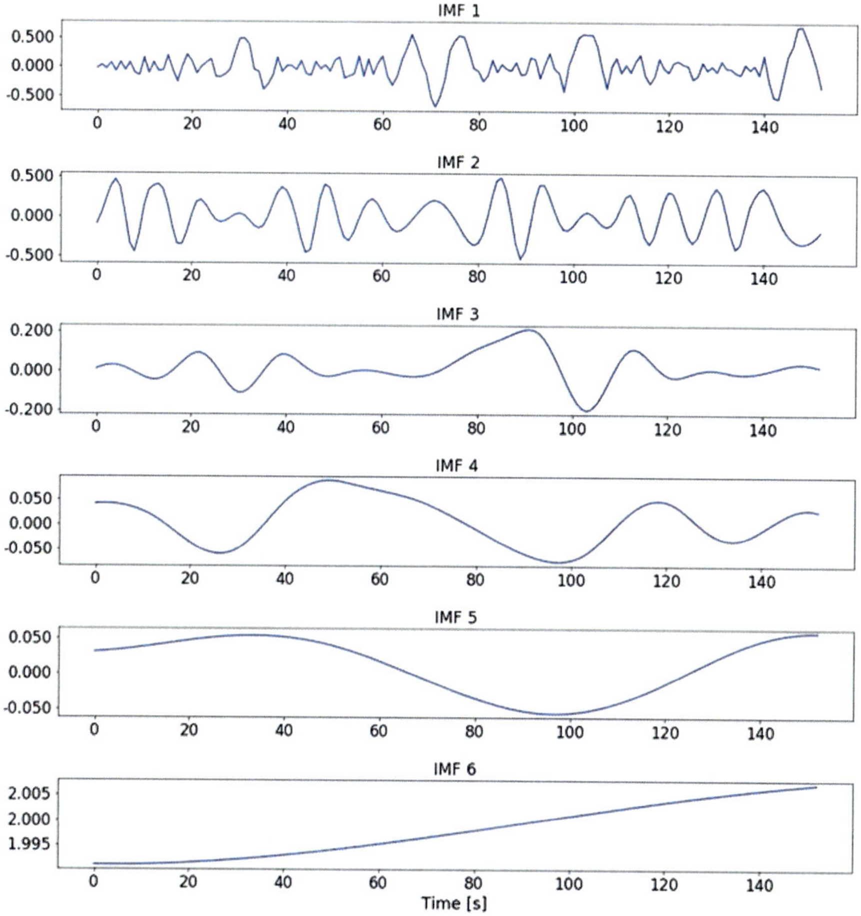

Fig. 3.8 EMD IMFs decompositions of temperature

find a hyperplane where the distance to any data point is the nearest. SVR achieves this goal by introducing two hyperparameters, ε (the maximum error) and ξ (the deviation from the margin).

The objective function of OLS is to minimize the following:

$$y = f(x) = wx_i + b$$

$$Minimize : \sum_{N}^{i=1} (y_i - w_i x_i)^2$$

where y_i is the target, w_i is the set of coefficients, and x_i is the vector of predictors.

SVR first introduces the maximum error term, ε, to minimize the regression coefficients:

$$Minimize : \frac{1}{2} w$$

$$\text{Subject to } \left| y_i - w_i x_i \right| \leq \varepsilon$$

The minimization is the $L2$ (L – Lagrangian) norm of the coefficient vector, reflected in the constraints (Cortes & Vapnik, 1995; Duan et al., 2016; Huang et al., 2014). In other words, the absolute error is set less than or equal to the specified margin, ε, called the maximum error. ε can be adjusted to gain the desired accuracy of a model. For example, two ε-derived margin lines (in green color in Fig. 3.9) explained the maximum errors of an SVR model. In our case study of searching for the best parameters for running the enhanced vegetation index SVR model, the value of ε for the first intrinsic mode function of EVI in Alax Right Banner is 4.2.

Many points still fall outside the maximum error lines ($w_i x_i \pm \varepsilon$). SVR introduces another hyperparameter, ξ (also called the deviation from the margin or the slack variable) (Cortes & Vapnik, 1995; Duan et al., 2016; Huang et al., 2014). For any value outside $w_i x_i \pm \varepsilon$, their deviations from the margins are denoted ξ. Now, the target is to solve the constrained optimization problem, which can be expressed as follows:

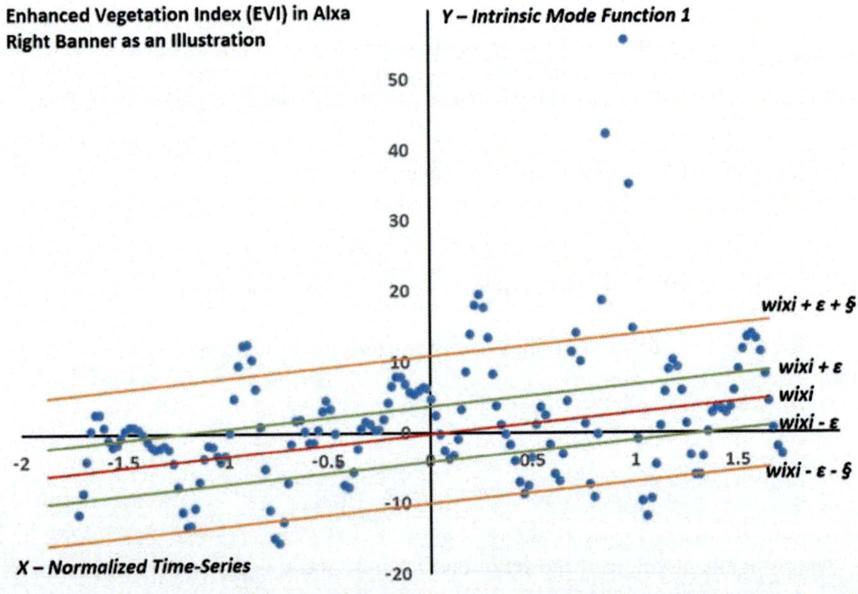

Fig. 3.9 Maximum errors and slack variables of support vector regression (SVR) model

$$Minimize: \frac{1}{2}w + C\sum_{N}^{i=1}(\| \xi_i \|)$$

$$\text{Subject to}: |y_i - wx_i| \leq \varepsilon + |\xi_i|$$

where C is the regularized constant that determines the degree of the empirical error in the optimization problem; ξ refers to the positive slack variable penalizing the training errors by the loss function within the error tolerance ε. Using the first intrinsic mode function of EVI in Alax Right Banner as the illustration, the value of ξ is 5.8. Two deviation lines ($w_i x_i \pm \varepsilon \pm \xi$) are the orange lines in Fig. 3.9.

The hybrid EMD and SVR method adopted in this chapter involves three processes. In other words, the complete procedures of the EMD–SVR model consist of three steps. In the first step, the original time-series data (EVI, precipitation, and temperature) were decomposed using the Rilling's EMD algorithm into stationary IMFs and a residual. During the second step, the SVR model was conducted individually for each IMF and residual curve to generate corresponding ε and ξ values, then applied to adjust the initial IMF and residual values (Duan et al., 2016). All adjusted IMFs and the residual curve were summarized to reconstruct new time-series data at the final step. It is worth pointing out that the adjusted residual curve can also be used directly as the adjusted trend change curve.

The time-series values of EVI in Alxa Right Banner were plotted in Fig. 3.10 in comparison with the EMD residual, the SVR-adjusted EMD residual, and the sums of SVR-adjusted all EMD IMFs. The curves of the SVR-adjusted EVI EMD residual and IMFs sums are smoother than the EVI original time series and its EMD residual curve. This finding visually confirms that the hybrid algorithm of SVR and EMD generates smoother curves than the EMD algorithm.

The input data, the Python codes for HCL filtering, and the HCL filtered data are stored in the folder, \Chapter_3\DataProcess\HCL. There is a readme text file that provides a brief explanation.

3.3 Stationarity Tests and Non-stationarity Removals

Stationarity is critical in judging if an autoregressive model can predict changes (Shumway & Stoffer, 2019). In other words, stationarity means a stable dependence structure exists within the time-series values. Otherwise, it is less meaningful to run autoregressive models if the dependence structure in the time-series data changes at any time point. Therefore, it is necessary to test if stationarity exists in time-series data before applying any time-series analysis. There is a good introductory book about stationarity in time-series data (Mills, 2019). In recent years, stationarity tests and non-stationarity removal have been essential data mining techniques in big data analytics. Srivastava (2015) has developed a complete tutorial on time-series modeling. Jain (2016) provided a comprehensive tutorial for beginners to create a

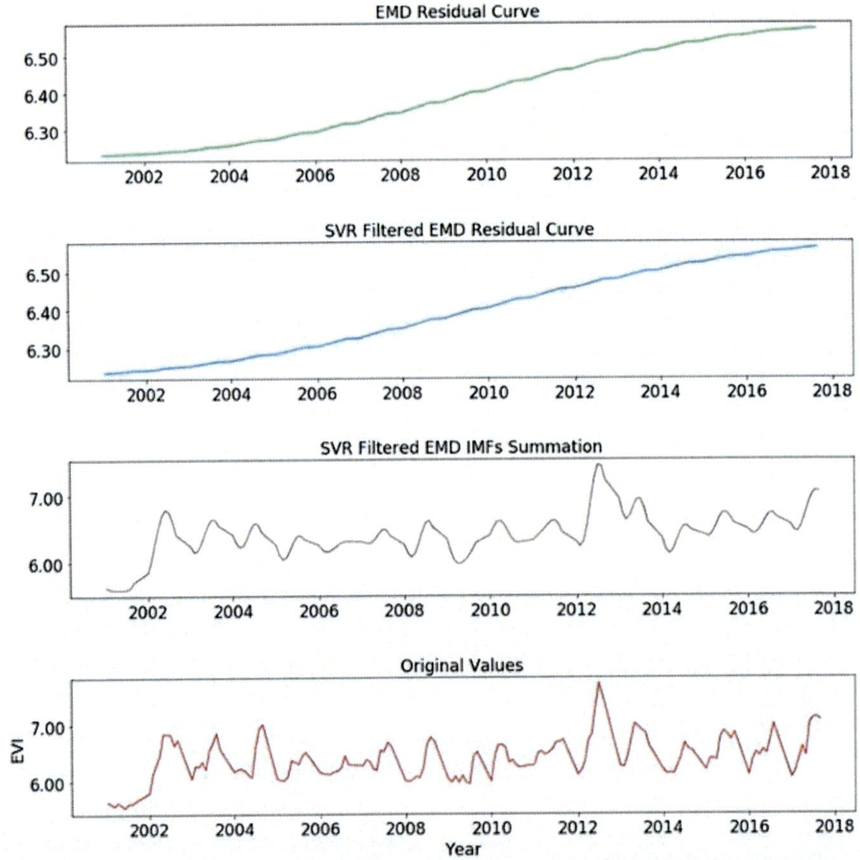

Fig. 3.10 EVI original time series in comparison with the EMD residual, the SVR-adjusted EMD residual, and EMD IMFs sums (IMAR Alxa Right Banner)

time-series forecast with Python. Singh (2018) illustrated how to handle non-stationary time-series data in Python. All these tutorials are hands-on, oriented for big data analytics, and easy to follow for implementing stationarity tests and non-stationarity removals.

There are two types of time-series data. One is a uni-section time-series data that contain daily, monthly, seasonal, periodic, or yearly changes, or combinations of some of those time-series changes, or simply yearly changes over one section (or one area). The other type is cross-sectional time-series data. This data structure is much more complicated because it includes time-series data collected over multiple sections, places, or subjects. This data structure is called panel data (Liu & Xie, 2013a, b). Different statistical procedures are developed to test stationarity embedded within these two different types of datasets.

3.3.1 Stationarity Test with Uni-section Time-Series Data

Many stationarity tests and non-stationarity removal methods are discussed in recent big-data mining. This section adopts the methods that Singh (2018) introduced to handle the stationarity in uni-section time-series data. The stationarity assessment is usually called the unit root stationarity test in statistics (Li et al., 2013). The existence of unit root in a time-series dataset indicates that the statistical properties of this given time series are not constant with time. The unit root can be explained by the following regression model equations (Singh, 2018). Assume an autocorrelation exists in the time series. The time-series value at the current time, t, in the time series can be expressed as

$$y_t = \alpha_{t-1} \times y_{t-1} + \varepsilon_t$$

where y_t is the time-series value at the present time, t, α is the coefficient, and ε_t is the error term. To calculate y_t, we need the value of y_{t-1}, which is

$$y_{t-1} = \alpha_{t-2} \times y_{t-2} + \varepsilon_{t-1}$$

For the entire time-series values, the value of y_{t-n-1} can be calculated as follows:

$$y_{t-n} = \alpha_{t-n-1} \times y_{t-n-1} + \Sigma\varepsilon_{t-n}$$

If the values of α_{t-n-1} are 1 (unit) in the above equations, the predictions of y_{t-n} will be equal to y_{t-n-1}, and the sum of all errors from $t - n$ will be t. This phenomenon means that the variance will increase with time. When this happens, this phenomenon is known as the existence of unit root in this time series. On the contrary, for a stationary time series, the variance must not be a function of time and the coefficient $\alpha \neq 1$. The two most commonly used unit root stationarity tests for the uni-section time-series data are ADF (Augmented Dickey-Fuller) test and KPSS (Kwiatkowski-Phillips-Schmidt-Shin) test.

ADF (Augmented Dickey-Fuller) Test

The null and alternate hypotheses of ADF are:

- *Null hypothesis*: The series has a unit root (value of $\alpha = 1$).
- *Alternate hypothesis*: The series has no unit root.

In other words, if the null hypothesis can be rejected, the ADF test confirms that this time series is stationary. The judgment if an ADF test is rejected depends on its probability value (i.e., p-value). If its p-value is less than the critical value (often 0.05), the null hypothesis is rejected (i.e., the time series is stationary). On the contrary, when the p-value is greater than 0.05, the null hypothesis cannot be rejected (which means the time series has a unit root and, thus, is not stationary).

KPSS (Kwiatkowski-Phillips-Schmidt-Shin) Test

KPSS is another commonly used test for checking the stationarity of a time series. The null and alternate hypotheses of KPSS are opposite to that of ADF. In other words, the null hypothesis of the KPSS test assumes that the time series is *trend stationary*, but the KPSS alternate hypothesis is a unit root series. Trend stationary indicates that a time series has no unit root but exhibits a trend, which is referred to as a trend stationary series (Singh, 2018). The term trend stationary separates the ADF test from the KPSS test. The ADF test generally identifies a strict stationary, but the KPSS test reveals a trend stationary.

It is strongly suggested that both tests be applied to exhibit whether a time series is stationary. Let us look at the results of using the ADF and KPSS tests on the three variables, enhanced vegetation index (EVI), precipitation (PRE), and temperature (TEM), in the IMAR dataset (Table 3.4).

There are three types of results:

- Type 1: Both tests conclude that the series is stationary. In the case of PRE, the ADF test rejects the null hypothesis (stationary) and the KPSS test accepts the null hypothesis (stationary). These two tests confirm that the PRE time series is strictly stationary.
- Type 2: Both tests conclude that the EVI time series is not stationary (accepting the ADF null hypothesis and rejecting the KPSS null hypothesis), indicating that the EVI time series is not stationary.
- Type 3: KPSS accepts its null hypothesis for TEM (trend stationary) and ADF also accepts its null hypothesis (a non-stationary time series). This difference stationary is trend stationary.
- Another type is the difference stationary. A time series is stationary with the ADF test but is non-stationary with the KPSS test. This variation between ADF and KPSS indicates a stationary difference, which can be made as a stationary time series using the differencing method.

Making a Non-stationary Time-Series Stationary

It is generally agreed that it is a prerequisite to transforming a non-stationary time series to a stationary series to conduct auto-regressive-based analysis. Four groups of methods can make a non-stationary series stationary, including differencing, seasonal differencing, transforming, and linear trend removing (Brownlee, 2020; Singh, 2018; Solanki, 2020a, b).

Table 3.4 The stationarity test results by ADF and KPSS

	Enhanced vegetation index – EVI		Precipitation – PRE		Temperature-TEM	
	ADF	KPSS	ADF	KPSS	ADF	KPSS
Statistic	−2.638	0.705	−6.935	0.052	−2.779	0.164
p-value	0.085[a]	0.013[a]	0.000[b]	0.100[b]	0.061[a]	0.100[b]

[a]Non-stationary based on its null hypothesis and the *p*-value of its statistic (0.05)
[b]Stationary based on its null hypothesis and the *p*-value of its statistic (0.05)

Differencing is a straightforward method of calculating the difference between the values at two consecutive times in the series. Differencing is often used to eliminate the varying mean. It can be expressed as the following equation:

$$Y_t' = Y_t - Y_{t-1}$$

where Y_t is the value at a time t.

Seasonal differencing (SD) calculates the difference between a current observation and a previous observation in the same season rather than calculating the difference between consecutive values. For example, in the IMAR dataset, an EVI observation taken in the first growing period of the current year can deduct the observation taken in the first season of the previous year. Mathematically, it can be written as:

$$Y_t' = Y_t - Y_{t-1}$$

where n is the number of seasons of a year (i.e., the number of growing periods – 9 – in the IMAR dataset).

Transforming includes several data standardization methods in statistical packages, such as log transformation, power transformation, and square root. Log transformation is the most popular among them, which will be illustrated here.

Linear trend removing (LTR) uses a linear regression model to remove the trend. In other words, a linear regression model is fitted to the time-series data. The fitted model is then used to predict time-series values. Afterward, the predicted values are subtracted from the original time series to remove the trend.

The enhanced vegetation index (EVI) in Alax Banner, IMAR, was used to illustrate these non-stationarity removal methods. The graphics of the original data series and four stationarity removed series by different methods are depicted in Fig. 3.11. The outcomes of the ADF and KPSS tests are reported in Table 3.5. The LTR removal method, in this case, did not work because it generated a noticeable trend curve. Moreover, both ADF and KPSS tests for LTR confirmed that the stationarity was not removed.

On the other hand, the remaining three methods, consecutive differencing, seasonal differencing, and log transformation, worked very well. The resultant curves of the consecutive differencing and log transformation removals looked quite analogous. The outcome curve of the seasonal differencing resembled the curve of the original series, although some amplifications happened to the peaks and valleys of the initial curve. Assessments on which non-stationarity removal methods perform better or worse are still missing.

The ADF and KPSS stationarity tests and four non-stationarity removal methods were coded in Python. The input data, the Python codes, and the stationarity removed data are stored in the folder, \Chapter_3\DataProcess\Stationarity. There is a readme text file that provides a brief explanation.

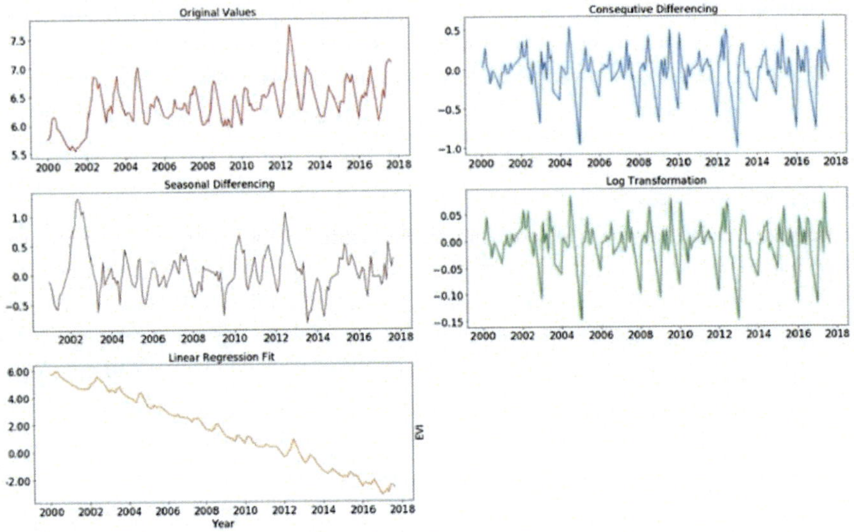

Fig. 3.11 Curves of the original series and the non-stationarity removed series

Table 3.5 The stationarity test results for non-stationarity removed series

	ADF	KPSS	ADF	KPSS
	Consecutive differencing		*Seasonal differencing*	
Statistic	−5.073	0.050	−4.297	0.050
p-value	0.000[a]	0.100[a]	0.000[a]	0.100[a]
	Log transform		*Linear trend removing*	
Statistic	−5.014	0.050	−0.092	1.182
p-value	0.000[a]	0.100[a]	0.950[b]	0.010[b]

[a]Stationary based on its null hypothesis and the *p*-value of its statistic (0.05)
[b]Non-stationary based on its null hypothesis and the *p*-value of its statistic (0.05)

3.3.2 *Stationarity Test with Panel Time-Series Data*

The ADF and KPSS stationarity tests only work with time series of a single object, or one case, or one place. More sophisticated unit root tests are needed for the panel time-series data consisting of both time series and multiple sections (or cases or places). Theoretical discussions of unit root tests for complex data can be found in the works by Patterson (2010, 2012). The software of Stata provides a variety of tests for unit roots or stationarity in panel datasets. For example, the Levin–Lin–Chu (Levin et al., 2002), Harris–Tzavalis (Harris and Tzavalis, 1999), Breitung (Breitung & Das, 2005), Im–Pesaran–Shin (Im et al., 2003), and Fisher-type (Choi, 2001) tests assume that the panel data contains a unit root as the null hypothesis.

The Hadri (2000) Lagrange multiplier (LM) test has the null hypothesis that all the panels are (trend) stationary.

We are adopting the Levin–Lin–Chu test to illustrate the unit root or stationarity test with the IMAR dataset. The widely used Levin, Lin, and Chu (LLC) unit root test assumes that the time series contain a common unit root, and the alternative is that this series is stationary (Levin et al., 2002). When a series rejects an LLC unit root test, this series is stationary and thus can be included for panel data analysis.

We have processed LLC tests for three variables containing seasonal changes, EVI, PRE, and TEM, and their variations using different filtering techniques (Table 3.5). We have also tested two sample variables only showing yearly changes, *DPop* (population density) and *gdppc* (per capita GPD). The LLC tests have rejected the null hypotheses of most original and filtered forms of EVI, PRE, and TEM, except TEMtse (TEM filtered by the TSE algorithm). Considering that TEM has nine seasonal readings per year, we have applied the method of nine seasonal differencing (S9.TEMtse in the terminology of the STATA software package) to achieve stationarity (Table 3.6). Moreover, Dpop contains the unit root based on the LLC test. We have used the consecutive differencing methods (1 differencing – D.Dpop in STATA terminology; Table 3.6).

The panel data stationarity test and removal methods were coded in the statistical package Stata. The input data, the Stata do files (i.e., the State scripts or macros), and the non-stationarity removed data are stored in the folder, \Chapter_3\DataProcess\Stationarity.

By now, we have introduced five time-series data standardization or transformation methods: trend-seasonality-error (TSE), which is also called trend-periodicity-error (TPE), empirical mode decomposition residual curve (EMD-R), hybrid Support Vector Regression-EMD-Residual (SEMD-R), hybrid Support Vector Regression-EMD-Additive (SEMD-A), and stationarity removal (SRM). We applied the six methods to EVI, PRE, and TEM. We also applied EMD-R, SEMD-R, and SEMD-A to the LULC and the socioeconomic data. Since most land uses have lots of missing records, only grassland (GRASS), agricultural land (Percent of Arable Land – PAA), and urban land (URBAN) are kept for further analysis.

- The original data are stored in the folder, \Chapter_3\DataProcess\IMAR_Data\Original.
- The filtered data and the Python codes used to filter the original data are stored in the folder, \Chapter_3\DataProcess\IMAR_Data\Filtered_step1.
- The filtered data are reorganized for creating panel datasets for further analysis and are stored in the folder, \Chapter_3\DataProcess\IMAR_Data\Filtered_step2.
- The final panel datasets and the Python codes for appending the filtered data into the final panels are stored in the folder, \Chapter_3\DataProcess\IMAR_Data\Filtered_Final.

Table 3.6 The results of LLC tests for selected variables in the IMAR dataset

Variable	LLC test	Statistic	*p*-value
EVIorg[a]	Unadjusted *t*	−88.107	
	Adjusted *t**	−94.853	0.000
PREorg[b]	Unadjusted *t*	−69.429	
	Adjusted *t**	−59.075	0.000
TEMorg[c]	Unadjusted *t*	−84.786	
	Adjusted *t**	−92.017	0.000
EVItse[d]	Unadjusted *t*	−33.920	
	Adjusted *t**	−18.335	0.000
PREtse	Unadjusted *t*	−24.722	
	Adjusted *t**	−6.141	0.000
TEMtse	Unadjusted *t*	−24.612	
	Adjusted *t**	−1.188	0.117
S9.TEMtse[e]	***Unadjusted t***	***−30.947***	
	Adjusted t*	***−12.570***	***0.000***
EVIemd[f]	Unadjusted *t*	−250.000	
	Adjusted *t**	−320.000	0.000
PREemd	Unadjusted *t*	−240.000	
	Adjusted *t**	−320.000	0.000
TEMemd	Unadjusted *t*	−240.000	
	Adjusted *t**	−310.000	0.000
EVIsvr_r[g]	Unadjusted *t*	−280.000	
	Adjusted *t**	−370.000	0.000
PREsvr_r	Unadjusted *t*	−280.000	
	Adjusted *t**	−360.000	0.000
TEMsvr_r	Unadjusted *t*	−50.699	
	Adjusted *t**	−66.313	0.000
EVIsur_a[h]	Unadjusted *t*	−280.000	
	Adjusted *t**	−370.000	0.000
PREsvr_a	Unadjusted *t*	−280.000	
	Adjusted *t**	−360.000	0.000
TEMsvr_a	Unadjusted *t*	−50.699	
	Adjusted *t**	−66.313	0.000
Dpop	Unadjusted *t*	−5.466	
	Adjusted *t**	9.930	1.000
D.Dpop[i]	***Unadjusted t***	***−82.730***	
	Adjusted t*	***−72.251***	***0.000***
gdppc[j]	Unadjusted *t*	−6.679	
	Adjusted *t**	−1.681	0.046

[a]*EVI* stands for enhanced vegetation index and *org* for the original data
[b]*PRE* stands for precipitation and *org* for the original data
[c]*TEM* stands for temperature and *org* for the original data
[d]tse stands for trend, seasonality, and error time-series data mining
[e]*S9* indicates the ninth seasonal differencing in the terminology of STATA software

(continued)

Table 3.6 (continued)

[f] *emd* stands for empirical mode decomposition
[g] *svr* stands for support vector regression, which indicates a hybrid data mining technique of *emd* and *svr*; *_r* implies the filtered *emd* residual values by the *svr* algorithm
[h] *a* implies that additive values (the sums) of the filtered *emd* IMFs by the *svr* algorithm
[i] *Dpop* represents population density and "*D*." implies the first consecutive differencing in the terminology of STATA software
[j] *gdppc* indicates per capita GDP

3.4 Validity of Time-Series Data Transformation

The most crucial purpose of data transformation is to augment causal relationships between the response and exploratory variables. Many statistical techniques have been developed to transform the data to achieve linear relationships in regression models. Data transformation is widespread in statistics. Variable rescaling, standardization, or transformation is a standard data preprocessing method in regression analysis. The question is whether different data standardizing techniques change the causal relationships between the initial response and exploratory variables because the standardized data differ from the original. Consequently, the model outcomes from different data standardizing techniques could reflect different causal relationships and hence cannot be compared.

There are some discussions in statistics about the potential impacts of data transformation. Standardizing the variables in a regression analysis can reveal statistically significant findings that might be missed otherwise (Newman & Browner, 1991). There are various techniques exist to standardize variables such as normalizing, scaling, standardizing, or transforming. These techniques often apply the statistics of mean, standard deviation, data range, or linear trend to filter out anomalies in the data. The common goals of variable standardizing are to remove multicollinearity and disclose causal relationships between the response variable and explanatory variables (Allen, 1997). In general, a linear transformation of a model does not statistically change the model itself (Marquardt, 1980, p. 87). Data standardizing methods rarely change the general causal relationships embedded in a regression model, although they may augment linear relations sought to examine (Fox, 2016).

In short, linear data transformation conceptually does not alter statistical relationships between the response variable and the explanatory variables. However, it is too early to confirm this notion before we fit the transformed datasets into linear regression models and compare the outcomes of the transformed linear models with the results of the same linear models fitted with the non-transformed dataset. We will compare 18 different panel models composed of six altered datasets from varied transformations and fitted with three panel models (conventional, dependent-lagged, and dynamic-lagged) in Chap. 4. We will also analyze how the different combinations of transformed datasets and panel models impact the physical and socioeconomic interpretations of the model outcomes.

3.5 Potential Applications of Time-Series Data Transformation

The new datasets derived from the time-series data transformation discussed in this chapter will be fitted into three different types of panel regression models in Chap. 4. Experimenting with which combination of time-series data transformation and panel regression model is to identify the best estimation or prediction of grassland productivity. Part of the results will be spatially interpolated in Chap. 8 to implement a two-level (town-county) panel regression model.

The time-series data transformation methods for environmental and socioeconomic data developed in this chapter can be applied to construct a temporally unified database to support integrated total human and nature system studies. The data mining methods and the accompanying Python codes are easily deployable to other integrated human–nature systems or study areas.

In addition, the stationarity tests for a single section and panel data are adaptable to time-series analysis of all environmental and socioeconomic variables. In summary, the temporal data mining methods are critical tools for building a temporally unified total socioenvironmental database.

References

Allen, M. P. (1997). Regression analysis with standardized variables. In *Understanding regression analysis*. Springer. https://doi.org/10.1007/978-0-585-25657-3_10

Bendre, M., & Manthalkar, R. (2019). Time series decomposition and predictive analytics using MapReduce framework. *Expert Systems with Applications, 116*, 108–120. https://doi.org/10.1016/j.eswa.2018.09.017

Bracewell, R. N. (1978). *The Fourier transform & its applications* (p. 496). McGraw-Hill Companies.

Breitung, J., & Das, S. (2005). Panel unit root tests under cross-sectional dependence. *Statistica Neerlandica, 59*(4), 414–433.

Brownlee, J. (2020). *How to remove trends and seasonality with a difference transform in python?* Retrieved from https://machinelearningmastery.com/remove-trends-seasonality-difference-transform-python/

Bruns, A. (2004). Fourier-, Hilbert- and wavelet-based signal analysis: Are they really different approaches? *Journal of Neuroscience Methods, 137*(2), 321–332. https://doi.org/10.1016/j.jneumeth.2004.03.002

Burrus, C., Gopinath, R., & Guo, H. (1998). *Introduction to wavelets and wavelet transforms: A primer* (1st ed., p. 268). Prentice Hall.

Chen, T. Y., Xie, Y. C., Liu, C., Bai, Y. F., Zhang, A. B., Mao, L. S., & Fan, S. Y. (2018). Trend analysis of relationship between primary productivity, precipitation and temperature in Inner Mongolia. *ISPRS International Journal of Geo-Information, 7*(6). https://doi.org/10.3390/ijgi7060214

Choi, I. (2001). Unit root tests for panel data. *Journal of International Money and Finance, 20*(2), 249–272.

Cortes, C., & Vapnik, V. (1995). Support-vector networks. *Machine Learning, 20*(3), 273–297. https://doi.org/10.1023/A:1022627411411

Demirel, H., & Anbarjafari, G. (2011). Discrete wavelet transform-based satellite image resolution enhancement. *IEEE Transactions on Geoscience and Remote Sensing, 49*(6), 1997–2004.

Duan, W. Y., Han, Y., Huang, L. M., Zhao, B. B., & Wang, M. H. (2016). A hybrid EMD-SVR model for the short-term prediction of significant wave height. *Ocean Engineering, 124*, 54–73. https://doi.org/10.1016/j.oceaneng.2016.05.049

Evans, J. P., & Geerken, R. (2006). Classifying rangeland vegetation type and coverage using a Fourier component based similarity measure. *Remote Sensing of Environment, 105*(1), 1–8. https://doi.org/10.1016/j.rse.2006.05.017

Fox, J. (2016). *Applied regression analysis and generalized linear models* (3rd ed., p. 816). SAGE Publications.

Fu, T. C. (2011). A review on time series data mining. *24*(1), 164–181. https://doi.org/10.1016/j.engappai.2010.09.007

Gaci, S. (2014). The use of wavelet-based denoising techniques to enhance the first-arrival picking on seismic traces. *IEEE Transactions on Geoscience and Remote Sensing, 52*(8), 4558–4563.

Gong, P., Li, X., & Zhang, W. (2019). 40-Year (1978–2017) human settlement changes in China reflected by impervious surfaces from satellite remote sensing. *Scientific Bulletin, 2019*(64), 756–763.

Hadri, K. (2000). Testing for stationarity in heterogeneous panel data. *The Econometrics Journal, 3*, 148–161.

Hafez, A. G., Ghamry, E., Yayama, H., & Yumoto, K. (2012). A wavelet spectral analysis technique for automatic detection of geomagnetic sudden commencements. *IEEE Transactions on Geoscience and Remote Sensing, 50*(11), 4503–4512.

Hao, L., Sun, G., Liu, Y., Gao, Z., He, J., Shi, T., & Wu, B. (2014). Effects of precipitation on grassland ecosystem restoration under grazing exclusion in Inner Mongolia. *China Landscape Ecology, 29*(10), 1657–1673. https://doi.org/10.1007/s10980-014-0092-1

Harris, R. D. F., & Tzavalis, E. (1999). Inference for unit roots in dynamic panels where the time dimension is fixed. *Journal of Econometrics, 91*, 201–226.

Hayes, S. (2021). *Finding seasonal trends in time-series data with python*. Retrieved from https://towardsdatascience.com/finding-seasonal-trends-in-time-series-data-with-python-ce10c37aa861

Huang, N. E., Shen, Z., Long, S. R., Wu, M. C., Shih, H. H., Zheng, Q., Yen, N.-C., Tung, C. C., & Liu, H. H. (1998). *The empirical mode decomposition and the Hilbert spectrum for nonlinear and non-stationary time series analysis.* Paper presented at the Proceedings of the Royal Society of London A: mathematical, physical and engineering sciences.

Huang, S. Z., Chang, J. X., Huang, Q., & Chen, Y. T. (2014). Monthly streamflow prediction using modified EMD-based support vector machine. *Journal of Hydrology, 511*, 764–775. https://doi.org/10.1016/j.jhydrol.2014.01.062

Im, K. S., Pesaran, M. H., & Shin, Y. (2003). Testing for unit roots in heterogeneous panels. *Journal of Econometrics, 115*, 53–74.

Jain, A. (2016). *A comprehensive beginner's guide to create a Time Series Forecast* (with Codes in Python and R). https://www.analyticsvidhya.com/blog/2016/02/time-series-forecasting-codes-python/. Accessed on 6 Feb 2022.

Juchem Neto, J. P., Claeyssen, J. C. R., & Pôrto Júnior, S. S. (2018). Economic agglomerations and spatio-temporal cycles in a spatial growth model with capital transport cost. *Physica A: Statistical Mechanics and its Applications, 494*, 76–86. https://doi.org/10.1016/j.physa.2017.12.036

Kong, Y. L., Meng, Y., Li, W., Yue, A. Z., & Yuan, Y. (2015). Satellite image time series decomposition based on EEMD. *Remote Sensing, 7*(11), 15583–15604. https://doi.org/10.3390/rs71115583

Körner, T. W. (1989). *Fourier analysis* (p. 604). Cambridge University Press.

Levin, A., Lin, C.-F., & Chu, C.-S. J. (2002). Unit root tests in panel data: Asymptotic and finite-sample properties. *Journal of Econometrics, 108*(1), 1–24.

Li, S., Xie, Y., Brown, D. G., Bai, Y., Hua, J., & Judd, K. (2013). Spatial variability of the adaptation of grassland vegetation to climatic change in Inner Mongolia of China. *Applied Geography, 43*, 1–12. https://doi.org/10.1016/j.apgeog.2013.05.008

Liu, Y., & Xie, Y. (2013a). Asymmetric adjustment of the dynamic relationship between energy intensity and urbanization in China. *Energy Economics, 36*, 43–54. https://doi.org/10.1016/j.eneco.2012.12.003

Liu, Y., & Xie, Y. (2013b). Measuring the dragging effect of natural resources on economic growth: Evidence from a space–time panel filter modeling in China. *Annals of the Association of American Geographers, 103*(6), 1539–1551. https://doi.org/10.1080/00045608.2013.784108

Liu, H., Chen, C., Tian, H. Q., & Li, Y. F. (2012). A hybrid model for wind speed prediction using empirical mode decomposition and artificial neural networks. *Renewable Energy, 48*, 545–556. https://doi.org/10.1016/j.renene.2012.06.012

Mallat, S. (2008). *A wavelet tour of signal processing: The sparse way* (3rd ed.). Academic Press.

Martínez, B., & Gilabert, M. A. (2009). Vegetation dynamics from NDVI time series analysis using the wavelet transform. *Remote Sensing of Environment, 113*(9), 1823–1842.

Marquardt, D. W. (1980). Comment, you should standardize the predictor variables in your regression models. *Journal of the American Statistical Association, 75*(369), 87–91. https://doi.org/10.1080/01621459.1980.10477430

Mills, T. C. (2019). *Applied time series analysis: A practical guide to modeling and forecasting*. Academic Press.

Montgomery, D. C., Jennings, C. L., & Kulahci, M. (2015). *Introduction to time series analysis and forecasting*. Wiley.

Moody, A., & Johnson, D. M. (2001). Land-surface Phenologies from AVHRR using the discrete Fourier transform. *Remote Sensing of Environment, 75*(3), 305–323. https://doi.org/10.1016/S0034-4257(00)00175-9

Newman, T. B., & Browner, W. S. (1991). In defense of standardized regression coefficients. *Epidemiology, 2*(5), 383–386.

Patterson, K. (2010). *A primer for unit root testing* (p. 277). Palgrave Macmillan Limited.

Patterson, K. (2012). *Unit root tests in time series volume 2: Extensions and developments* (p. 550). Palgrave Macmillan Limited.

Proisy, C., Couteron, P., & Fromard, F. (2007). Predicting and mapping mangrove biomass from canopy grain analysis using Fourier-based textural ordination of IKONOS images. *Remote Sensing of Environment, 109*(3), 379–392. https://doi.org/10.1016/j.rse.2007.01.009

Propastin, P. A., Kappas, M., & Muratova, N. R. (2008). Inter-annual changes in vegetation activities and their relationship to temperature and precipitation in Central Asia from 1982 to 2003. *Journal of Environmental Informatics, 12*(2), 75–87. https://doi.org/10.3808/jei.200800126

Ren, Y., Suganthan, P. N., & Srikanth, N. (2015). Ensemble methods for wind and solar power forecasting—A state-of-the-art review. *Renewable and Sustainable Energy Reviews, 50*(82-91), 1364–0321. https://doi.org/10.1016/j.rser.2015.04.081

Rilling, G., Flandrin, P., & Goncalves, P. (2003). *On empirical mode decomposition and its algorithms.* Paper presented at the IEEE-EURASIP workshop on nonlinear signal and image processing.

Roman, P., & Thiry, G. (2016). The inclusive wealth index. A critical appraisal. *Ecological Economics, 124*, 185–192.

Shumway, R., & Stoffer, D. (2019). *Time series: A data analysis approach using R*. Chapman and Hall/CRC.

Singh, A. (2018). *A gentle introduction to handling a non-stationary time series in python*. Retrieved from https://www.analyticsvidhya.com/blog/2018/09/non-stationary-time-series-python/

Solanki, S. (2020a). *Time series – How to remove trend & seasonality from time-series data using pandas [Python].* Retrieved from https://coderzcolumn.com/tutorials/data-science/how-to-remove-trend-and-seasonality-from-time-series-data-using-python-pandas

Solanki, S. (2020b). *Time series: Resampling & moving window functions in python using pandas*. Retrieved from https://coderzcolumn.com/tutorials/data-science/time-series-resampling-and-moving-window-functions

Srivastava, T. (2015). *A complete tutorial on time series modeling in R. Time series analysis*. https://www.analyticsvidhya.com/blog/2015/12/complete-tutorial-time-series-modeling/. Accessed on 6 Feb 2022.

Valenzuela, O., Rojas, I., Rojas, F., Pomares, H., Herrera, L. J., Guillen, A., et al. (2008). Hybridization of intelligent techniques and ARIMA models for time series prediction. *Fuzzy Sets and Systems, 159*(7), 821–845. https://doi.org/10.1016/j.fss.2007.11.003

Verbesselt, J., Hyndman, R., Newnham, G., & Culvenor, D. (2010). Detecting trend and seasonal changes in satellite image time series. *Remote Sensing of Environment, 114*(1), 106–115. https://doi.org/10.1016/j.rse.2009.08.014

Wang, X., Smith, K., & Hyndman, R. (2006). Characteristic-based clustering for time series data. *Data Mining and Knowledge Discovery*. https://doi.org/10.1007/s10618-005-0039-x

Wang, C. D., Yuan, J. J., Zhang, J., Deng, N., Zhou, Z. H., & Gao, F. (2020). Multi-criteria comprehensive study on predictive algorithm of heating energy consumption of district heating station based on timeseries processing. *Energy, 202*. https://doi.org/10.1016/j.energy.2020.117714

Watson, S. J., Luck, G. W., Spooner, P. G., & Watson, D. M. (2014). Land-use change: Incorporating the frequency, sequence, time span, and magnitude of changes into ecological research. *Frontiers in Ecology and the Environment, 12*(4), 241–249. https://doi.org/10.1890/130097

Wilcox, D., & Xie, Y. (2007). Predicting wetland plant community responses to proposed water-level-regulation plans for Lake Ontario: GIS-based modeling. *International Journal of Great Lakes Research, 33*(4), 751–773.

Wu, Z., & Huang, N. E. (2009). Ensemble empirical mode decomposition: A noise-assisted data analysis method. *Advances in Adaptive Data Analysis, 1*(1), 1–41.

Wu, Z., Huang, N. E., Long, S. R., & Peng, C.-K. (2007). On the trend, detrending, and variability of nonlinear and non-stationary time series. *Proceedings of the National Academy of Sciences, 104*(38), 14889–14894.

Xie, Y., Crary, D., Bai, Y., Cui, X., & Zhang, A. (2019). Modeling grassland ecosystem responses to coupled climate and socioeconomic influences in multi-spatial-and-temporal scales. *Journal of Environmental Informatics, 33*(1), 37–46. https://doi.org/10.3808/jei.201600337

Yan, X., Li, J., Shao, Y., Hu, Z., Yang, Z., Yin, S., & Cui, L. (2020). Driving forces of grassland vegetation changes in Chen Barag Banner, Inner Mongolia. *GIScience & Remote Sensing, 57*(6), 753–769. https://doi.org/10.1080/15481603.2020.1794395

Yang, H. F., & Chen, Y. P. P. (2019). Hybrid deep learning and empirical mode decomposition model for time series applications. *Expert Systems with Applications, 120*, 128–138. https://doi.org/10.1016/j.eswa.2018.11.019

Zhang, W., Liu, F., Zheng, X., & Li, Y. (2015a). *A hybrid EMD-SVM based short-term wind power forecasting model*. Paper presented at the 2015 IEEE PES Asia-Pacific power and energy engineering conference (APPEEC).

Zhang, X., Miller-Hooks, E., & Denny, K. (2015b). Assessing the role of network topology in transportation network resilience. *Journal of Transport Geography, 46*, 35–45.

Zhou, C., Liang, J., & Xie, Y. (2022). Investigating causal relationships between grassland deterioration and climate and socioeconomic changes through time-series computational learning. *Journal of Cleaner Production, 366*, 132963. https://doi.org/10.1016/j.jclepro.2022.132963

Chapter 4
Empirically Investigating Time-Series Causality Between Grassland Productivity and Socioenvironmental Factors: Inner Mongolia as a Case Study

The Inner Mongolia Autonomous Region (IMAR) of China locates between 37°24′–53°23′N and 97°12′–126°04′E. IMAR is on the northern border of China and the southern portion of the Mongolia Plateau. It has a total area of 1.18 million km^2 (Fig. 3.1), extending diagonally from the northeast to the southwest in a long and narrow shape. The terrain is flat, with the Greater Khingan Range in the east and Yinshan and Henan Mountains in the south. The average elevation of IMAR is between 1000 and 1200 m and is dominantly covered by the temperate steppe. The climate in the steppe area is a typical temperate continental climate, with annual precipitation of 50–450 mm and an annual average temperature of −1.01 °C to 9.15 °C. The weather gradually transitions from humid and semi-humid in the east to semiarid and arid in the west. The average precipitation decreases from the northeast to the southwest, but the temperature increases. The IMAR temperate grasslands account for about 67% of the region's total area and 22% of the grassland area of China.

IMAR and Mongolia consist of one of the most wide-ranging remaining grasslands in the world. There are 89 banners (the Mongolian term for counties) in IMAR. Grasslands in IMAR have suffered severe degradation from excessive population growth and economic development in recent decades. For example, IMAR saw dramatic economic development and population growth from 2000 to 2017. The GDP increased by 1040.87% from 14.26 billion Chinese Yuan in 2000 to 16.28 trillion in 2017 (IMAR Statistical Bureau, 2001–2018). The livestock increased 127.9% from 49.36 million sheep units in 2000 to 112.50 million in 2017. The urban construction area increased by 114.23% from 2000 to 2017. The total population increased by almost 9.35% from 23.10 million to 25.26 million. The excessive socioeconomic activities and urban expansion have threatened the grassland's

---

Supplementary Information The online version contains supplementary material available at https://doi.org/10.1007/978-3-031-39594-9_4.

Y. Xie, *Total Socioenvironmental Systems*,
https://doi.org/10.1007/978-3-031-39594-9_4

ecological security and urban growth sustainability. Therefore, IMAR is one of the best sites to study integrated socioenvironmental systems because of its fragile semiarid environment and excessive human activities (Brown et al., 2013).

4.1 Causal Structures of Grassland Deterioration

This chapter enriches literature in several ways. First, it develops a causal framework to examine grassland deterioration from four major components of a grassland ecosystem, global climate change, grassland deterioration, ecological/environmental consequences, and socioeconomic adaptations. The chapter comprises a comprehensive database reported in the current literature, describing these elements of Inner Mongolia grasslands with 32 variables. Second, the chapter applies the time-series trend filtering methods developed in Chap. 3 to construct a unified socioenvironmental database. It uses the unified database to investigate the interactions and trade-offs between grassland deterioration, diving climate, and socioeconomic factors. Third, it designs a most sophisticated experiment to explore which time-series trend filtering and panel regression model combination can effectively augment the causal relationships between grassland deterioration and other socioenvironmental driving factors. Fourth, the chapter integrates three panel regression models and a block-entered regression approach to distinguish the contributions of various climate, agricultural, socioeconomic, land-use, education, and industrial factors to grassland deterioration. Although three-panel regression models (Zhou et al., 2022) and the block-entered regression approach (Zhou et al., 2022) were applied in recent literature, this chapter confirms that the integration of them effectively differentiates how various driving factors impact grassland deterioration in Inner Mongolia.

Another note is that a subset of the database (13 variables over 54 counties in Inner Mongolia) was analyzed to explore temporal relationships between grassland deterioration and socioeconomic factors (Zhou et al., 2022). The work reported in 2022 (Zhou et al., 2022) was a pilot. This chapter provides a complete dataset including 33 variables in all 89 counties available now. The chapter conducts a comprehensive and full-blown experiment with detailed methodological discussions, result validations, and finding discussions. More importantly, this extensive experiment obtains several new insights, which will be discussed later in this chapter.

The causal relationships between grassland deterioration (using grassland productivity change as a proxy) and other socioeconomic and environmental factors are very complex (Fig. 4.1). The studies of grassland deterioration involve four interconnected perspectives: (1) the direct causes of grassland deterioration; (2) the global climate change, which usually intensifies grassland deterioration; (3) the socioeconomic driving effect and adaptation; and (4) the ecological/environmental response. Population explosion, overgrazing, excessive agricultural plantation, and urban construction are direct or tangible causes of grassland deterioration (Robinson et al., 2017). Another important cause of grassland deterioration is land use and land cover (LULC) change. LULC includes the area changes or expansions of irrigated

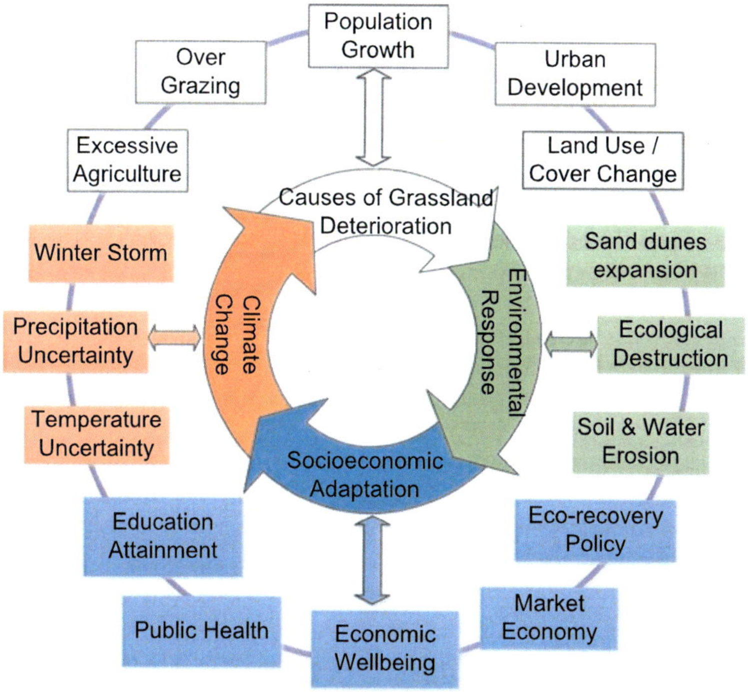

Fig. 4.1 The causal structure of grassland deterioration in IMAR

agriculture, commercial crops, urban build-ups, commercial and industrial establishments, urban constructions, and transportation networks. In the predominant grassland area of Inner Mongolia, LULC expansions indicate grassland conversion to other types of land uses (Hu & Nacun, 2018; Sha & Li, 2020).

The study of the climate change impact on grassland deterioration focuses on the uncertainty of climate change. A large section of the study area is in the semiarid and arid regions. Precipitation, temperature, and their variabilities are critical factors affecting grassland productivity (Hao et al., 2014). Moreover, the long and cold winter often causes snowstorms, posing significant risks to human and animal lives (Brown et al., 2013).

Rapid economic growth since the economic reforms in IMAR has fostered a greater desire for better living standards and financial well-being (Li et al., 2013). Increased income for farmers and herders, improved government revenues, better educational and public health facilities, raised purchasing power, and more available consumer goods and services have dramatically expanded demands for agricultural, animal husbandry, industrial, and mining productions (Shang et al., 2019). As a result, all these socioeconomic improvements have led to the accelerated conversion of grasslands to other land uses, provoking vicious cycles of grassland deterioration.

Grassland deterioration has induced many ecological and environmental responses, which, in turn, have been accelerating grassland deterioration (Deng et al., 2017). Grassland deterioration often accelerates negative plant community succession (Steffens et al., 2008), decreases plant quality and nutrients, degrades grassland grazing-bearing capacity, worsens soil and water erosion, and leads to sand dune development (Wu et al., 2015). Human adaptation strategies and intervention policies also affect the direction and strength of ecological and environmental responses to grassland deterioration. For example, several large-scale ecological compensations and recovery programs have been launched in IMAR since the 1990s (Chen et al., 2017). However, there are no credible enforcement mechanisms or too many abstractors prevent effective implementation of these policies (Srinivasan et al., 2012; Chen et al., 2017).

However, it has been a challenging research question regarding which factors are the primary causes of grassland deterioration, population growth, economic development, climate change, or LULC. In other words, the challenge is quantitatively examining the interactions and feedbacks between grassland deterioration and its various driving factors and responses from socioeconomic, ecological, environmental, and policy dimensions. Grassland ecosystems are one example of complex socioenvironmental systems (SESs). Numerous actors and processes exist in these complex SESs. They coexist and interact to form solid path dependencies or create causal structures to determine positive or negative trad-offs between them (Brandon & Lombardi, 2009; Timon et al., 2016). It is commonly recognized that statistical regression modeling based on empirical data is a practical approach to exploring causes and effects in complex socioenvironmental systems (Liu et al., 2018).

Regression techniques are one of the most popular statistical techniques used for causal relation exploration and predictive analysis based on the causal relation. Many regression analysis techniques exist, such as multivariate regression and logistic regression. However, the causal relationships and feedbacks between socioeconomic drivers and natural phenomena are much more complicated and have not been thoroughly examined in the current literature. As we explained in Chap. 3, many natural phenomena and processes show notable variations in time series (Propastin et al., 2008; Watsonet al., 2014). Conventional regression analysis cannot handle time-series causal relations. Several statistical methods have been developed to analyze time-series causal relationships. For instance, autocorrelation, autoregressive integrated moving average (ARIMA), and seasonal ARIMA (SARIMA) are standard statistical methods (Brown et al., 2013; Metcalfe & Cowpertwait, 2009; Stoffer, 2016; Xie et al., 2019).

However, human beings and biophysical systems interact over time and across space (Liu et al., 2007; Brown et al., 2013; Xie et al., 2019). The panel regression model (PRM) can simultaneously examine causal relationships across locations and over time. PRM is a standard analytical model in econometrics (Parent and LeSage, 2011). PRM has been increasingly applied in environmental research (Demeke et al., 2011; (Li et al., 2013) and geographical studies (Oud et al., 2012; Li & Xie, 2013; Fan, 2017). PRM can detect change trends and identify spatial variations hidden in complex cross-sections and time-series data. Thus, this chapter chooses conventional, dependent-lagged, and dynamic-lagged PRM models to analyze the causal relationships between grassland productivity, climate change, and socioeconomic transformation variables.

Moreover, PRM supports examining differentiated contributions to the causal explanation relationship of the response variable (grassland deterioration in our case study) just like other regression models by entering explanatory variables by blocks (Allison, 1999). In other words, we can enter the interconnected components of the grassland deterioration framework (Fig. 4.1) as blocks in PRM to analyze their impacts on grassland deterioration, respectively. It is commonly agreed that the block-based regression can differentiate specific contribution of a unique group of explanatory variables to the variance explanation of the response variable (Zhou et al., 2022). We integrate this block–input–variable approach with three PRM models to identify how each driving factor affects grassland deterioration.

4.2 Regional Grassland Deterioration Modeling Framework

The analytical framework for investigating time-series causality between grassland productivity and socioenvironmental driving factors based on the total socioenvironmental system approach is depicted in Fig. 4.2. This framework consists of six steps: (1) constructing a total socioenvironmental system (TSES) dataset; (2) conducting time-series data transformation or standardization through various data

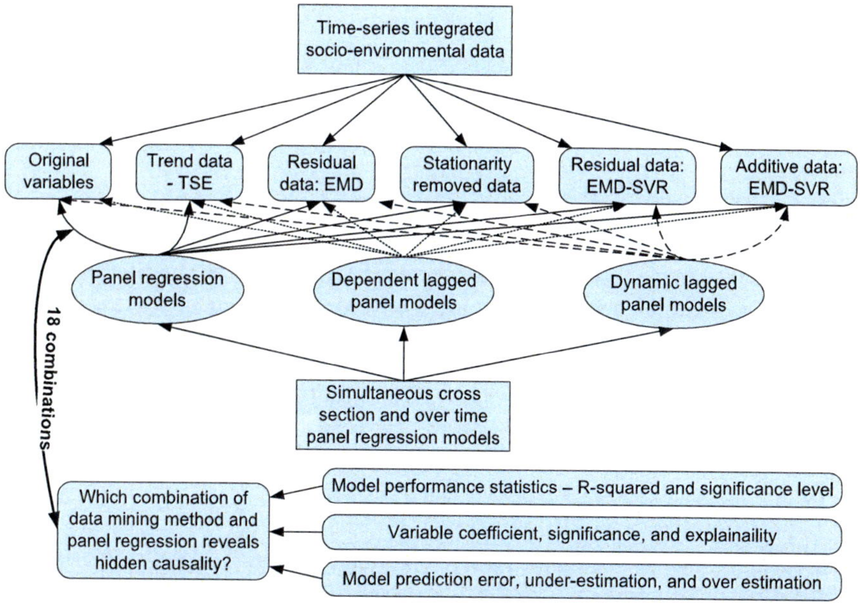

Fig. 4.2 Analytical framework for investigating causes of grassland deterioration through time-series data mining and panel regression modeling

mining techniques; (3) choosing suitable regression models; (4) fitting the chosen regression models with the transformed data; (5) assessing the model performance from the model performance statistics and the model coefficient contributions; and (6) evaluating the model estimation errors and the under- or overestimations of the model errors.

We have compiled a TSES dataset that consists of 32 variables such as grassland productivity (EVI as the proxy), climate (precipitation and temperature), 20 socio-economic factors, 8 land-use-land-cover (LULC) change variables, and one categorical grazing variable (Table 3.1). Several time-series data mining methods have been applied to transform the data into six TSES datasets (see details in Chap. 3):

1. the original observed variables (ORI),
2. the non-stationarity removed variables (STA),
3. the trends of change transformed through trend, seasonality, and error (TSE) time-series data mining,
4. the trends of change preprocessed as the residuals by empirical mode decomposition (EMD) algorithm,
5. the trends of change preprocessed as the residuals by hybrid EMD and support vector regression (SVR-R) method, and,
6. the additive intrinsic mode functions (IMFs) initially generated by EMD and then smoothed by SVR (SVR-A).

These datasets are processed in Chap. 3 and the processing details are also provided in Chap. 3. However, for the reader's convenience, the final transformed data in panel data format are provided in the folder,\Chapter_4\DataProcess.

The chapter applies conventional, dependent-lagged, and dynamic panel regression models to examine causal relationships between grassland deterioration, climate change, and socioeconomic drivers.

The chapter formulates the conventional panel regression model (Coven) as:

$$Y_{it} = \alpha + \beta_{it} X_{it} + \delta_i + \gamma_t + \varepsilon_{it} \tag{4.1}$$

where Y_{it} is the response variable EVI (enhanced vegetation index) for panel i (=1,…, N) at time t (=1,…, T); X_{it} is a vector of explanatory variables; β_{it} is the vector of slope coefficients of X_{it}; α is the intercept; δ_i represents the cross-sectional effect; γ_t is the overtime effect; and ε_{it} represents the errors that vary across panels and over times.

The chapter defines the dependent-lagged panel regression model (Lagged) in the following equation:

$$Y_{it} = \alpha + \lambda Y_{i(t-n)} + \beta_{it} X_{it} + \delta_i + \gamma_t + \varepsilon_{it} \tag{4.2}$$

The difference between Eq. (4.1) and Eq. (4.2) is the addition of $\lambda Y_{i(t-n)}$. $Y_{i(t-n)}$ is the nth lagged value of Y_{it} and correlated with the temporal-specific effect, γ_t. For instance, n is often taken a value 1 and $Y_{i(t-1)}$ is the one-time lagged value of Y_{it}.

The chapter expresses the dynamic panel regression model (Dynamic) as follows:

$$Y_{it} = \alpha + \lambda Y_{i(t-j)!} + \beta_{it} X_{it} + \delta_i + \gamma_t + \varepsilon_{it} \tag{4.3}$$

where $\lambda Y_{i(t-j)!}$ represents the dynamic-lagged effect of the response variable. This dynamic-lagged effect is not simply a one-time influence but a combined effect of all possible lagged differences (Arellano & Bond, 1991). The combined effect means that the response variable's dynamic-lagged effect combines the dependent variable's second lagged value with all the feasible lags after that. The combination creates a set of moments, which are called instruments in terms of STATA statistical software package (Pinzon, 2015). For instance, when a dynamic panel has 6 time periods, each panel will include 10 moments: $t = 6$, $Y_{i(t-4)}$, $Y_{i(t-3)}$, $Y_{i(t-2)}$, $Y_{i(t-1)}$; $t = 5$, $Y_{i(t-3)}$, $Y_{i(t-2)}$, $Y_{i(t-1)}$; $t = 4$, $Y_{i(t-2)}$, $Y_{i(t-1)}$; $t = 3$, $Y_{i(t-1)}$.

The block-entry method is also adopted to differentiate how each group of explanatory variables affects grassland deterioration. Taking Eq. (4.1) as an example, its block-entry form looks as follows:

$$Y_{it} = \alpha + \sum_{b,j}^{b=1,j=1} X_{itbj} \beta_{itbj} + \delta_i + \gamma_t + \varepsilon_{it} \tag{4.4}$$

The main difference between Eq. (4.1) and Eq. (4.4) is that the list of explanatory variables is broken into several blocks (or groups). These blocks of explanatory variables are entered into the PRM model in an order determined by the analysts. The 31 variables are grouped into 11 blocks (Table 4.1).

Six datasets and three models lead to 18 combinations of the fitted PRM models. To thoroughly discriminate the performances of the 18 models, we further analyze these various combinations of the datasets and the models from the following angles: (1) the entire PRM model performance statistics, such as R-squared (R-sq.) values and Wald Chi-squared (Chi-sq.) values, (2) the total predicted errors, (3) the regional variations of the predicted errors for the selected models, and (4) the physical and socioeconomic interpretations of the impacts of the explanatory variables on grassland productivity for the selected models.

Furthermore, we compute the model residuals to assess the model accuracy levels. The model residual is the difference between an observed value and a predicted value and hence is a measure of model accuracy (Pillonetto et al., 2011). The model predicted values in panel data analysis include two parts: the model fixed or random effect ($\alpha + \beta X_{it} + \mu_i$) and the prediction from the model coefficients. The chapter uses the statistic of root mean squared error (RMSE) to assess the accuracy degrees of the 18 models. In addition, the panel dataset includes 89 counties (banners) engaged in grazing or semi-grazing–semi-agricultural activities as the cross sections. It also covers 18 years and nine periods per year as the overtime. Therefore, two sets of RMSE will be computed, the total RMSE of the entire dataset as a measure of the total model accuracy and the RMSE values for each county over the 18 × 9 periods to disclose regional or geographical disparities of the model performance.

Table 4.1 Eleven blocks of 31 variables in panel regression models

Block ID	Block name	Variables by block	Remark
	Dependent	EVI	Response variable
1.	Population	DPop, prpop	Population growth
2.	Grazing	Lssk	Grazing density
3.	Agriculture	Paa, CropPG, gpsk	Arable land, grain production
4.	Land use and land cover	Forest, Grass, Sand, Shrub, Urban, Water, Wetland, Lohsk	Land changes
5.	Climate	PRE, TEM	Climate changes
6.	Socioeconomic	gdppc, gova, fai, lgr, TCRV	Economic development
7.	Education	MHT, PTE	Education endowment
8.	Health	HMP, HPB	Health development
9.	Income	PcUDI, iofp	Residents' income
10.	Industrial	PIO, SIO, TIO	Industrial development
11.	Grazing zone	A categorical variable: 1: Agricultural, 2: Mixed agriculture and grazing, and 3: Grazing	Grazing policy

The 18 models are implemented in the statistical software package, Stata. The Stata data files and DO (the script) files are stored in the folder, \Chapter_4\DataProcess. The analysis results are saved in the folder, \Chapter_4\DataProcess\Outcomes. Python codes are developed to compute RMSE values. The Python codes are provided in the folder, \Chapter_4\DataProcess, while the results are saved in the folder, \Chapter_4\DataProcess\Outcomes. The RMSE maps are generated with ArcGIS Desktop 10.8. The ArcGIS project file and the shapefiles are stored in the folder, \Chapter_4\DataProcess\RMSE_Models.

4.3 Outcomes of the 18 PRM Models

Running a PRM model requires two statistical tests: (1) the unit root test to confirm if the selected variables are time-series stationary; and (2) the Hausmann test to determine if a PRM model is a fixed-effect model or a random-effect model. The stationarity tests and their results for the case study datasets were discussed in Chap. 3.

The second test is the Hausmann test that assesses which type of estimation a PRM model should apply, a fixed-effect model or a random-effect model (Li et al., 2013). A fixed-effect model means either a cross-section-fixed model or an overtime-fixed mode. The former indicates that the model outcomes do not vary across sections, while the latter means that the model estimations do not change overtimes.

Table 4.2 The statistics of Hausman test

chi^2 =	190.05
Prob > chi^2	0.000

A random-effect model assumes that the model outcomes vary across sections or overtimes, or cross sections and overtimes. One standard method for testing whether a PRM model is a fixed-effect or a random-effect model is the Hausman test (Baltagi, 2005; Li et al., 2013). The null hypothesis of the Hausman test is that the differences in estimated coefficients from the PRM model are not systematic, which means the random effect. The alternative is that the differences of estimated coefficients from the PRM model are systematic, which indicates the fixed-effect model.

The IMAR grassland deterioration panel model was executed through the software Stata 16. The Hausman test got a relatively small chi^2 value for the PRM model with more than 30 independent variables (Table 4.2). The chi^2 statistic was also highly significant (Prob > chi^2 = 0.000). Therefore, we determined that the model estimation coefficients were not systematic and hence exhibited a random effect. In other words, the estimation method for Eq. (4.1) ought to be the random-effect approach.

4.3.1 The Performance Statistics of the 18 Models

The model performance statistics of the 18 models are reported in Table 4.3. To assess the PRM model performance, three sets of statistics, Wald Chi-sq. and its significance test, Prob > Chi-sq., and three R-sq. values were critical to examine. Wald Chi-sq. revealed whether the predictors were significantly explaining the dependent variable. The higher the Wald Chi-sq. value, the more significant the predictors were explaining the dependent variable. Prob > Chi-sq. confirmed if Wald Chi-sq was statistically confident at a 0.05 or 0.01 level. There were three R-sq. values. Within R-sq. indicated how much of the variation in EVI (the dependent variable) within counties (sections) was explained by the PRM model. Between R-sq. specified how much of the variation in EVI between counties was captured. Overall R-sq. was the weighted average of the two. The calculation of each one can be obtained from the user manual of the Stata software package (StataCorp, 2019). Overall R-sq. indicated an overall percentage of the variance of EVI was explained by the PRM model.

Based on these statistics, the six conventional models showed relatively poor performances. Among them, the TSE-fitted conventional panel model had the highest overall R-sq. The OROGIN and SVR-A fitted models had relatively higher overall R-sq.

Strikingly, the Lagged (dependent-lagged) models all achieved much higher overall R-sq. values as well as enormously high Wald Chi-sq. values. The overall R-sq. values of the EMD-R, SVR-R, and SVR-A fitted Lagged models were perfect

Table 4.3 The model performance statistics of the 18 models

	ORIGIN	TSE	STAT	EMD-R	SVR-R	SVR-A
Conventional panel models						
Within R-sq.	0.291	0.290	0.296	0.280	0.278	0.224
Between R-sq.	0.203	0.309	0.155	0.161	0.212	0.234
Overall R-sq.	0.239	0.306	0.212	0.163	0.212	0.233
Wald Chi-sq.	5508	5263	5595	5081	4346	3839
Prob > Chi-sq.	0.000	0.000	0.000	0.000	0.000	0.000
Dependent-lagged panel models						
Within R-sq.	0.729	0.950	0.732	0.999	0.999	0.999
Between R-sq.	0.987	1.000	0.983	1.000	1.000	1.000
Overall R-sq.	0.874	0.998	0.871	1.000	1.000	1.000
Wald Chi-sq.	93,792	5,090,000	91,151	57,700,000	61,700,000	72,100,000
Prob > Chi-sq.	0.000	0.000	0.000	0.000	0.000	0.000
Dynamic-lagged panel models						
Wald Chi-sq.	261,998	778,935	1207	2,710,000	1,210,000	429,799
Prob > Chi-sq.	0.000	0.000	0.000	0.000	0.000	0.000
Arellano-Bond Test-$z(1)$	−7.605	−5.733	n/a	1.138	0.572	2.421
Prob > $z(1)$	0.000	0.000	n/a	0.255	0.567	0.016
Arellano-Bond Test-$z(2)$	−1.280	−2.818	n/a	1.113	0.549	2.412
Prob > $z(2)$	0.201	0.000	n/a	0.265	0.583	0.016

1.0. Even the TSE-fitted model had an overall R-sq. value, 0.998. The remaining two overall R-sq. values were above 0.87. Therefore, the Lagged models displayed very high performances in general.

The dynamic panel model is a unique statistical analysis. It does not simply add a lagged dependent variable. Still, it adds a large set of subsequently lagged instruments within the entire time series to reveal their impacts on the dependent variable. The dynamic model does not compute R-sq. statistic. Instead, it provides two critical statistics, Wald Chi-sq. and Arellano-Bond Test-z, and their associated significance tests. The Wald Chi-sq. test determines if the predictors are significantly explaining the dependent variable. The higher it is, the bigger the predictors contribute to the dependent variable. The Arellano-Bond Test is used to test no-autocorrelation in first-differenced errors (Arellano & Bond, 1991). The Arellano-Bond Test assumption (or the null hypothesis) is that not all error terms are autocorrelated. In this case, either the first-step test (Arellano-Bond Test-$z(1)$) or the second-step test (Arellano-Bond Test-$z(2)$) was not rejected because the probability test (Prob > $z(1)$ or Prob > $z(2)$) failed at the 0.05% significance level. As Table 4.3 reported, the six dynamic models, except the STAT-fitted model, performed well. Especially, the EMD, SVR-R, and SVR-A fitted models had very high and significant Wald Chi-sq. Moreover, the STAT-fitted model failed the Arellano-Bond Test, because their variance–covariance matrices of the two-step estimators were not fully filled (in other words, the two-step estimators were not available). As a result, the STAT-fitted dynamic Model was not reliable. The remaining five dynamic models pass the Arellano-Bond Test, indicating that the models meet the model assumption.

4.3.2 *The Predicted Errors of the 18 Models as a Whole*

We computed the root mean square errors (RMSE) between the initial and the predicted values derived from the three panel models and six datasets to assess the prediction errors (Table 4.4). We also listed the overall R-sq. values for the conventional and lagged panel models, or Wald Chi-sq, values for the dynamic models. We examined two statistics, RMSE and R-sq. or Wald Chi-sq., for the 18 models and required that both statistics had to be high. As a result, we found that four models performed much better than the other 14 models: (1) the TSE-Lagged Model, (2) the EMD-Lagged Model, (3) the SVR-A-Lagged Model, and (4) the EMD-Dynamic Model. We did not select the TSE-Dynamic Model because its Wald Chi-sq. statistic was not very high although it had a lower RMSE. Furthermore, for comparison, we included the ORIGIN-Conventional Model for further analysis. This model was the original data-fitted conventional panel model, indicating that no data transformation and no lagged effect were considered. This model was the baseline model for comparison.

4.3.3 *The Regional Variations of the Predicted Errors for the Four Selected Models*

In the previous section, five models – the TSE-Lagged Model, the EMD-Lagged Model, the SVR-A-Lagged Model, the EMD-Dynamic Model, and the ORIGIN-Conventional Model – were selected for further analysis. There were two purposes for in-depth analyses: (1) how the model performances varied over space (i.e., the spatial disparities of the model performances) and (2) what the model performances meant to physical and socioeconomic implications of causal relationships between grassland deterioration and environmental and socioeconomic driving factors. This section will answer the first question, while the next section will address the second inquiry.

Table 4.4 The overall predicted errors of the 18 models

	Conventional panel		Lagged panel		Dynamic panel	
Dataset	RMSE[a]	R_Sq.[b]	RMSE	R_Sq.	RMSE	Wald Chi-sq.
ORGIN	**6.710**[c]	**0.239**	4.184	0.874	5.436	261,998
TSE	1.660	0.306	**2.019**	**0.998**	2.110	778,935
STAT	6.170	0.212	4.230	0.871	6.179	1207
EMD-R	2.581	0.163	**3.163**	**1.000**	**3.180**	**2,710,000**
SVR-R	10.663	0.212	10.797	1.000	10.820	1,210,000
SVR-A	2.527	0.233	**3.162**	**1.000**	3.172	429,799

[a]RMSE: root mean squared errors
[b]R_Sq.: R-squared values
[c]Bold font: indicating the selected models based on RMSE and R_Sq

We calculated the RMSE for each county over the 18 × 9 periods for the five selected models. We then investigated the RMSE values of the five models and designed thematic legends by taking the Jenkin breaks of the RMSEs of the five models. This method of legend classification was able to provide a consistent assessment of the regional patterns of RMSEs among the five models. We mapped the RMSE values of the five models in Fig. 4.3.

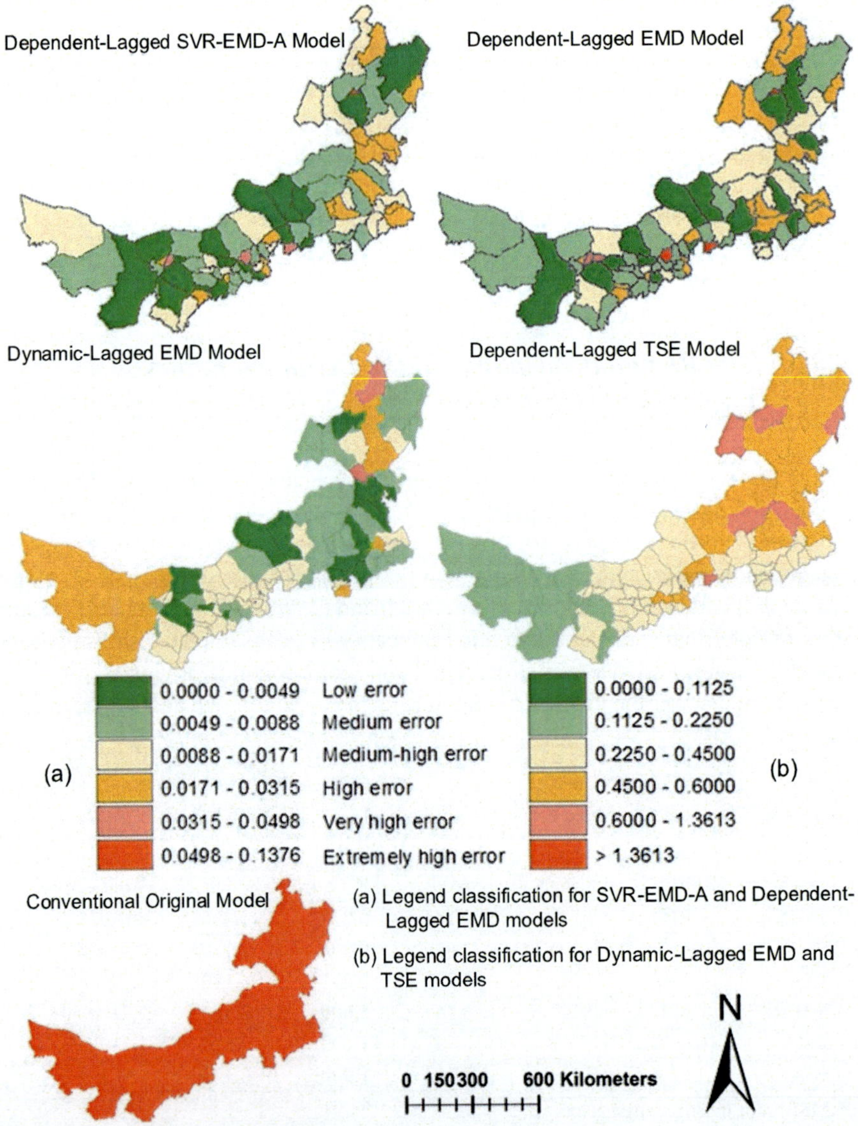

Fig. 4.3 Regional RMSE disparities between five selected models

The highest RMSE values were 0.138 for the Dependent-Lagged SVR-A Model and the Dependent-Lagged EMD Model, respectively, 0.71 for the Dependent-Lagged TSE Model, and 1.361 for the Dynamic-Lagged EMD Model. However, the RMSE values of the ORIGIN-Conventional Model ranged between 3.024 and 11.535. If a unified legend classification method was applied, the maps of the SVR-A-Lagged Model and Dependent-Lagged EMD Model were all dark green, indicating the lowest RMSE values. In contrast, the Conventional Original Model displayed an overall red color, meaning extremely high RMSE values. To distinguish spatial patterns of the RMSE changes, two different legends with varied value ranges were designed for comparing the RMSE spatial variations between the five models. Legend A depicted the RMSE variations for the Dependent-Lagged SVR-A Model and Dependent-Lagged EMD Model. Legend B showed the spatial patterns of the Dynamic-Lagged EMD, Dependent-Lagged TSE, and Conventional Original models (Fig. 4.3).

The maps of the model prediction errors displayed contrasting spatial patterns among these models. The study area is mainly covered by various grasslands, which are affected by precipitation and temperature gradients. The precipitation decreases from the semi-humid zone in the east and northeast to the semiarid in the middle and to the arid in the west. The temperature gradient is reversed, colder in the east and warmer in the west. As a result, the plant communities are meadow steppe and forests in the east and northeast, typical steppe in the near east and the middle, and desert steppe and deserts in the west. The grassland productivity decreases from the east to the west.

The Dynamic-Lagged EMD Model displayed a general spatial pattern of RMSE values – lower in the east, medium in the middle, and higher in the west. In other words, the RMSE values were lower in the high yield meadow steppe and typical steppe, medium in the less fertile desert steppe, and higher in the low yield deserts. There were some exceptions. For example, the Hulun-Buir grasslands in the northeast showed high RMSE values because of overgrazing (Zhou et al., 2022). The Da-Xian-Ling mountainous areas in the northeast displayed high RMSEs because of excessive lumping and deforestation (Chang et al., 2007). There were several counties in the middle west, showing some lower RMSE values. The Yellow River flows through these areas and has an irrigated fertile agricultural plantation, which offers high EVI values because of high-yield agrarian plants. In brief, the spatial patterns of the Dynamic EMD Model were consistent with the underlying plant communities. The higher the plant productivities, the lower the RMSE values. This consistency could indicate that the Dynamic EMD Model revealed the correct causal relationships.

However, the Dependent-Lagged TSE Model displayed a divergent RMSE spatial pattern. Although the spatial pattern was spatially well structured or configurated, the RMSE values were not aligning with the ground grasslands. This inconsistency between the RMSE and the underlying vegetation is questionable. Furthermore, the RMSE spatial pattern disclosed by the Dependent-Lagged EMD Model was also unpredictable. Although the total RMSE of the Dependent-Lagged EMD Model was lower than that of the TSE model, the RMSE displayed an irregular spatial pattern. The lower RMSE appeared in the western and central sections, but the higher RMSE interspersed with lower RMSE patches. The RMSE of the Dependent-Lagged SVR-A EMD Model revealed a unique pattern, which showed an obvious local interference to the general east-toward-west decline tendency. The following section will provide a more thorough analysis and discussion.

4.3.4 The Block-Based EMD-Logged Model

This chapter adopts a block-based panel data modeling approach. The EMD-Lagged Model was a good illustration among the five selected models – the TSE-Lagged Model, the EMD-Lagged Model, the SVR-A-Lagged Model, the EMD-Dynamic Model, and the ORIGIN-Conventional Model. The panel modeling was implemented in the software, Stata 16 (StataCorp, 2019). We discussed possible blocks that affected the health of the grasslands in Fig. 4.1 and Table 4.1. The realization in the Stata coding language is provided as Fig. 4.4. The outcomes of the block-entered EMD-Lagged Model are reported in Table 4.5.

```
xtset CID TimeStamp
xtreg EVIemd DPop prpop, re (# Population Growth*)
xtreg EVIemd DPop prpop lssk, re (# adding Grazing)
xtreg EVIemd DPop prpop lssk Paa CropPG gpsk, re (# adding
Agriculture)
xtreg EVIemd DPop prpop lssk Paa CropPG gpsk Grass Urban lohsk, re
(# adding Urban Development and Land-use Changes)
xtreg EVIemd DPop prpop lssk Paa CropPG gpsk Grass Urban lohsk
PREemd TEMemd, re (# adding Climate Change)
xtreg EVIemd DPop prpop lssk Paa CropPG gpsk Grass Urban lohsk
PREemd TEMemd gdppc gova fai lgr TCRV, re (# adding Economic
Wellbeing)
xtreg EVIemd DPop prpop lssk Paa CropPG gpsk Grass Urban lohsk
PREemd TEMemd gdppc gova fai lgr TCRV MHT PTE, re (# adding
Education Attachments)
xtreg EVIemd DPop prpop lssk Paa CropPG gpsk Grass Urban lohsk
PREemd TEMemd gdppc gova fai lgr TCRV MHT PTE HMP HPB, re (# adding
Public Health)
xtreg EVIemd DPop prpop lssk Paa CropPG gpsk Grass Urban lohsk
PREemd TEMemd gdppc gova fai lgr TCRV MHT PTE HMP HPB PcUDI iofp,
re (# adding Market Economy)
xtreg EVIemd DPop prpop lssk Paa CropPG gpsk Grass Urban lohsk
PREemd TEMemd gdppc gova fai lgr TCRV MHT PTE HMP HPB PcUDI iofp
PIO SIO TIO, re (# adding Industrial Development)
```

xtset CID TimeStamp
xtreg EVIemd DPop prpop, re (# Population Growth*)
xtreg EVIemd DPop prpop lssk, re (# adding Grazing)
xtreg EVIemd DPop prpop lssk Paa CropPG gpsk, re (# adding Agriculture)
xtreg EVIemd DPop prpop lssk Paa CropPG gpsk Grass Urban lohsk, re (# adding Urban Development and Land-use Changes)
xtreg EVIemd DPop prpop lssk Paa CropPG gpsk Grass Urban lohsk PREemd TEMemd, re (# adding Climate Change)
xtreg EVIemd DPop prpop lssk Paa CropPG gpsk Grass Urban lohsk PREemd TEMemd gdppc gova fai lgr TCRV, re (# adding Economic Wellbeing)
xtreg EVIemd DPop prpop lssk Paa CropPG gpsk Grass Urban lohsk PREemd TEMemd gdppc gova fai lgr TCRV MHT PTE, re (# adding Education Attachments)
xtreg EVIemd DPop prpop lssk Paa CropPG gpsk Grass Urban lohsk PREemd TEMemd gdppc gova fai lgr TCRV MHT PTE HMP HPB, re (# adding Public Health)
xtreg EVIemd DPop prpop lssk Paa CropPG gpsk Grass Urban lohsk PREemd TEMemd gdppc gova fai lgr TCRV MHT PTE HMP HPB PcUDI iofp, re (# adding Market Economy)
xtreg EVIemd DPop prpop lssk Paa CropPG gpsk Grass Urban lohsk PREemd TEMemd gdppc gova fai lgr TCRV MHT PTE HMP HPB PcUDI iofp PIO SIO TIO, re (# adding Industrial Development)
xtreg EVIemd DPop prpop lssk Paa CropPG gpsk Grass Urban lohsk PREemd TEMemd gdppc gova fai lgr TCRV MHT PTE HMP HPB PcUDI iofp PIO SIO TIO i.Graze, re (# adding Grazing Policy)
xtreg EVIemd DPop prpop lssk Paa CropPG gpsk Grass Urban lohsk PREemd TEMemd gdppc gova fai lgr TCRV MHT PTE HMP HPB PcUDI iofp PIO SIO TIO i.Graze L.EVIemd, re (# adding the EVI Lagged Effect)
* *Referring Figure 4.1 for the blocks entered in the Stata Implementation*

Fig. 4.4 The Stata code for implementing the block-entered EMD-Lagged panel model

```
xtreg EVIemd DPop prpop lssk Paa CropPG gpsk Grass Urban lohsk
PREemd TEMemd gdppc gova fai lgr TCRV MHT PTE HMP HPB PcUDI iofp
PIO SIO TIO i.Graze, re (# adding Grazing Policy)
xtreg EVIemd DPop prpop lssk Paa CropPG gpsk Grass Urban lohsk
PREemd TEMemd gdppc gova fai lgr TCRV MHT PTE HMP HPB PcUDI iofp
PIO SIO TIO i.Graze L.EVIemd, re (# adding the EVI Lagged Effect)
*Referring Figure 4.1 for the blocks entered in the Stata
Implementation
```

Table 4.5 points out that lagged effects of previous grassland conditions (L.EVIemd) had the most significant impact on current grassland health. However, the effect of previous EVI values on current values was treated as the legacy or lagged effect but not an external driving factor. Here, we focused on the driving factors of grasslands deterioration.

Table 4.5 The model performance statistics of the block-entered EMD-Lagged Model

	B-1[a]	B-2	B-3	B-4	B-5	B-6
Within R-sq.	0.0036	0.0077	0.0825	0.1022	0.1836	0.2316
Between R-sq.	0.0175	0.0206	0.1774	0.0023	0.2151	0.1667
Overall R-sq.	0.0168	0.0193	0.1646	0.0030	0.2032	0.1529
Within R-sq. change		0.0041	0.0748	0.0197	0.0814	0.0480
Wald Chi-sq.	49	106	1195	1499	2967	3946
Prob > Chi-sq.	0.0000	0.0000	0.0000	0.0000	0.0000	0.0000
Wald Chi-sq. change		57	1089	303	1469	978
	B-7	**B-8**	**B-9**	**B-10**	**B-11**	**L.EVIemd**
Within R-sq.	0.2586	0.2600	0.2676	0.2796	0.2797	0.9992
Between R-sq.	0.1080	0.1244	0.1037	0.0799	0.1613	1.0000
Overall R-sq.[2]	0.1048	0.1174	0.0992	0.0769	0.1628	1.0000
Within R-sq. change	0.0270	0.0014	0.0076	0.0120	0.0001	0.7195
Wald Chi-sq.	4555	4568	4744	5029	5081	57,700,000
Prob > Chi-sq.	0.0000	0.0000	0.0000	0.0000	0.0000	0.0000
Wald Chi-sq. change	610	13	175	285	53	57,694,919

[a]The variables in each block are listed in Table 4.1

The population block was entered first. Its Within R-sq. and Wald Chi-sq. statistics were significant (Prob > Chi-sq. = 0.000) and strong. Ten more blocks were entered afterward. There were several interesting findings. First, the climate block was the strongest driving factor of EVI change based on the absolute changes of the Within R-sq. and Wald Chi-sq. statistics. Second, the agriculture block was the second strongest driving factor of EVI variation. Third, the socioeconomic block was the third strongest driving factor. The education block was ranked fourth and the land-use change block was fifth. The blocks of industrial development, residential income, grazing pressure, grazing zone, and public health followed in declining order. Fourth, the climate block was the strongest but was far surpassed jointly by the agricultural and socioeconomic blocks. In addition, three blocks of education, land-use change, and industrial development had relatively high values of Wald Chi-sq. change. This finding suggests a nuance about the effects of climate change on grassland health. It is a common notion that climate is the biggest driving factor of grassland degradation (Guo et al., 2021; Li & Xie, 2013; Su et al., 2020). The block-entered model revealed that there were so many factors affecting the health of the grasslands. Their combined effects far exceeded the climate change influence. Furthermore, the agricultural and socioeconomic blocks were strong enough to balance the effects of climate change.

Agriculture had complex relationships with EVI. The agricultural block included three variables, *Paa*, *CropPG*, and *gpsk*. The first two variables positively affected EVI, but the last one negatively impacted EVI. *Paa* is the percentage of arable land over the total land area, while *CropPG* is the percentage of crop plantation area over the grasslands area. The variable of *gpsk* indicated grain production per square kilometers. It is critical to point out that *Paa*, *CropPG*, and *gpsk* measure different aspects of agricultural activities. *Paa* and *CropPG* reflect the concentration degrees of the arable land and crop planation. However, *gpsk* is a measurement of agriculture yield by land area. From the physics of remote sensing, higher *Paa* and *CropPG* indicate a large area covered by agriculture plants, which should correspond to higher greenness or higher *EVI* values on the remotely sensed images. The physical characteristics explained their positive relationships. The following section will provide more discussions on the effects of *gpsk* on EVI in comparison with four other models.

One issue should be pointed out is whether the order of entering the blocks into the EMD-Lagged panel model would affect the model outcomes. The purpose of entering the independent variables by blocks is to differentiate how a specific set of variables affects the dependent variable. Based on the studies in the same area, no significant difference was identified when the order of the blocks of independent variables was entered differently (Zhou et al., 2022).

4.4 The Physical and Socioeconomic Interpretations of the Selected Models

4.4.1 The ORIGIN-Fitted Conventional Model

The ORIGIN-fitted Conventional Model had the worst performance in terms of the R-sq. statistic, the overall RMSE, and the regional (county-based) RMSE. However, this model was the baseline from the panel regression model and the original dataset. It is worth taking an examination. Table 4.6 reported the coefficients of this model. Table 4.6 revealed four findings. First, two climate factors, TEMorg and PREorg, were the most vital driving forces of grassland productivity. The coefficient *z* values reached 54.6 and 32.2, respectively, and were 15–25 times higher than the other coefficients except for *gpsk*. Second, four socioeconomic variables, *gpsk*, *gdppc*, *fai*, and *lssk*, had positive and significant impacts on EVI. The *z* value of *gpsk* was almost 9.0, and the other three coefficients were less than 3.0. The statistic judged the term of significance, $P > z < 0.05$. Third, five factors showed substantial but negative impacts on EVI, *Paa*, *iofp*, *Grass*, *lohsk*, *PTE*, and *TIO*. Fourth, GRAZE, (1 agricultural, 2 mixed agriculture and grazing, and 3 grazing) showed significant negative impacts. Compared with the agricultural production, whole grazing and mixed grazing negatively affected EVI. Fifth, 11 variables did not significantly impact EVI.

4.4.2 The EMD-Lagged Model

The EMD-Lagged Model got dramatically different coefficients from the ORIGIN-Conventional Model (Table 4.7). The most significant difference was the addition of a new variable, the dependent-lagged variable, L1.EVIemd. In the terminology of

Table 4.6 Coefficients of the ORIGIN-Conventional Model

EVI	Coef.	Std. Err.	Z	P > z
DPop	−0.032	0.137	−0.230	0.818
prpop	0.011	0.006	1.790	0.073
lssk	**0.029**	**0.013**	**2.270**	**0.023**
Paa	*−0.234*	*0.035*	*−6.780*	*0.000*
CropPG	−0.001	0.000	−1.330	0.184
gpsk	**0.023**	**0.003**	**8.960**	**0.000**
Grass	*−0.022*	*0.008*	*−2.830*	*0.005*
Urban	−0.029	0.019	−1.500	0.134
lohsk	*−0.860*	*0.336*	*−2.560*	*0.011*
PREorg	**0.781**	**0.024**	**32.160**	**0.000**
TEMorg	**8.676**	**0.159**	**54.630**	**0.000**
gdppc	**0.004**	**0.002**	**2.650**	**0.008**
Gova	0.000	0.004	−0.030	0.979
fai	**0.005**	**0.002**	**2.390**	**0.017**
lgr	0.019	0.034	0.570	0.567
TCRV	0.012	0.011	1.070	0.283
MHT	0.011	0.008	1.290	0.198
PTE	*−0.002*	*0.001*	*−2.420*	*0.015*
HMP	−0.009	0.006	−1.630	0.104
HPB	0.011	0.006	1.850	0.064
PcUDI	0.000	0.000	−1.200	0.230
iofp	*−0.189*	*0.063*	*−2.990*	*0.003*
PIO	0.040	0.010	3.770	0.000
SIO	−0.001	0.001	−0.740	0.462
TIO	*−0.018*	*0.009*	*−2.010*	*0.044*
Graze				
2	*−2.350*	*1.128*	*−2.080*	*0.037*
3	*−8.368*	*1.006*	*−8.320*	*0.000*
_cons	10.882	1.078	10.090	0.000

Stata software, L1 indicates one-time lagged and hence L1.EVIemd was a new variable by taking the one-time lag of the original dependent variable, EVIemd. The most important finding was that L1.EVIemd exhibited an enormous impact on the original dependent variable, EVIemd. Its z value was 6108.35, 300–600 times higher than the other z values. The second difference was that almost all socioeconomic and land-use variables had significant impacts on EVIemd, except for three variables, *fai*, *SIO*, and *TIO*. In addition, many coefficients and their associate z values were substantially high. The third difference was that influence magnitudes of the climate factors, PREemd and TEMemd, dramatically diminished although they were still significant. Moreover, the impact of PREemd on EVIemd was turned negative.

Table 4.7 Coefficients of the EMD-Lagged Model

EVIemd	Coef.	Std. Err.	z	P > z
L1.EVIemd	**1.00233**	**0.00016**	**6108.35**	**0.00000**
DPop	*−0.01005*	*0.00059*	*−17.14000*	*0.00000*
prpop	*−0.00008*	*0.00003*	*−2.33000*	*0.02000*
lssk	*−0.00019*	*0.00006*	*−2.88000*	*0.00400*
Paa	**0.00170**	**0.00023**	**7.30000**	**0.00000**
CropPG	**0.00002**	**0.00000**	**6.71000**	**0.00000**
gpsk	*−0.00032*	*0.00001*	*−23.50000*	*0.00000*
Grass	**0.00041**	**0.00004**	**10.40000**	**0.00000**
Urban	**0.00128**	**0.00009**	**13.84000**	**0.00000**
lohsk	**0.03330**	**0.00171**	**19.51000**	**0.00000**
PREemd	*−0.00416*	*0.00077*	*−5.41000*	*0.00000*
TEMemd	**0.02736**	**0.00652**	**4.20000**	**0.00000**
gdppc	*−0.00007*	*0.00001*	*−8.59000*	*0.00000*
gova	*−0.00063*	*0.00006*	*−10.03000*	*0.00000*
fai	0.00001	0.00001	1.16000	0.24700
lgr	**0.00055**	**0.00020**	**2.71000**	**0.00700**
TCRV	−0.00030	0.00005	−6.48000	0.00000
MHT	**0.00012**	**0.00005**	**2.46000**	**0.01400**
PTE	**0.00002**	**0.00000**	**3.98000**	**0.00000**
HMP	*−0.00021*	*0.00003*	*−7.72000*	*0.00000*
HPB	*−0.00005*	*0.00003*	*−1.81000*	*0.07000*
PcUDI	*−0.00000*	*0.00000*	*−9.04000*	*0.00000*
iofp	*−0.00186*	*0.00034*	*−5.46000*	*0.00000*
PIO	**0.00047**	**0.00009**	**5.11000**	**0.00000**
SIO	−0.00001	0.00001	−0.89000	0.37600
TIO	0.00007	0.00005	1.35000	0.17800
Graze				
2.00000	0.00679	0.00353	1.92000	0.05500
3.00000	0.00470	0.00352	1.34000	**0.18100**
_cons	−0.09101	0.01391	−6.54000	0.00000

4.4.3 The TSE-Lagged Model

The coefficients of the TSE-Lagged Model were reported in Table 4.8. The lagged model generated a one-time lagged new variable, L1.EVItse, which was the one-time lag of the original dependent variable, EVItse. L1.EVItse exhibited a massive impact on the original dependent variable, EVItse. Its z value was 1302, 40–400 times higher than the other z values. Although the lagged EVItse (L1.EVItse) coefficient was not as enormous as L1.EVIemd, this overweighted L1.EVItse made other variables in the TSE-Lagged Model less significant and became trivial. For example, many of the Coef. values were minimal. We had to apply the four decimals to keep them from 0.000. The second finding was that two climate factors, PREtse

Table 4.8 Coefficients of the TSE-Lagged panel model

EVItse	Coef.	Std. Err.	z	P > z
L1.EVItse	**0.9864**	**0.0008**	**1301.5900**	**0.0000**
DPop	*−0.0082*	*0.0034*	*−2.4400*	*0.0150*
prpop	*−0.0020*	*0.0003*	*−6.9600*	*0.0000*
lssk	0.0010	0.0009	1.1000	0.2700
Paa	0.0024	0.0016	1.5600	0.1190
CropPG	0.0000	0.0000	−0.8500	0.3930
gpsk	*−0.0003*	*0.0001*	*−2.7700*	*0.0060*
Grass	0.0002	0.0002	1.3200	0.1880
Urban	−0.0019	0.0011	−1.6800	0.0930
lohsk	**0.0607**	**0.0211**	**2.8800**	**0.0040**
PREtse	**0.1783**	**0.0053**	**33.9500**	**0.0000**
TEMtse	**0.2609**	**0.0345**	**7.5700**	**0.0000**
gdppc	*−0.0004*	*0.0001*	*−3.9700*	*0.0000*
gova	0.0002	0.0003	0.6600	0.5080
fai	−0.0003	0.0001	−2.8400	0.0040
lgr	0.0005	0.0021	0.2600	0.7930
TCRV	−0.0008	0.0005	−1.7500	0.0810
MHT	−0.0006	0.0004	−1.4000	0.1610
PTE	*−0.0001*	*0.0000*	*−2.3300*	*0.0200*
HMP	0.0000	0.0003	−0.1100	0.9150
HPB	−0.0001	0.0003	−0.3600	0.7190
PcUDI	0.0000	0.0000	−1.1200	0.2610
iofp	**0.0079**	**0.0029**	**2.6800**	**0.0070**
PIO	**0.0025**	**0.0006**	**3.9100**	**0.0000**
SIO	*−0.0003*	*0.0001*	*−2.8400*	*0.0050*
TIO	−0.0004	0.0005	−0.8100	0.4170
Graze				
2	−0.0627	0.0129	−4.8800	0.0000
3	−0.0377	0.0152	−2.4800	0.0130
_cons	−0.4740	0.0727	−6.5200	0.0000

and TEMtse, were still critically affecting EVI. Third, the strength and direction of the causal relations between other socioeconomic and land-use variables and EVI differed from those reflected in the ORIGIN-Conventional and the EMD-Lagged models. We will now compare and analyze their differences after introducing the other two models.

4.4.4 The SVR-A-Lagged Model

The SVR-A-Lagged Model illustrated some commonalities with the EMD-Lagged Model (Table 4.9). First, the SVR-A-Lagged Model had a common characteristic of the Dependent-Lagged Model, a new addition of a one-time lagged variable,

Table 4.9 Coefficients of the SVR-A-Lagged Model

EVIsvr_a	Coef.	Std. Err.	Z	P > z
L1.EVIsvr_a	**1.00169**	**0.00015**	**6535.81000**	**0.00000**
DPop	**0.00143**	**0.00049**	**2.92000**	**0.00400**
prpop	**0.00012**	**0.00003**	**4.25000**	**0.00000**
lssk	*−0.00020*	*0.00008*	*−2.34000*	*0.01900*
Paa	*−0.00393*	*0.00016*	*−23.88000*	*0.00000*
CropPG	0.00000	0.00000	0.75000	0.45300
gpsk	*−0.00015*	*0.00001*	*−9.86000*	*0.00000*
Grass	*−0.00010*	*0.00003*	*−3.02000*	*0.00300*
Urban	**0.00059**	**0.00009**	**6.93000**	**0.00000**
lohsk	**0.00896**	**0.00283**	**3.16000**	**0.00200**
PREsvr_a	*−0.00710*	*0.00070*	*−10.11000*	*0.00000*
TEMsvr_a	**0.02208**	**0.00587**	**3.76000**	**0.00000**
gdppc	*−0.00004*	*0.00001*	*−4.87000*	*0.00000*
gova	0.00004	0.00005	0.89000	0.37500
fai	0.00001	0.00001	0.66000	0.50800
lgr	*−0.00058*	*0.00020*	*−2.89000*	*0.00400*
TCRV	*−0.00032*	*0.00005*	*−5.82000*	*0.00000*
MHT	0.00000	0.00004	0.03000	0.97500
PTE	0.00000	0.00000	−0.95000	0.34000
HMP	−0.00004	0.00003	−1.52000	0.12900
HPB	0.00002	0.00003	0.77000	0.44000
PcUDI	**0.00000**	**0.00000**	**4.15000**	**0.00000**
iofp	*−0.00275*	*0.00034*	*−8.00000*	*0.00000*
PIO	−0.00003	0.00007	−0.47000	0.63700
SIO	0.00001	0.00001	0.79000	0.42800
TIO	*−0.00025*	*0.00005*	*−5.61000*	*0.00000*
Graze				
2	−0.00075	0.00303	−0.25000	0.80400
3	−0.02783	0.00305	−9.12000	0.00000
_cons	0.00925	0.01256	0.74000	0.46100

L1.EVIsvr_a. This lagged variable's coefficient and associated z statistic were significantly higher, reflecting its dominant impact on EVIsvr_a. Second, the magnitudes and directionalities of two climate factors, PREsvr_a and TEMsvr_a, were analogous to PREemd and TEMemd in the EMD-Lagged Model. Third, many socioeconomic and land-use variables had sizable coefficients and associated z values. However, there were some differences between the SVR-A-Lagged and the EMD-Lagged models. For example, ten variables in the SVR-A-Lagged Model did not significantly impact EVIsvr_a (while the EMD-Lagged Model only had three nonsignificant variables). Furthermore, several variables showed different magnitudes and directionality in their impacts on EVI between the two models. We will carry out more discussions in the following sections.

4.4.5 The EMD-Dynamic Model

The Dynamic-Lagged EMD Model is the last selected one. This model contained a set of sequentially lagged instruments, called dynamically lagged, within the time series to model the lagged effects of the dependent variable. DL.EVIemd still had the most significant coefficient and *z* statistic (Table 4.10). This model came to several surprising findings. The first one was that the climate factors were no longer significant influencers of grassland productivity. The second one was that only two socioeconomic variables (*DPop* and *gpsk*) negatively affected EVIemd, and all other socioeconomic variables became nonsignificant factors to EVI. The third one was that all three land uses positively and significantly impacted EVIemd.

Table 4.10 Coefficients of the EMD-Dynamic Model

EVIemd	Coef.	Std. Err.	Z	P > z
DL.EVIemd	**0.9925**	**0.0048**	**207.47**	**0.0000**
DPop	*−0.0886*	*0.0237*	*−3.7400*	0.0000
prpop	−0.0002	0.0014	−0.1200	0.9060
lssk	0.0024	0.0067	0.3500	0.7240
Paa	0.0073	0.0103	0.7100	0.4780
CropPG	0.0005	0.0003	1.7300	0.0840
gpsk	*−0.0017*	*0.0007*	*−2.5300*	*0.0110*
Grass	**0.0095**	**0.0028**	**3.3500**	**0.0010**
Urban	**0.0068**	**0.0035**	**1.9400**	**0.0530**
lohsk	**0.1906**	**0.0603**	**3.1600**	**0.0020**
PREemd	−0.0236	0.0179	−1.3200	0.1880
TEMemd	0.0044	0.3465	0.0100	0.9900
gdppc	0.0001	0.0002	0.3600	0.7210
gova	−0.0021	0.0021	−1.0100	0.3110
fai	0.0008	0.0009	−0.8800	0.3810
lgr	−0.0103	0.0078	−1.3300	0.1840
TCRV	−0.0014	0.0019	−0.7600	0.4490
MHT	−0.0019	0.0041	−0.4500	0.6510
PTE	0.0000	0.0002	0.0300	0.9740
HMP	0.0003	0.0009	0.3200	0.7460
HPB	−0.0002	0.0027	−0.0700	0.9430
PcUDI	0.0000	0.0000	−0.4600	0.6430
iofp	−0.0112	0.0108	−1.1000	0.2700
PIO	0.0034	0.0018	1.8600	0.0630
SIO	0.0001	0.0002	0.6600	0.5080
TIO	0.0029	0.0023	1.2400	0.2130
_cons	−0.2882	0.7352	−0.3900	0.6950

4.5 Discussions of the Physical and Socioeconomic Interpretations of the Selected Models

It is important to point out that the strength (the coefficient magnitudes and associated z values) and direction (the sign) of the causal relations between the explanatory variables and EVI varied between the different combinations of data transformations and panel models. Consequently, these explanatory variables may have different physical and socioeconomic meanings to EVI. We need to understand how different combinations of data transformation and panel models affected these variables' physical and socioeconomic interpretations. We found the following commonalities and discords by delving into the differences between the five selected models.

The ORIGIN-Conventional Model had the lowest R-sq. value but the highest overall and regional RMSE values. However, this model signified the strong and positive influences of the climate factors (Precipitation – PRE and Temperature – TEM) on EVI compared with the other four models. This finding is consistent with the long-held common notion that climate factors are the most significant influencers on grassland productivity in IMAR (Guo et al., 2021; Su et al., 2020). However, the EMD-Lagged Model and the SVR-A-Lagged Model challenged this notion. As we explained, the SVR-A data transformation is a hybrid data transformation that applies EMD to decompose the original data into intrinsic mode functions (IMFs) and then smooth these IMFs by using support vector regression (SVR). Finally, all SVR smoothed IMFs are added together as the transformed data. The SVR-A transformation is an improved variation of the EMD decomposition. Therefore, the two dependent lagged models fitted with the EMD and the SRV-A datasets revealed similar causal relationships with PRE and TEM. PRE displayed weak negative impacts on EVI, but TEM exhibited weak positive effects on EVI.

PRE and TEM showed weak impacts because many variables in these two lagged models had high magnitudes of coefficients. In particular, the lagged variables, EVIemd and EVIsvr_a, had coefficients higher than 1.0 and z values larger than 6000. Compared with other variables, these coefficients and z values were significantly higher, indicating that grassland productivity (EVI) showed a strong legacy (historical) effect on its current state. In other words, the past grassland productivity dramatically impacted its current grow condition. Moreover, the past climate impacts on grassland productivity were also reflected in the past grassland productivity.

Consequently, the addition of lagged grassland productivity implicitly considered the past impact of climate factors. As a result, the direct impacts of PRE and TEM on current EVI were considerably reduced. More importantly, since the study area is mainly within the semiarid and arid zones, PRE became a limiting factor to EVI from the perspective of time-series changes. Therefore, the panel models with the time-series transformed data and the dependent-lagged variables correctly modeled the climate impacts on grassland productivity. This new insight is obtained from this new modeling approach with time-series transformed data.

There is a controversy about the impact of transportation development (measured by the highway length per square kilometer, *lohsk*). It is commonly agreed that rapid urbanization and industrialization accelerated large-scale road construction. As a result, high road density caused direct damage to grassland and worsened grassland deterioration (Sha et al., 2020). The coefficient of *lohsk* in the ORIGIN-Conventional Model was significant and negative, which supports this common notion. However, all four other models (EMD-Lagged, TSE-Lagged, SRV-A-Lagged, and EMD-Dynamic) rejected this notion. The *lohsk* coefficients in these four models were highly positive. These models were the lagged models, which considered the EVI legacy effects and revealed a historical perspective. As a result, these models suggested that highway development positively impacted EVI in the long run, which has support from some literature (Deng et al., 2011). Good transportation infrastructure can provide diversified economic opportunities to herders and improve accessibility to markets. As a result, transportation development helps alleviate pressure on grasslands (Gao et al., 2016).

A third difference was that grain production density (*gpsk*). The *gpsk* coefficient in the ORIGIN-Conventional Model was positive. The explanation was physical. High grain production density indicated the plantation of high-yield agriculture plants, which would increase the greenness of the vegetation. Consequently, this would significantly increase the vegetation index from the physical meaning of remote sensing (Xie et al., 2021).

On the other hand, the *gpsk* coefficients in the other four models were negative, disclosing a negative impact. These models' findings indicated that grassland productivity would eventually decrease when grain production kept increasing over time. This finding suggested that increasing grain production would lead to a larger area of grassland converted to agricultural land, causing grassland deterioration. Some current literature supports this finding (Tang et al., 2016).

A fourth variation was the per capita GDP (*gdppc*). The variable of *gdppc* in the ORIGIN-Conventional Model positively affected EVI. Still, it disclosed adverse effects on EVI in the EMD-Lagged, TSE-Lagged, and SVR-A-Lagged models. This variable became a nonsignificant factor in the EMD-Dynamic Model. The common notion is that the increase of *gpdpc* in IMAR is driven by the fast-growing urban economy, commercial and industrial growth (especially mining industries), and grain production. These developments foster the conversion of large areas of grasslands and lead to grasslands deterioration (Hu & Nacun, 2018). However, three models in our study provide nuance. These models suggest that the relationship between GDP and grassland health is complex. Healthy economic development can mean more available financial resources supporting sophisticated grassland management and utilization (Deng et al., 2011; Xie et al., 2019).

Another difference that should be discussed is the inclusion of Graze in these models. Graze is a categorical variable, "1" indicating agricultural county, "2" meaning a mixed agricultural and grazing county, and "3" belonging to a grazing county. The categorical variable was not permitted in a Dynamic-Lagged Model. Therefore, only four models reported the results. Because Graze=1 was held as the comparison base in our analyses. The ORIGIN-Conventional and the TSE-Lagged models revealed differences between the agricultural and the mixed counties and the agricultural and grazing counties. The SVR-A-Lagged Model only disclosed the difference between the agricultural and the grazing counties. However, the EMD-Lagged Model did not unveil significant differences among the rural production modes. The explanation was that so many socioeconomic, environmental, and land-use factors disclosed their significant effects on EVI, and the impacts of pastoral production modes became less critical.

Among the four models (EMD-Lagged, EMD-Dynamic, SRV-A-Lagged, and TSE-Lagged), the land use and population variables existed apparent differences as well. First, urban land-use change displayed positive impacts on grassland productivity in three of those four models except the TSE-Lagged Model, in which urban growth was a nonsignificant factor. This finding also distinguishes us from our current understanding of the role of urbanization in grassland health. Urbanization is often perceived as a cause of grassland loss because of the conversion of grassland to other types of land uses for development (Seto, 2011; Xie et al., 2005). Our analyses of the impact of urban growth on grasslands showed that the effects of urbanization on grasslands might be more complicated. This finding supports one recent study (Allington et al., 2017). They suggest that the migration toward urban centers is shifting grazing pressures spatially. The urbanization in Inner Mongolia and Mongolia is decreasing pressure in grazing areas but creating increased pressure in peri-urban areas.

Second, the EMD-Lagged and the EMD-Dynamic models disclosed that the coefficients of grasslands positively affected EVI. However, the *Grass* coefficients in the TSE-Lagged, SVR-A-Lagged, and ORIGIN-Conventional models negatively impacted EVI. The common notion is that grasslands increase and restoration should improve grasslands' health. One possible explanation is that the study area is covered predominantly with grasslands. The grassland area is no longer a critical factor compared with other environmental and socioeconomic variables.

Third, the population density (*DPop*) and the rural population concentration (*prpop*) showed adverse effects on EVI in the EMD-Lagged, TSE-Lagged, and EMD-Dynamic models. These findings could lead to controversial explanations. The rapid growth of rural population worsened the burden on natural resources and the desire for socioeconomic growth, causing intensified grazing. Eventually, these economic activities would accelerate the conversion of grasslands into residential and agricultural usages (Hu & Nacun, 2018). On the other hand, a high rural population percentage could indicate less urban and commercial development. As a result, grasslands witness less disturbance and retain higher productivity.

Finally, it is worth noting if the data transformation is statistically valid. However, it deserves further discussion whether or how data transformation techniques impact causal relationships between the response and exploratory variables. The question is whether, when different data standardizing procedures transform the initial data into other standardized datasets, the causal relationships between the initial response and exploratory variables will change or not after the standardization. Subsequently, the model outcomes from various data transformations could reflect different causal relationships. In other words, the model outcomes cannot be compared when the input data changed.

From the point of view of statistics, prior data preprocessing before running regression analysis is expected. Various techniques exist to standardize variables, such as normalizing, scaling, standardizing, or transforming. Standardizing the variables in a regression analysis aims to remove outliers or noises in these variables' observations. As a result, variable standardization can remove multicollinearity and disclose causal relationships between the dependent and independent variables (Allen, 1997). Thus, variable standardization in a regression analysis can reveal statistically significant findings that might be missed otherwise (Newman & Browner, 1991). These techniques commonly employ the statistics of mean, standard deviation, data range, or linear trend to filter out anomalies in the data. Because of these normalized statistics, though data standardization methods intend to augment linear relations sought to examine, they rarely change general causal relationships embedded in a regression model (Fox, 2015). Therefore, a linear transformation of a model does not statistically change the model itself (Marquardt, 1980, p. 7).

This chapter uses the mean statistic to extract trend, seasonality, and error (TSE) and remove stationarity (Stat). A stepwise linear filtering of noises is adapted to conduct empirical mode decomposition (EMD). Sub-data-range linear filtering and reconstruction are combined to implement EMD-SVR. Statistically, these techniques are the extensions of the common variable standardization techniques for time-series data. In the case studies in this chapter, all four data transformations (TSE, Stat, EMD, and EMD-SVR) effectively and significantly increased the R-squared values. The panel regression models fitted with these transformed datasets significantly augmented the causal relationships between grassland deterioration, climate change, and socioeconomic transformation (Zhou et al., 2022). Therefore, the time-series data transformations in this chapter confirm similar findings for the conventional statistical data standardization in a regression analysis.

4.6 Conclusions

Several conclusions are based on the model performances (R-squared and Wald Chi-sq. values), block-entered sequences, prediction errors, causalities, and physical and socioeconomic meanings.

First, the lagged panel models fitted with the transformed time-series data have suggested seven nuances:

1. The climate factors (Precipitation and Temperature) are no longer the dominant driving factors determining grasslands' health in IMAR. Joint interactions of other socioenvironmental factors were leading causes of grassland deterioration.
2. Transportation infrastructure improvement in IMAR begins to show positive effects on grasslands' health.

3. Grain production has a complex relationship with grasslands' health and shows noticeable adverse effects.
4. GDP is not an absolute negative factor regarding grassland health and can provide financial resources for better grassland management practices.
5. Urbanization can be a positive factor for grassland health. Urbanization causes spatial reorganization of grazing pressure, affecting the functioning of the grassland system as a whole, and needs more in-depth analysis.
6. The impacts of rural population density and percentage on grassland productivity vary between different models, which will be further investigated in Chap. 8.
7. Education shows a noticeable impact on grassland productivity.

Second, the conventional panel model with the original (non-transformed) data exemplifies the causal relationships between the climate factors and EVI but fails to disclose causal relationships between grasslands' health with other socioeconomic and land-use factors.

Third, the block-entered modeling approach effectively distinguishes various climate, agricultural, socioeconomic, land-use, education, and industrial factors contributions to grassland productivity.

Fourth, the causal directions revealed by the transformed data varied between the panel models and the transformed data. In particular, the SVR-A-Lagged and EMD-Lagged models generated very low RMSE values. In addition, based on the regional RMSE patterns, the SVR-A-Lagged Model captured the general linear relationships between grassland productivity and beneath grassland types. Moreover, the SVR-A-Lagged Model revealed important causalities between grassland productivity and several socioeconomic driving factors. This model attributed grassland health to various factors, including agricultural, economic, industrial, education, and land-use. As a result, its prediction error spatial pattern did not simply reflect climate zones but reflected the modified impacts of these socioeconomic factors on climate zones. Thus, this model in general overestimated the predictions in the highly fertile grasslands and underestimated the projections in the low-productivity grasslands. The matches between predictions and productivities were consistent with the assumption of linear regression analysis. As a result, this model improved the prediction accuracy and achieved the highest performance among the panel model and input dataset combinations.

It is worth pointing out that the EMD-Lagged Model generated similarly low total RMSE values as the SVR-A-Lagged Model did. However, the spatial patterns of prediction errors of the EMD-Lagged Model were less conforming with the linear relationships between grassland productivity and grassland types than the SVR-A-Lagged Model. Furthermore, the EMD-Lagged Model was the sole model among the 18 models that revealed some causal relationships between grassland deterioration and almost all socioeconomic and LULC variables (except for 3 out of 28). In other words, this model significantly augmented the causalities between grassland deterioration and socioeconomic and LULC variables. Therefore, the spatial errors of this model were less conforming with the regional patterns of climate changes. Furthermore, the Dynamic-Lagged EMD Model correctly revealed the spatial patterns of prediction errors, although its error range was relatively higher. In summary, the SVR-A-Lagged and EMD-Lagged models are the best to choose from when examining the causal relationships underlying grassland deterioration.

On the other hand, the pilot experiment conducted in 2022 by this book author and his colleagues (Zhou et al., 2022) led to different conclusions and suggested that the SVR-A-Lagged EMD Model was not a preferred choice. Moreover, all the above conclusions in this chapter were refined compared with the findings from its pilot experiment (Zhou et al., 2022). The dataset in this book covered all 89 counties in the study area and included a completed list of 28 socioeconomic and LULC variables available at this time, compared with 54 counties and 10 socioeconomic variables in the pilot experiment. The data coverage improvement in this chapter apparently contributed to its new findings and insights.

In brief, the 18 experiments demonstrate that the conceptual framework and the analytical methods developed in this chapter help reveal time-series causal relationships embedded in TSES systems. The experiments can differentiate different causal relationships and directions among distinctive combinations of regression models and transformed datasets. They also confirm that the time-series preprocessing ecological/environmental and socioeconomic data to construct temporally consistent or unified socioenvironmental data has an important implication in total social-environmental system (TSES) studies. These techniques are meaningful applications of big-data analytics and data mining algorithms in integrated socioenvironmental studies (Carpenter et al., 2015; Kalantari et al., 2019; Zhou et al., 2022). Furthermore, the 18 experiments confirm that incorporating lagged effects is essential to disclose time-series causal relationships between environmental and socioeconomic factors. Integrating time-series data mining techniques and temporal statistical analysis approaches is an emerging research approach for examining TSES or coupled human–nature systems. These tools and methods have the potential to be explored in urban informatics and added to planning support technologies (Pan et al., 2020).

Finally, it should be cautioned that this chapter's selection of temporal data preprocessing methods and regression models is not exhaustive. The chapter only chose one data mining method to represent time-series data mining, computational data mining, and hybrid computational learning, respectively. Moreover, the selection of environmental and socioeconomic variables is limited to published statistical data and publicly available image sources. In other words, more socioenvironmental data and more time-series data mining algorithms are warranted in the near future to conduct more experiments for further exploration.

References

Allen, M. P. (1997). Regression analysis with standardized variables. *Understanding Regression Analysis*, 46–50.

Allington, G. R. H., Li, W., & Brown, D. G. (2017). Urbanization and environmental policy effects on the future availability of grazing resources on the Mongolian plateau: Modeling socio-environmental system dynamics. *Environmental Science & Policy, 68*, 35–46.

Allison, P. (1999). *Multiple regression: A primer* (Undergraduate Research Methods & Statistics in the Social Sciences) (p. 202). Pine Forge Press.

Arellano, M., & Bond, S. (1991). Some tests of specification for panel data: Monte Carlo evidence and an application to employment equations. *The Review of Economic Studies, 58*(2), 277–297. https://doi.org/10.2307/2297968

Baltagi, B. H. (2005). Solutions to problems posed in Volume 20 (1) and 20 (2): 04.1. 1. A Hausman test based on the difference between fixed effects two-stage least squares and error components two-stage least Squares—Solution. *Econometric Theory, 21*(2), 483–484.

Brandon, P. S., & Lombardi, P. (2009). *Evaluating sustainable development: In the built environment*. Wiley.

Brown, D. G., Agrawal, A., Sass, D. A., Wang, J., Hua, J., & Xie, Y. (2013). Responses to climate and economic risks and opportunities across national and ecological boundaries: Changing household strategies on the Mongolian plateau. *Environmental Research Letters, 8*(4), 045011.

Carpenter, S. R., Booth, E. G., Gillon, S., Kucharik, C. J., Loheide, S., Mase, A. S., & Seifert, J. (2015). Plausible futures of a social-ecological system: Yahara watershed, Wisconsin, USA. *Ecology and Society, 20*(2).

Chang, Y., He, H. S., Bishop, I., Hu, Y., Bu, R., Xu, C., & Li, X. (2007). Long-term forest landscape responses to fire exclusion in the great Xing'an mountains, China. *International Journal of Wildland Fire, 16*(1), 34–44.

Chen, H., Shao, L., Zhao, M., Zhang, X., & Zhang, D. (2017). Grassland conservation programs, vegetation rehabilitation and spatial dependency in Inner Mongolia, China. *Land Use Policy, 64*, 429–439.

Demeke, A. B., Keil, A., & Zeller, M. (2011). Using panel data to estimate the effect of rainfall shocks on smallholders food security and vulnerability in rural Ethiopia. *Climatic Change, 108*(1–2), 185–206. https://doi.org/10.1007/s10584-010-9994-3

Deng, X., Huang, J., Huang, Q., Rozelle, S., & Gibson, J. (2011). Do roads lead to grassland degradation or restoration? A case study in Inner Mongolia,China. *Environment and Development Economics, 16*(6), 751–773.

Deng, L., Shangguan, Z.-P., Wu, G.-L., & Chang, X.-F. (2017). Effects of grazing exclusion on carbon sequestration in China's grassland. *Earth-Science Reviews, 173*, 84–95.

Fan, S. (2017). *An analysis of long-term effects of climate change and socioeconomic activities on grassland productivity of inner Mongolia*. Ph.D. Dissertation, Eastern Michigan University.

Fox, J. (2015). *Applied regression analysis and generalized linear models*. Sage Publications.

Gao, L., Kinnucan, H. W., Zhang, Y., & Qiao, G. (2016). The effects of a subsidy for grassland protection on livestock numbers, grazing intensity, and herders' income in Inner Mongolia. *Land Use Policy, 54*, 302–312.

Guo, D., Song, X., Hu, R., Cai, S., Zhu, X., & Hao, Y. (2021). Grassland type-dependent spatiotemporal characteristics of productivity in Inner Mongolia and its response to climate factors. *Science of the Total Environment, 775*, 145644.

Hao, L., Sun, G., Liu, Y., Gao, Z., He, J., Shi, T., & Wu, B. (2014). Effects of precipitation on grassland ecosystem restoration under grazing exclusion in Inner Mongolia, China. *Landscape Ecology, 29*(10), 1657–1673.

Hu, Y., & Nacun, B. (2018). An analysis of land-use change and grassland degradation from a policy perspective in Inner Mongolia, China, 1990–2015. *Sustainability, 10*(11), 4048.

Kalantari, Z., Ferreira, C. S. S., Page, J., Goldenberg, R., Olsson, J., & Destouni, G. (2019). Meeting sustainable development challenges in growing cities: Coupled social-ecological systems modeling of land use and water changes. *Journal of Environmental Management, 245*, 471–480.

Li, S., & Xie, Y. (2013). Investigating coupled impacts of climate change and socioeconomic transformation on desertification by using multitemporal Landsat images: A case study in central Xilingol, China. *IEEE Geoscience and Remote Sensing Letters, 10*(5), 1244–1248.

Li, S., Xie, Y., Brown, D. G., Bai, Y., Hua, J., & Judd, K. (2013). Spatial variability of the adaptation of grassland vegetation to climatic change in Inner Mongolia of China. *Applied Geography, 43*, 1–12.

Liu, J., Dietz, T., Carpenter, S. R., Alberti, M., Folke, C., Moran, E., & Tayler, W. (2007). Complexity of coupled human and natural systems. *Science, 317*(5844), 1513–1516.

Liu, J., Dou, Y., Batistella, M., Challies, E., Connor, T., Friis, C., et al. (2018). Spillover systems in a telecoupled Anthropocene: Typology, methods, and governance for global sustainability. *Current Opinion in Environmental Sustainability, 33*, 58–69.

Marquardt, D. W. (1980). Comment: You should standardize the predictor variables in your regression models. *Journal of the American Statistical Association, 75*(369), 87–91.

Metcalfe, A. V., & Cowpertwait, P. S. P. (2009). *Introductory time series with R*. Springer.

Newman, T. B., & Browner, W. S. (1991). In defense of standardized regression coefficients. *Epidemiology*, 383–386.

Oud, J. H. L., Folmer, H., Patuelli, R., & Nijkamp, P. (2012). Continuous-time modeling with spatial dependence. *Geographical Analysis, 44*(1), 29–46. https://doi.org/10.1111/j.1538-4632.2011.00834.x

Pan, H., Geertman, S., & Deal, B. (2020). *What does urban informatics add to planning support technology?* (Vol. 47, pp. 1317–1325). SAGE Publications.

Parent, O., & LeSage, J. P. (2011). A space–time filter for panel data models containing random effects. *Computational Statistics and Data Analysis, 55*(1), 475–490.

Pillonetto, G., Chiuso, A., & De Nicolao, G. (2011). Prediction error identification of linear systems: a nonparametric Gaussian regression approach. *Automatica, 47*(2), 291–305.

Pinzon, E. (2015, November 12). Xtabond cheat sheet, the DTATA Blog. https://blog.stata.com/author/epinzon/

Propastin, P. A., Kappas, M., & Muratova, N. R. (2008). Inter-annual changes in vegetation activities and their relationship to temperature and precipitation in Central Asia from 1982 to 2003. *Journal of Environmental Informatics, 12*(2).

Robinson, B. E., Li, P., & Hou, X. (2017). Institutional change in social-ecological systems: The evolution of grassland management in Inner Mongolia. *Global Environmental Change, 47*, 64–75.

Seto, K. C. (2011). Exploring the dynamics of migration to mega-delta cities in Asia and Africa: Contemporary drivers and future scenarios. *Global Environmental Change, 21*, S94–S107.

Sha, Z., & Li, R. (2020). Assessment of human-related driving forces for reduced carbon uptake using neighborhood analysis and geographically weighted regression: A case study in the grassland of Inner Mongolia, China. *Applied Sciences, 10*(21), 7787.

Sha, Z., Li, R., Li, J., & Xie, Y. (2020). Estimating carbon sequestration potential in vegetation by distance-constrained zonal analysis. *IEEE Geoscience and Remote Sensing Letters, 18*(8), 1352–1356.

Shang, C., Wu, T., Huang, G., & Wu, J. (2019). Weak sustainability is not sustainable: Socioeconomic and environmental assessment of Inner Mongolia for the past three decades. *Resources, Conservation and Recycling, 141*, 243–252.

Srinivasan, V., Lambin, E. F., Gorelick, S. M., Thompson, B. H., & Rozelle, S. (2012). The nature and causes of the global water crisis: Syndromes from a meta-analysis of coupled human-water studies. *Water Resources Research, 48*, W10516. https://doi.org/10.1029/2011WR011087

StataCorp, L. L. C. (2019). *Stata statistical Software: Release* (Vol. 16).

Steffens, M., Kölbl, A., Totsche, K. U., & Kögel-Knabner, I. (2008). Grazing effects on soil chemical and physical properties in a semiarid steppe of Inner Mongolia (PR China). *Geoderma, 143*(1–2), 63–72.

Stoffer, D. (2016). Applied statistical time series analysis. In *Package 'astsa.*

Su, R., Yu, T., Dayananda, B., Bu, R., Su, J., & Fan, Q. (2020). Impact of climate change on primary production of inner Mongolian grasslands. *Global Ecology and Conservation, 22*, e00928.

Tang, J., Davy, A. J., Jiang, D., Musa, A., Wu, D., Wang, Y., & Miao, C. (2016). Effects of excluding grazing on the vegetation and soils of degraded sparse-elm grassland in the Horqin Sandy Land, China. *Agriculture, Ecosystems & Environment, 235*, 340–348.

Timon, M., Pickett, S. T. A., Grimm, N. B., Niemelä, J., Alberti, M., Elmqvist, T., et al. (2016). Advancing urban ecology toward a science of cities. *Bioscience, 66*(3).

Watson, S. J., Luck, G. W., Spooner, P. G., & Watson, D. M. (2014). Land-use change: Incorporating the frequency, sequence, time span, and magnitude of changes into ecological research. *Frontiers in Ecology and the Environment, 12*(4), 241–249.

Wu, J., Zhang, Q., Li, A., & Liang, C. (2015). Historical landscape dynamics of Inner Mongolia: patterns, drivers, and impacts. *Landscape Ecology, 30*(9), 1579–1598. https://doi.org/10.1007/s10980-015-0209-1

Xie, Y., Mei, Y., Guangjin, T., & Xuerong, X. (2005). Socio-economic driving forces of arable land conversion: A case study of Wuxian City, China. *Global Environmental Change, 15*(3), 238–252.

Xie, Y., Crary, D., Bai, Y., Cui, X., & Zhang, A. (2019). Modeling grassland ecosystem responses to coupled climate and socioeconomic influences in multi-spatial-and-temporal scales. *Journal of Environmental Informatics, 33*(1).

Xie, Y., Fan, S., & Zhou, C. (2021). Examining ecosystem deterioration using a total socioenvironmental system approach. *Science of the Total Environment, 784*, 147171.

Zhou, C., Xie, Y., Zhang, A., Liu, C., & Yang, J. (2022). Spatiotemporal analysis of interactions between seasonal water, climate, land use, policy, and socioeconomic changes: Hulun-Buir steppe as a case study. *Water Research, 209*, 117937.

Chapter 5
Empirical Evaluation of Urban Sustainability from Underlying Causal Structures and Legacy Effects: The Prefecture Cities in China as a Case Study

There is an increasing and convincing call that the interactions among various components of an urban system shall be structured as a complex causal chain to examine its sustainability (Liu et al., 2015). Identifying and interpreting causal chains or structures underlying an urban system is critical to measure its sustainability level or status. In the recent 40 years, China has made significant progress in urbanization and economic development. The urban population increased from 170 million to 730 million and the urbanization rate from 17.9% to 53.7%. On the other hand, rapid growth costs are high and most reflected in the environmental problems, which threaten sustainable development in China. Therefore, China's prefecture cities have become a giant laboratory for studying urban sustainability, where various economic, environmental, social, cultural, and political policies were implemented at many spatial scales (Zhang et al., 2020).

5.1 Introduction

The world's population reached 8.0 billion in mid-November 2022 from an estimated 2.5 billion people in 1950 (UN, 2022). About 55% of the world's population – 4.2 billion inhabitants – live in cities, and nearly 7 of 10 people in the world will live in the cities in 30 years (World Bank, 2020). The speed and scale of urbanization bring numerous challenges, including deterioration of natural resources, air and water pollution, climate change, and unprecedented demand for affordable housing, essential services, and jobs. Therefore, sustainable urban systems (SUS) studies have attracted a broad spectrum of audiences and generated an increasingly growing literature volume in recent years (Ali-Toudert & Ji, 2017; Pupphachai &

Supplementary Information The online version contains supplementary material available at https://doi.org/10.1007/978-3-031-39594-9_5.

Y. Xie, *Total Socioenvironmental Systems*,
https://doi.org/10.1007/978-3-031-39594-9_5

Zuidema, 2017). For example, more than 300 articles dealing with SUS assessment were published annually in recent years in the Web of Science (Sharifi, 2021). These publications illustrate that SUS assessment studies have diversified from the initial few themes to multidimensional contents (Pupphachai & Zuidema, 2017).

Current urban sustainability measurements heavily focus on evaluating and choosing sustainable urban system (SUS) indicators. SUS indicators are quantifiable indicators for assessing progress toward sustainable growth in the context of sustainable development goals and targets (Liu et al., 2015; Bai et al., 2016; Pupphachai & Zuidema, 2017). These measurable sustainable growth indicators are called sustainability indicators (SIs) (Cutaia, 2016). SIs can be simple socioeconomic indicators such as gross domestic product (GDP) or highly composite measurements. In general, there are three common types of indicators. The first type is the global sustainability measurement with an emphasis on measuring comprehensive urban and socioeconomic development. For example, the City Prosperity Index (CPI) has been proposed by the United Nations to measure the overall achievement of a city (UN-HABITAT, 2015). CPI is a composite index consisting of six dimensions of city prosperity: productivity, environmental sustainability, equity and social inclusion, infrastructure, quality of life, and governance and legislation. Each dimension includes two to four measurable indicators. CPI can be measured at four scales: (1) basic CPI as an initial diagnosis that is internationally comparable; (2) global city ranking for global and regional monitoring; (3) extended CPI for in-depth diagnosis that is comparable within a specific country; and (4) contextual CPI that is policy implementation oriented as an urban monitoring tool. CPI advocates and also serves as a flexible and robust assessment framework that provides conceptual guidance and methodological solutions for developing a comprehensive index and connecting healthy cities' indicators with the needs of policy and governance (Wong, 2015).

The inclusive wealth index (IWI) and the genuine progress indicator (GPI) are other examples. IWI measures the wealth of nations, including all of the assets from which human well-being is derived, including manufactured, human, and natural capital (Roman & Thiry, 2016). IWI serves as a synthetic indicator to supersede or complement the iconic *Human Development Index (HDI)* (Dasgupta, 2009). GPI is developed as a national-level measure of economic growth and prosperity. GPI accounts for externalities, such as environmental and carbon footprints, resource depletion, pollution, and long-term environmental damage (Kubiszewski et al., 2013).

The second type includes a variety of measurements focusing on urban construction sustainability and inner city-built environments. For example, architects, civil and environmental engineers, and urban planners have developed many well-known systems to rate and certify sustainable building construction and surrounding urban development. For instance, Building Research Establishment Environmental Assessment Method (BREEAM) (Cole & Valdebenito, 2013), Comprehensive Assessment Method for Sustainable Urban Development (CAMSUD) (Sharifi & Murayama, 2013; Ali-Toudert et al., 2019), Leadership in Energy and Environmental Design for Neighborhood Development (LEED-ND) (Benson & Bereitschaft, 2020), and the German Sustainable Building Council – New City District (the German abbreviation is DGNB-NSQ) are good representatives. Many of these urban sustainability rating systems use multicriteria-based or indicator-based approaches.

The third group of SI indicators consists of broad scopes of research and policy-oriented SUS measurements, representing the most dynamic development of SI studies. A recent review based on multi- and cross-typological systematic analysis has examined 67 indicator sets (for a total of 2847 indicators) from academic research to policy implication (Merino-Saum et al., 2020). Numerous processes have been developed in multiple disciplines to compute SI (Kaur & Garg, 2019). GIS and remote sensing techniques are also integrated to calculate or estimate SI. One example is to calculate dynamic ecosystem service values (DESVs) through GIS and remote sensing (Liang et al., 2020). Urban sustainability is the square root of the sum squares of DESVs, gross domestic product (GDP), and per capita net income (PCNI) (Fu et al., 2016; Xue & Luo, 2015). Although the estimation of DESVs is relatively sophisticated, the ecosystem service value-based SI has firm roots in the concepts of ecosystem sustainability and ecological economics.

The formation of an SI often involves six essential elements (Merino-Saum et al., 2020): (1) a title clearly conveys what this indicator is going to measure, (2) a definition explains how to understand this indicator; (3) a conceptual framework supports the formation of this indicator or justifies this indicator; (4) a dataset is required to derive the indicator; (5) a unit of measurement makes this indicator meaningful or understood to a potentially interested audience; and (6) an expectation or target or benchmark is critical to this indicator. Based on this SI assessment template, current SI measurements have encountered many criticisms. First, many SIs are exclusively dependent on available and easily collectable data independently from the context, concept, and expectation behind their use (Sebastien et al., 2014). Unfortunately, many of them are based on statistical data available at the national scale and involve a large set of macroeconomic variables that are not enumerated at local and regional scales. It is impossible to calculate these SIs to assess urban sustainability at a regional or state/provincial scale. As a result, these SIs are only suitable for global assessment but not for meaningful policy guidelines in a nation or region to improve sustainability.

Second, the large body of current SIs lack well-defined and integrated conceptual frameworks. Many essential elements of urban sustainability such as cultural, ecological, economic, environmental, health, management, policy, and social contents are measured or modeled relatively independently from each other like silos (Gibson et al., 2005; von Edmund, 2012). In many SI conceptual frameworks, these subthemes of urban sustainability are relatively loosely coupled rather than closely integrated as a whole system (Spindler, 2011; Ali-Toudert & Ji, 2017). Quantitatively, complex interactions between these subthemes are modeled as inputs or outputs between them (Dawson et al., 2010). Various subthemes are coupled through procedural modeling to link as an entire system. Sometimes the subthemes are linked through state variables of these subthemes (Ness et al., 2010). To a large degree, these subthemes of urban sustainability are relatively loosely coupled rather than closely integrated as a whole system. In other words, many current SI conceptual frameworks are flawed to describe complex interactions and feedbacks between various urban sustainability elements.

Third, no matter what types of quantitative measurement techniques (such as multicriteria assessment or statistical analysis) are adopted, most SIs make their final judgments upon a summative score. It is difficult to preserve the measurable complexity of urban sustainability through a single score or quantity (Xie et al., 2022). Fourth, many SIs often have ambiguous sustainability goals, targets, and thresholds (Verma & Raghubanshi, 2018). In addition, some of these goals, targets, and thresholds are neither scientifically determined nor policy goal-based (Mori & Christodoulou, 2012). Fifth, many SIs disclose a general ignorance about national and regional differences in understanding the contextual meanings and interpretations of sustainability assessment (Xie et al., 2022). In other words, these SIs fail to reveal regional variations of sustainability within a nation. Finally, current SIs miss many other critical traits for systematically measuring sustainability, which will be discussed in the following sections.

In this chapter, we are developing six complementary approaches for a comprehensive measurement of sustainable urban system (SUS) level: (1) developing a conceptual framework based on cross-lagged panel model for urban sustainability (CLPM-US) to measure SUS level; (2) examining simultaneously SUS spatial variations and temporal changes; (3) using CLPM-US to reveal how legacy effects of a sustainable urban system (SUS) and its underlying causal constructs affect the present conditions and future directions of urban sustainability; (4) using CLPM-US to compute summative SUS scores over times to measure SUS levels and their time-series changes; (5) supplementing the summative scores with their underlying causal structures to assess unique contributions of urban causal constructs to sustainability scores; and (6) validating and comparing the urban sustainability rankings of CLPM-US with the rankings from the multi-objective optimization problems algorithm. The chapter illustrates these new urban sustainability measurements through a case study of Chinese 278 prefecture cities.

5.2 The Data and Research Methods

5.2.1 The Data and Explanation

There are 662 cities in China, including four municipalities directly under the central government (Fig. 5.1), 333 prefecture-level cities, and 329 county-level cities. However, comprehensive fine-scale city statistical data in China are only available at the prefecture-level cities. Therefore, this chapter examines urban sustainability in prefecture-level cities. In addition, it should be pointed out that there are a large number of autonomous prefectures and leagues in western and northern China. Although these units are equivalent to the prefecture-level cities in eastern China administratively, they have low urbanization rates, less developed economies and

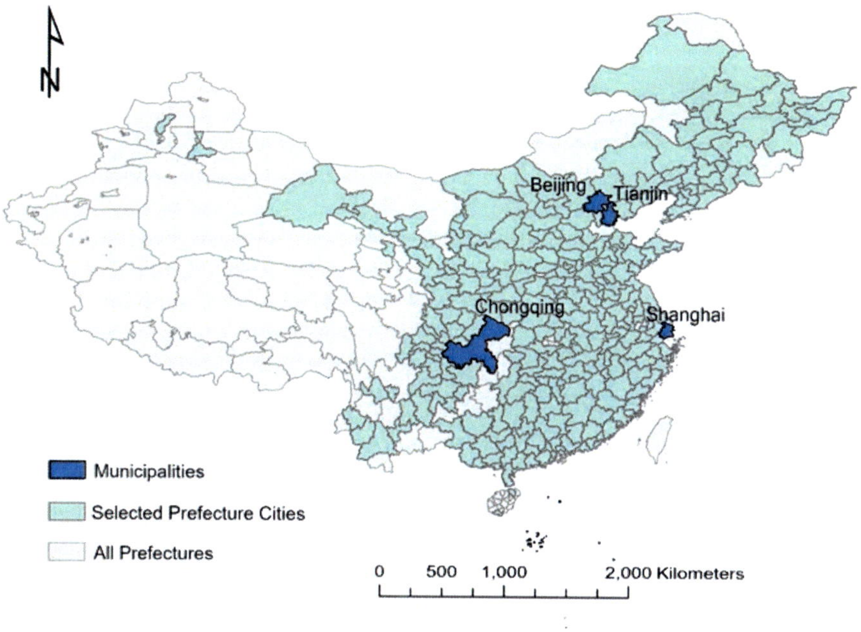

Fig. 5.1 The selected prefecture cities in China

industries, and sparse population density. They lack statistical data appropriate for measuring urban sustainability. Thus, they are excluded from the case study. As a result, the final dataset includes 278 prefecture-level cities, including 251 in 2000, 267 in 2005, 276 in 2010, and 278 in 2015. We composed a time-series panel dataset of 29 observed variables in four periods (2000, 2005, 2010, and 2015) over the 278 prefecture cities (Fig. 5.1). This panel dataset is one of the most comprehensive Chinese city-data reported so far (Table 5.1). Based on the complex urban systems theory, the literature review (Zhang et al., 2020; Xie et al., 2022), and statistical analyses, we proposed that the sustainability measurement of the Chinese prefecture cities consisted of ten causal constructs. The causal constructs were also called latent variables. The term "causal constructs" was used in this chapter for consistency. The ten causal constructs were: cleaner production (CP), economic performance (ECON), economic vitality (ECOV), natural resources consumption (NAT), patents and R & D status (PAT), population resources (POPU), quality of life (QOL), state investment (STA), transportation development (TRAN), and urban sustainability status (SUS) (Table 5.1). Each construct contained three to four measured variables.

The primary research design is to develop CLPM-US to measure SUS level. This framework innovatively integrates three modeling paradigms, panel data analysis (PDA), structural equation model (SEM), and longitudinal causal relations (cross-lagged effects), into a coherent framework to measure SUS levels.

Table 5.1 The list of measured variables and latent variables[a]

Latent	Indicators[b]	Data sources[c]
Economic performance (***ECON***)	Per capita GDP (**gdppc**)	China City Statistical Yearbook
	Workers' wage (**wage**)	China City Statistical Yearbook
	Average land GDP (**gdp_l**)	China City Statistical Yearbook
Economic vitality (***ECOV***)	Per capita retail sales of consumer goods (**cons**)	China City Statistical Yearbook
	Per capita wholesale and retail sales (**whole**)	China City Statistical Yearbook
	Practitioners in the wholesale and retail industry (**w_emp**)	China City Statistical Yearbook
	Practitioners in the hotel and catering industry (**h_emp**)	China City Statistical Yearbook
Quality of life (***QOL***)	Per capita universities (**uni_p**)	China City Statistical Yearbook
	Per capita university teachers (**unitea**)	China City Statistical Yearbook
	Per capita hospitals (**hos_p**)	China City Statistical Yearbook
	Per capita doctors (**doc_p**)	China City Statistical Yearbook
Transportation development (***TRAN***)	Per capita buses (**bus**)	China City Statistical Yearbook
	Per capita cars (**car**)	China City Statistical Yearbook
	Per capita road area (**road**)	China City Statistical Yearbook
Cleaner production (***CP***)	Industrial output per unit of sulfur dioxide emissions (**so2**)	China City Statistical Yearbook
	Industrial output produced per unit of industrial dust emissions (**dust**)	China City Statistical Yearbook
	Industrial output produced per unit of industrial dust emissions (**no**)	China City Statistical Yearbook
	Industrial output produced per unit of sewage discharge (**sewage**)	China City Statistical Yearbook
Natural resources (***NAT***)	Percentage of crop area over the total land area (**cro_p**)	Remote sensing monitoring datab ase of land use status in China
	Percentage of water area over the total land area (**wat_p**)	Remote sensing monitoring database of land use status in China
	Percentage of wetland area over the total land area (**wet_p**)	Remote sensing monitoring database of land use status in China
State investment (***STA***)	Per capita investment in fixed assets (**fix_acc**)	China City Statistical Yearbook
	Net value of fixed assets per capita (**fix_net**)	China City Statistical Yearbook
	Per capita fiscal revenue (**gover_re**)	China City Statistical Yearbook

(continued)

Table 5.1 (continued)

Latent	Indicators[b]	Data sources[c]
Patents and R & D (*PAT*)	Education expenditure per capita **(edu)**	China City Statistical Yearbook
	Per capita funding for scientific undertakings **(sci)**	China City Statistical Yearbook
	Patent applications per capita **(patent)**	Patent cloud database
Population resources (*POPU*)	All population **(popu_all)**	China City Statistical Yearbook
	Urban population **(popu_urban)**	China City Statistical Yearbook
	Employees **(employee)**	China City Statistical Yearbook

[a]The dataset is shared by Professor Xueliang Zhang and Dr. Yang Yang in Shanghai University of Finance and Economics, Shanghai, China. The data is part of the database developed by the Institute for Yangtze River Delta Development Research, Shanghai University of Finance and Economics for supporting urban and regional development strategy studies in China

[b]The data used in structural equation models is in the form of a Likert scale. In order to make the macro data more consistent with the basic assumptions of the structural equation model, while eliminating the influence of dimensions and units, we use the Likert scale to convert the original data. The actual data of each year is divided into five categories: 1, 2, 3, 4, and 5 according to the quantile from small to large

[c]The data in this chapter are mainly from online China's urban statistical yearbook (https://data.cnki.net/yearbook/Single/N2020050229), China's land use remote sensing monitoring database (http://www.resdc.cn/data.aspx?DATAID=184), and China's patent cloud database (https://app.patentcloud.com/). Some indicators in 2000 were missing, and we replaced these indicators with the data first available in a later year, ranging from 2001 to 2003. Specifically, the per capita retail sales of social consumer goods, per capita buses, per capita cars, per capita road area, per capita financial income, per capita education funds, and per capita science funds were replaced by the data of 2001. Employees of hotel and catering industry, industrial output per unit of sulfur dioxide emission, industrial output per unit of industrial dust emission, and industrial output per unit of wastewater emission were replaced by the 2003 data. The description of latent variables, corresponding indicators, and data sources are shown in Table 5.1

5.2.2 *The New Sustainability Measurement Framework: The Cross-lagged Panel Model for Measuring Urban Sustainability (CLPM-US)*

Cities are complex systems growing like biological organisms evolving in a Darwinian fitness landscape (Batty, 2013; Szell, 2014). Cities are networked spatial configurations where different sets of interactions, trade-offs, and transactions take place. These interconnected complexities are better interpreted through urban models built on complexity theories and supported with "big data" (Szell, 2014; Thrift, 2013). Big data and new data sources through mass media have increasingly played critical roles in examining spatial and temporal dynamics of socioenvironmental systems (Goodchild, 2016). Moreover, cities as complex human–natural systems consist of numerous demographical, ecological, environmental, socioeconomic, and political (policy) processes that are interconnected to form various levels of causal chains (Batty, 2013; Batty, 2018; Brandon & Lombardi, 2010; Liu et al.,

2015). These processes interact to create a causal structure to determine positive or negative trad-offs between them and hence their sustainability levels (Bai et al., 2016; McPhearson et al., 2016).

Moreover, urban sustainability is an evolving concept involving history, representing the current, and predicting the future. Lagged and cross-lagged effects of an urban system and its underlying causal structures are critical traits of urban system evolutionary dynamics. Furthermore, the measurement of urban sustainability is not a single composite score. Still, it involves a large set of trade-offs between sub-aims of urban sustainability and an empirical interpretation of its underlying urban system causal structures.

Hence, urban sustainability measurement from the perspectives of causal relationships among various components of urban systems and big data analytics needs a new research paradigm. This new analytical framework shall integrate functions of three sophisticated statistical methods, PDA (panel data analysis), SEM (structural equation modeling), and autoregressive modeling of lagged effects of causal structures. Phycologists have developed a cross-lagged panel model (CLPM) to analyze longitudinal causal relationships between repeatedly measured variables embedded in panel data (Rogosa, 1980). CLPM is not simply a panel data analysis but a discrete time structural equation model (SEM) that analyzes panel data in which some variables are repeatedly measured over different times. Therefore, CLPM is an integrated modeling paradigm combining PDA and SEM, called a cross-lagged structural equation model (CLSEM).

CLPM has been commonly recognized by many phycologists as a powerful statistical method to examine cross-lagged correlations as a solution to explore causal influences in longitudinal panel data (Hamaker et al., 2015). A CLPM was applied to examine protective behavioral strategies under alcohol abuse (Napper et al., 2014). Several CLPM models were developed to investigate causal relationships between weight stigma, weight gain, reduced psychological well-being, and sexual activities (Austen & Griffiths, 2022). The empirical applications of CLPM have rapidly expanded into the studies in medicine, health, crime, education, and mass media addiction during the current decade. CLPM models have been used to analyze interdependencies across the life course (Mund & Nestler, 2019), the association directions between radiation risk perception and relocation (Murakami et al., 2020), the relationship between symptoms of obsessive-compulsive disorder and depression during therapy (Simkin et al., 2022), and the causal correlations between the anxiety and depression symptoms and personality and individual differences (Sorjonen et al., 2023). CLPM models are also applied to analyze the attitudes about aggression and perpetration of Adolescent Dating Aggression (Sabol et al., 2020), the temporal directional relationship between problematic internet use and depressive symptoms (Kojima et al., 2021), and the relations between Adolescent girls' Instagram and TikTok use with body image-related constructs over time (Maes & Vandenbosch, 2022). In addition, CLPM models are deployed to examine the longitudinal relationship between teacher self-efficacy and work engagement (Burić et al., 2022) and to estimate the co-development of executive functions and math achievement throughout the elementary grades (Zhang et al., 2023).

However, the adaptation or application of CLPM in econometric analyses is sporadic (Allison et al., 2017; Williams et al., 2018) and still rare in urban sustainability assessment and integrated human–nature systems studies. This chapter systematically introduces CLPM and related modeling paradigms, such as structural equation modeling (SEM), panel data analysis (PDA), and longitudinal autocorrelation, into urban sustainability studies. The chapter discusses several essential lagged effects in measuring urban sustainability. It innovatively adopts the three components of CLPM (SEM, PDA, and lagged effects) in the arena of measuring urban sustainability to build a CLPM-US. It then applies CLPM-US to 278 Chinese prefecture cities to measure urban sustainability, unveil its underlying causal structures, and investigate cross-lagged effects of past sustainability levels and causal structures on current sustainability prospects. The chapter provides a comprehensive illustration of applying CLPM-US in measuring urban sustainability in China and serves as an experimental application of CLPM in urban sustainability and integrated socioenvironmental system studies.

5.2.3 The Structural Equation Modeling (SEM)

SEM is an essential foundation of CLPM. SEM is more fitted to interpret an urban system's causal structure than other statistical methods do (Grace, 2006). SEM has been applied to understand coupled human and natural systems and enhance empirical interpretation of their causal relationships (Brown et al., 2013). SEM is an analytic framework that encompasses more traditional analysis, such as multiple regression and analysis of variance, but it has substantial advantages over conventional methods because it allows for simultaneous testing of variables and hypotheses that result from the complex social–ecological systems (Grace, 2006; Kline, 2011). SEM also incorporates path analysis that places a heavier focus on interpretive causal structures than regression analysis does (Grace, 2006).

The crucial element of SEM is its inclusion of latent variables (LVs). LVs are not directly measured variables but hypothetical causal constructs (Fernandes et al., 2019). A causal construct (a latent variable) is designed to explain observed covariations of several measured variables interacting in one specific domain or reflecting one unique behavior. For instance, "Natural Resource Consumption Change (NAT)" in Table 5.1 is a causal construct. NAT explains what variance of natural resource consumptions could be explained by three covariates, *crop_p* (percentage of crop area over the total land area), *wat_p* (the percentage of water area over the total land area), and *wet_p* (the percentage of wetland area over the total land area). A SEM model consists of many measurement models and a structural model (Halkos et al., 2021). For example, NAT responds to *crop_p*, *wat_p*, and *wet_p*, comprising a measurement model. On the other hand, the structural model describes the relationships between all causal constructs (latent variables), which is also called a path diagram (Byrne, 2016).

5.2.4 Examining SUS Spatial and Temporal Changes and Variations Simultaneously

Urban sustainability levels and their underlying causal constructs change over times and vary across regions. It is imperative to examine urban growth dynamics simultaneously between places and over time. Panel data analysis (PDA), also called econometric panel data modeling, provides a flexible mechanism to address both spatial and temporal changes and disparities (Xie et al., 2021). PDA is a regression technique that can examine multi-scaled spatiotemporal causal relationships between ecological functions and coupled natural and human systems. PDA explicitly considers both time and space simultaneously and has been widely applied in econometrics (Baltagi et al., 2007; Elhorst, 2010; Parent & LeSage, 2011) and has started to emerge in recent ecological (Scrieciu, 2007; Miller-Rushing et al., 2008; Demeke et al., 2011; Li et al., 2013) and geographical studies (Oud et al., 2012; Liu & Xie, 2013a). PDA has the advantage of capturing complex data trends in multiple dimensions of cross-sections and time series concurrently.

Moreover, PDA can be asymmetrically adjusted to examine how the variables in time-series respond to the deviations from the equilibrium through an integration of threshold vector error correction model (TVECM) (Esso, 2010; Liu & Xie, 2013b). The asymmetric adjusted PDA can account for structural breaks in the time-series caused by disastrous natural disturbances or abrupt policy change. Moreover, pass-through variables can be added into PDA to examine sequential (lagged) effects, whereas dummy variables can be included to investigate unexpected (extreme) events (Petrie et al., 2012).

A panel regression model is formulated as follows:

$$Y_{it} = \alpha + \beta_{it} X_{it} + \delta_i + \gamma_t + \varepsilon_{it} \tag{5.1}$$

where Y_{it} is the dependent variable for panel i (=1,…, N) at time t (=1,…, T), α is the intercept, X_{it} is a vector of independent variables, β_{it} is the vector of slope coefficients of X_{it}, δ_i represents the cross-sectional effect, γ_t is the time-series effect, and ε_{it} represents the errors that vary across panels and over times.

5.2.5 Investigating Legacy Effects of SUS and Their Underlying Causal Constructs on Urban Sustainability

Legacy effects are critical characteristics of environmental sustainability and societal stability (Hampton et al., 2013; Chelleri et al., 2015; Shumway & Stoffer, 2017). Particularly, time-autocorrelation or inclusion of longitudinal lagged effect is a practical approach for examining temporal dynamics of an integrated socioenvironmental system (De Blander, 2020; Filho et al., 2014; Reed & Zhu, 2017; Xie et al., 2019). The time-lagged SUS levels (the past conditions of SUS) affect the present conditions and future directions of SUS (Hull et al., 2015). In addition, the underlying causal structures of SUS also have their legacy effects. In other words, the causal constructs of SUS change over times. They have direct effects on the SUS levels during the same period and indirect impacts (legacy effects) on their future structures and the future SUS levels as well.

Based on our understanding of urban sustainability and its underlying urban system dynamics, we argue that there are at least four types of legacy effects in the context of Chinese prefecture city's sustainability case study. Type 1 lagged effects are the direct lagged effects between the super constructs, SUS, at any two consecutive periods. In other words, the SUS level at a previous period will only impact the SUS level at the next period (Fig. 5.2). Type 2 lagged effects are the cross-lagged effects between the latent variables (other constructs) in a previous period and the super construct, SUS, in the current period. Type 3 lagged effects are the straight lagged effects between the latent variables during any two consecutive periods. A cross-lagged panel model can be a combination of Types 1, 2, and/or 3 lagged effects. For example, Fig. 5.2 depicts the combinations of these three types. In other

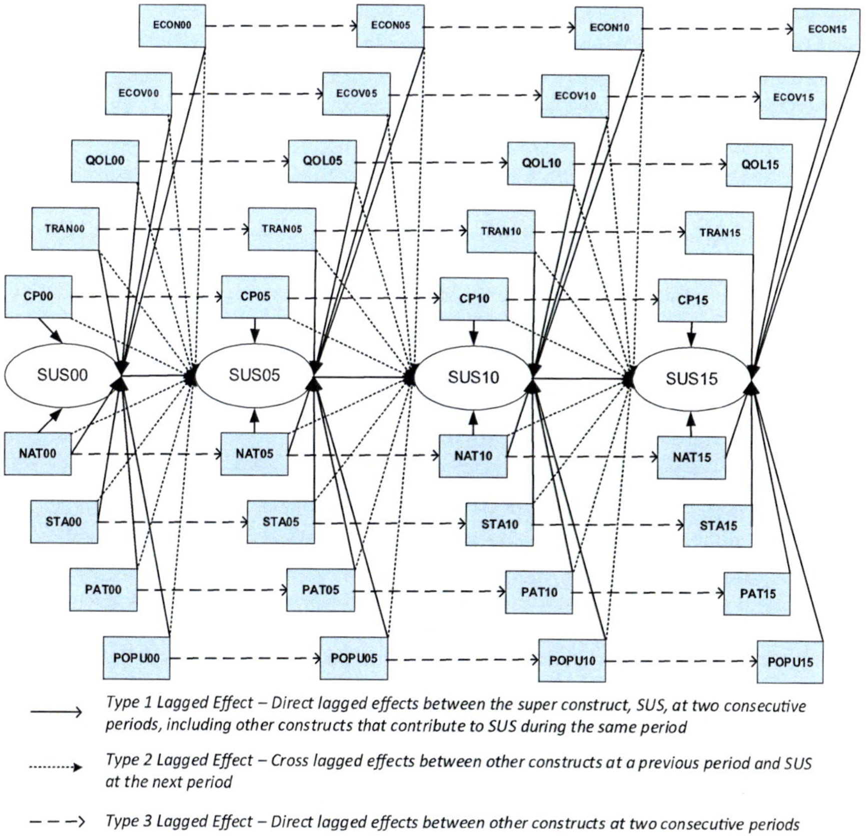

Fig. 5.2 A cross-lagged SUS measurement model

words, this model considers the lagged effects of the latent variables in a previous period on the latent variables and the SUS level in the current period, and the lagged effects of the SUS level in a precious period on the SUS level at the current period (Fig. 5.2).

Type 4 lagged effects are complex multi-consecutive lagging effects. In other words, the current three types of lagged effects are confined within two consecutive years. The lagging effects can be considered to happen between multi-consecutive periods. For instance, the SUS level in 2015 (SUS15) can have the lagged effects from SUS10, SUS05, and SUS00; SUS10 inherits the lagged effects from SUS05 and SUS00; SUS05 shows the lagged effects of SUS00. The same approach also applies to all other constructs (latent variables). As a result, Type 4 lagged effects can be named "fully lagged effects". A fully lagged effect model adds more complexity and considers the lagged effects of the latent variables on the latent variables and on the SUS levels over all previous periods. An application of Type 4 lagged effects will be explained and analyzed in Sect. 5.3.3.

A CLPM-US model attempts to describe how urban sustainability measurements respond to historical conditions, reflect their lagged impacts on current developments, and unveil evolutionary dynamics in the near future. Current SEM models can reveal various types of causal structures and associated covariations at a single time point only. PDA models can simultaneously study socioenvironmental interactions cross-sections and overtimes. Autoregressive PDA models (including lagged dependent variables) can examine simple lagged effects of either response variables or explanatory variables but cannot handle latent variables (causal structures). Cross-lagged panel models (CLPM) originated in phycological studies combine the capabilities of SEM and PDA models to disclose longitudinal causal correlations embedded in panel data, which provide a solution for us to investigate causal structures underlying urban sustainability. Therefore, this chapter adopts the concept of CLPM to develop a fully-lagged CLPM-US model to examine lagged effects of SUS levels, causal structures, and measured variables over time and measure SUS levels as a set of evolving quantities.

5.2.6 *Using CLPM-US to Compute Summative SUS Scores Over Time to Measure SUS Levels and Their Time-Series Changes*

Each construct in SEM is equivalent to a confirmatory factor analysis (CFA), which is a significant advantage of SEM. We can carry out CFA to test a measurement model. In other words, we can use CFA to quantify conceptual constructs by estimating their corresponding factor loadings (or scores) (Schumacker, 2002; Joreskog et al., 2006). We can further use the factor scores of these causal constructs to predict the scores of the super construct (Zyphur et al., 2020), the SUS levels in this case study. Thus, we can apply CLPM-US to quantify the causal constructs and estimate the SUS levels.

5.2.7 *Supplementing the Summative Scores with Their Underlying Causal Structures to Assess Unique Contributions of Urban Causal Constructs to Sustainability Scores*

The summative scores (the SUS levels) are linear estimates of the underlying causal constructs. Although we know how well each of these causal constructs correlates with the SUS levels, the scores themselves cannot disclose complex interactions between the SUS levels and the underlying causal constructs. Therefore, we need to examine the relationship between the SUS levels and the construct loadings. We will select the top 30 lists from the SUS ranking and the primary underlying constructs, respectively. We will apply two statistics, *not-matched* and *unique*, to reveal the relationships between these top 30 rankings. The number of *not-matched* is the number of the top 30 cities with the highest urban sustainability estimates that were not within the top 30 cities based on the factor scores of these causal constructs. The *not-matched* number indicated the ranking difference between the urban sustainability estimates and the factor scores of these causal constructs that underlined urban sustainability. The number of *unique* is the number of the top 30 cities only appeared once in one construct's top 30 scores but did not appear in the top 30 cities of the remaining causal constructs. This *unique* number indicated heterogeneity of this latent variable in comparison with the others.

5.2.8 *An Alternative Urban Sustainability Ranking Based on Multi-objective Optimization Problems Algorithm*

The SUS levels are the linear estimates of the nine constructs underlying the urban sustainability of the Chinese prefecture cities. The linear estimation is obtained through the correlation coefficients in the path diagrams in four periods, 2000, 2005, 2010, and 2015. We can use the SUS estimates to rank the SUS levels. We can also compare the variations between the SUS rankings and the factor score rankings of the underlying constructs to assess how these constructs contribute to the SUS rankings. We did these analyses in the above sections. In recent years, a new approach has been developed to look into the relationship between the SUS rankings and those factors that determined the rankings by evaluating their multi-dimensional trade-offs based on multi-objective optimization (MOO) (Xie et al., 2022).

MOO deals with multicriteria decision-making. When there exist multiple decisions, these decisions can complement or conflict with each other. Subsequently, optimizing (or minimizing) each decision candidate could generate a different solution. Therefore, the choices for a set of decision candidates involve a group of equally important or optimal trade-offs. In a specific case, Solution A outperforms Solution B considering one criterion, but Solution B is better than Solution A according to another standard. As a result, the outputs of multi-objective optimization will include a set of optimal but non-dominant solutions. Obtaining such a solution solves a multi-objective optimization problem (MOOP), and all answers are called MOOPs (Reed & Zhu, 2017).

Moreover, MOOPs constitute a Pareto front encompassing a set of solutions (Han et al., 2017). One solution outperforms all other solutions for at least one criterion, but will not surpass all other solutions for all criteria (Longo et al., 2019). In the case of the SUS assessment and its underlying constructs in this chapter, they compose a vector of decision candidates affecting urban sustainability. When each construct is treated as one solution and not a single construct dominates the others, the SUS assessment becomes a MOO to search for MOOPs.

There are many multi-objective evolutionary algorithms for solving MOOPs (Deb, 2011). The weighted sum method, the Tchebycheff approach, and the boundary intersection method are typical techniques (Zadeh, 1963; Geoffrion, 1968). Early algorithms solve MOOPs by transforming multi-objective problems into single-objective problems. An evolutionary optimization algorithm was developed to solve MOOPs soon (Hwang & Masud, 1979). Afterward, multi-objective optimization began to grow rapidly, resulting in many methods for solving MOOPs. These methods can be roughly divided into two categories. The first category mainly focuses on the Pareto-dominant individual selection and fitness value sharing. The multi-objective genetic algorithm (MOGA) (Fonseca & Fleming, 1993), the non-dominated sorting genetic algorithm (NSGA) (Srinivas & De, 1994), and the Niched Pareto genetic algorithm (NPGA) (Horn et al., 1994) are representatives. The other category mainly emphasizes the non-dominated solutions or individuals in the evolution process to preserve the explicit diversity. The Strength Pareto evolutionary algorithm II (SPEA-II) (Zitzler et al., 2001), the non-dominated sorting genetic algorithm II (NSGA-II) (Deb et al., 2002), the non-dominated sorting genetic algorithm III (NSGA-III) (Deb & Jain, 2014), and the Pareto envelop-based selection algorithm-II (PESA-II) (Corne et al., 2001) are typical examples.

The NSGA-III algorithm was adopted in this chapter (Xie et al., 2022). Annual Pareto fronts were calculated with nine SEM constructs and the super construct SUS in 4 years, 2000, 2005, 2010, and 2015. We designed two sets of Pareto fronts. The first set was to maximize all ten constructs: the super construct, SUS (sustainable urban status), cleaner production (CP), economic performance (ECON), economic vitality (ECOV), natural resources consumption (NAT), patents and R & D status (PAT), population resources (POPU), quality of life (QOL), state investment (STA), and transportation development (TRAN). The second set was to maximize SUS, CP, ECON, QOL, PAT, and STA but to minimize ECOV, NAT, TRAN, and POPU. The consideration of maximizing all ten constructs was focusing on the priority of comprehensive socioeconomic and urban development (Tan et al., 2018).

The rationale of choosing the second vector set was prioritizing ecological consequences of sustainable urban development (Song et al., 2021, 2022).

5.3 The Findings

5.3.1 Important Data Validation and Model Goodness-of-Fit Tests for Running an SEM Model

We composed a time-series panel dataset of 278 prefecture cities and 29 observed variables in four periods (2000, 2005, 2010, and 2015). This panel dataset is one of few comprehensive Chinese city-datasets reported so far (Table-1). Based on the social and biophysical coevolution theory, the literature review, and the principal component analysis, we proposed that Chinese prefecture cities' urban sustainability causal chain consisted of nine measurement models (causal constructs) and one structural model (the super construct) (Table 5.2). The causal constructs are also called latent variables. The term "causal constructs" was used in this chapter for consistency. The nine causal constructs were cleaner production (CP), economic performance (ECON), economic vitality (ECOV), natural resources consumption (NAT), patents and R & D status (PAT), population resources (POPU), quality of life (QOL), state investment (STA), and transportation development (TRAN). The super construct was the sustainable urban status (SUS). Each construct except SUS responded to three or four measured variables. We confirmed these measurement models through the overall Alpha statistic and the overall KMO statistics. Cronbach's alpha is a coefficient of reliability (or consistency) of the variables contained in the dataset. The *Kaiser-Meyer-Olkin* (*KMO*) is a test of sample adequacy for the factor analysis (i.e., the measurement models). Both overall Alpha and KMO values were higher than 0.90, confirming that the data was reliable and suitable for confirmatory factor analysis (structural equation model analysis) (Table 5.2).

However, the test statistics for the SEM structural model were controversial. For instance, three statistics under the population error revealed the model fit was moderate but non-satisfactory (Table 5.3). In general, a good fit model prefers having a RMSEA (root mean squared error of approximation) less than 0.08 (Xia & Yang, 2019), the lower bound of the 90% CI (confidence interval) below 0.05, and the upper bound less than 0.10. (Browne and Cudeck (1993). Our model did not achieve these statistical thresholds although they were statistically significant. Moreover, the Akaike information criterion (AIC) and Bayesian (or Schwarz) information criterion (BIC) are used to compare the fit of different models. Smaller values indicate a better fit. Since no comparison was done in our case study, AIC and BIC were not meaningful. The other group of statistics include the standardized root mean squared residual (SRMR), the comparative fit index (CFI), and the Tucker-Lewis index (TLI.) Usually, when SRMR is less than 0.10 and CFI and TLI are close to 1.0, the model shows an excellent fit. In our case study, SRMR = 0.143, CFI = 0.680, and

Table 5.2 The Cronbach's alpha and *Kaiser-Meyer-Olkin* (*KMO*) tests

Latent variable	Measured variable	Internal correlation (Alpha)	Overall Alpha	Sampling adequacy(KMO)	Overall KMO
NAT	cro_p	0.6561	0.9374	0.7682	0.9293
	wat_p			0.8867	
	wet_p			0.6819	
ECON	gdppc	0.8026		0.9533	
	wage			0.9516	
	gdp_l			0.9100	
POPU	popu_all	0.7879		0.7963	
	p_urban			0.8599	
	employee			0.9064	
EDU	edu	0.8062		0.9653	
	sci			0.9573	
	pat			0.9671	
CP	dust	0.7086		0.8564	
	so2			0.8202	
	sewage			0.8368	
STA	fixnet	0.9143		0.9613	
	fixacc			0.9818	
	gov_re			0.9624	
ECOV	cons	0.8340		0.9615	
	whole			0.9759	
	h_emp			0.8904	
	w_emp			0.9525	
TRANS	bus	0.7299		0.9364	
	road			0.9477	
	car			0.9027	
QOL	uni_p	0.7251		0.9127	
	unitea			0.9218	
	hos_p			0.8366	
	doc_p			0.9620	

TLI = 0.647, indicated that the SEM model was not a well-fit model. On the other hand, CD (coefficient of determination) is like R^2 for the whole model. CD = 0.997 in this case study was very high. Finally, the model got a warning that the maximization process was not converged.

The common challenge faced by a SEM model is a limited data size but contains a large set of parameters. Our model includes 10 latent variables and 29 measured variables. In addition, the model examines around 278 cities from 2000 to 2015, but only has 4 years of data. The measurement models and the structural model jointly generate an enormous number of parameters in the modeling process. However, many of these parameters cannot be identified in the small dataset the case study provides. This challenge is a typical SEM model identification problem, which

Table 5.3 The SEM structural model goodness-of-fit statistics

Fit statistic	Value	Description
Likelihood ratio		
chi^2_ms(368)	8224	Model vs. saturated
$p > chi^2$	0.000	
chi^2_bs(406)	24,953	Baseline vs. saturated
$p > chi^2$	0.000	
Population error		
RMSEA	0.145	Root mean squared error of approximation
90% CI, lower bound	0.142	
90% CI, upper bound	0.147	
pclose	0.000	Probability RMSEA $\leq$ 0.05
Information criteria		
AIC	87,579	Akaike's information criterion
BIC	88,053	Bayesian information criterion
Baseline comparison		
CFI	0.680	Comparative fit index
TLI	0.647	Tucker-Lewis index
Size of residuals		
SRMR	0.143	Standardized root mean squared residual
CD	0.997	Coefficient of determination

leads to non-convergency and non-satisfactory model-fit statistics (Kelly, 2011; Schechter, 2019). The solution is to simplify the model and add more data.

Nevertheless, it is interesting to point out that the SEM model designed in this chapter disclosed meaningful findings from the socioeconomic and environmental implications of urban development in China (see the following sections). Although the model fit was just moderate and the convergency was not achieved, the application and policy implementations of the case study are very informative. Finally, this chapter has a methodology focus, and it presents a comprehensive introduction of the CLPM model in urban sustainability studies. It also discusses the advantages and limitations of the CLPM-US modeling method.

5.3.2 *Causal Constructs of SUS of Chinese Prefecture Cities*

The structural path diagram in 2015 (Fig. 5.3) depicted the relationships between the ten constructs. SUS in 2015 (SUS15) was the super construct and responded to five constructs, NAT15, ECON15, CP15, QOL15, and STA15. In other words, SUS 15 was dependent on NAT15, ECON15, CP15, QOL15, and STA15. On the other hand, SUS15 explained two constructs, POPU15 and PAT15. We could also say SUS15 served as an explanatory construct to POPU15 and PAT15. The socioeconomic explanation was that the SUS level depended on the status or levels of NAT15,

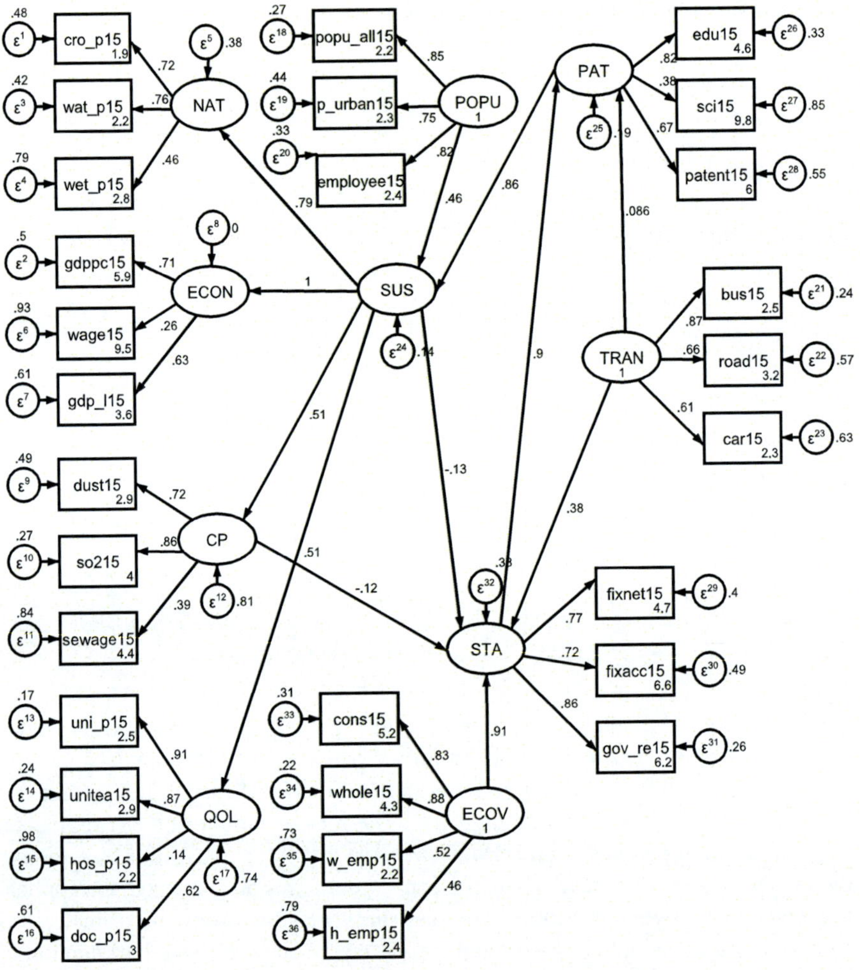

Fig. 5.3 The structural path diagram of urban sustainability status of 278 Chinese prefecture cities in 2015. *The full description of the 29 measured variables is provided in S1-C5

ECON15, CP15, QOL15, and STA15. SUS15 could affect POPU15 and PAT15. PAT15, in turn, explained two constructs, ECOV15 and TRAN15.

Another critical construct was STA15, which responded to PAT15 but explained SUS15, CP15, and ECOV15. The socioeconomic implication was that STA15 was a vital independent construct to affect SUS15, CP15, and ECOV15. China is a centralized state. The state investment policy decides the governmental intervention on SUS levels, clean production, and market development (Zhang et al., 2020). On the other hand, this structural model indicated that STA15 depended on innovation and research development. Moreover, TRAN15 responded to PAT15 and STA15. This meant that transportation development heavily relied on innovation and state investment. This structural path diagram described complex interconnected causal relationships between urban sustainability status and nine physical, economic, environmental, and social constructs in the Chinese prefecture cities.

The 2015 SEM model found out that ECON15 was the most significant contributor to SUS15, and one unit of ECON15 increase would lead to a one-unit increase of SUS15. NAT15 was the second largest contributor to SUS15, and one-unit change of NAT15 would cause a 0.79-unit change of SUS15. CP15 and QOL15 were significant and positive contributors to SUS15 as well. However, STA15 was a negative and nonsignificant actor to SUS15.

The structural path diagrams in 2000, 2005, and 2010 were provided in Figs. 5.4, 5.5, and 5.6. The coefficients and their associated statistics between the ten causal constructs were summarized in Table 5.4. The coefficients and associated statistics

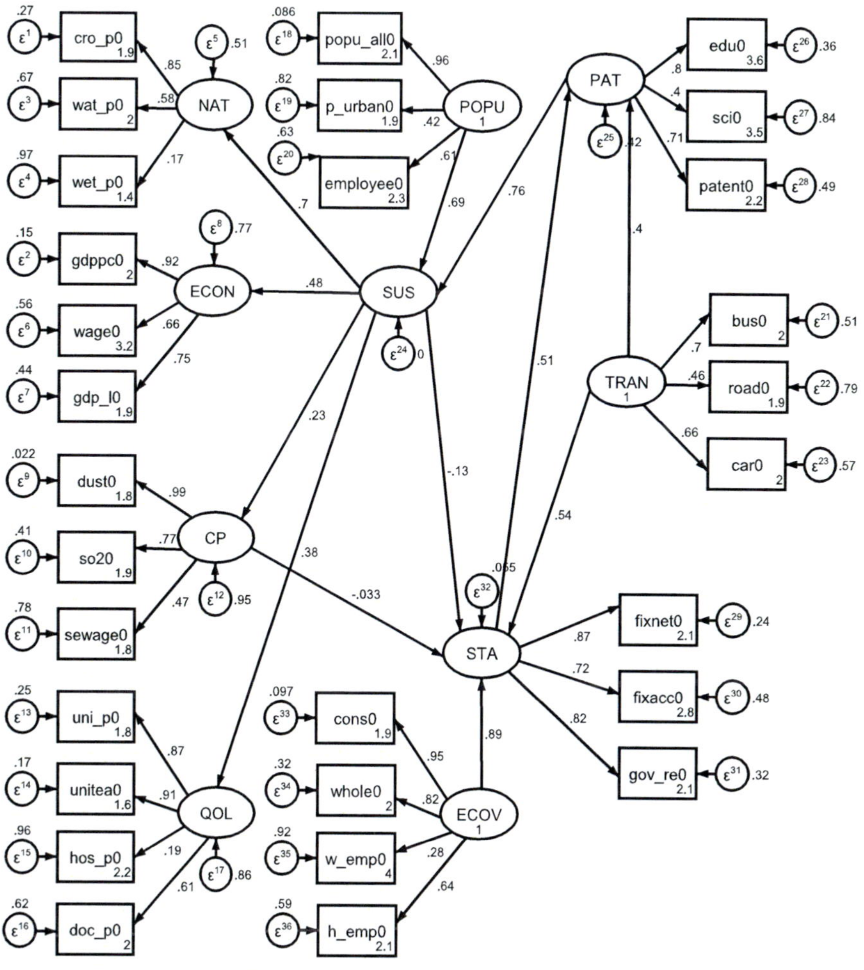

Fig. 5.4 The structural path diagram of urban sustainability status of Chinese prefecture cities in 2000

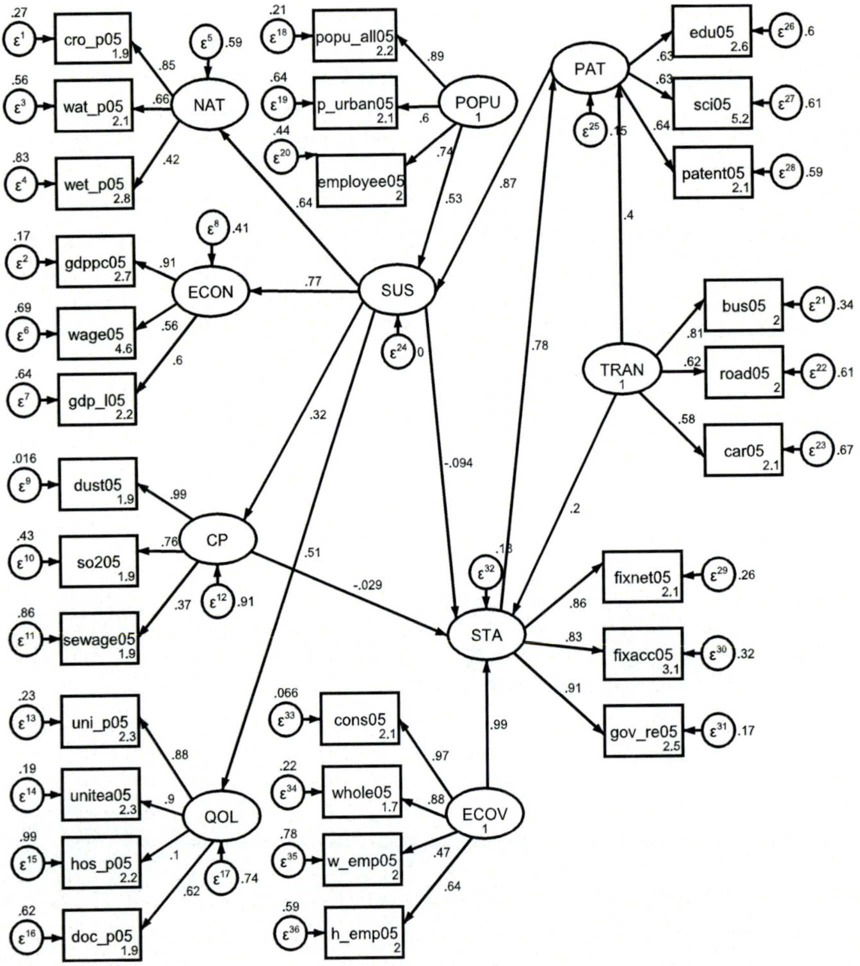

Fig. 5.5 The structural path diagram of urban sustainability status of Chinese prefecture cities in 2005

between the causal constructs and their corresponding measured variables in 2000, 2005, 2010, and 2015 were reported in this chapter – Supplementary Information.

When examining the causal constructs at each time point, we found that no causal constructs (or latent variables) were significantly affecting the urban sustainability status in 2000 (SUS00) (Table 5.4). By 2005, the constructs of economic performance (ECON05), natural resources (NAT05), quality of life (QOL05), and clean production (CP05) were making statistically significant contributions to the urban sustainability status in 2005 (SUS05) (based on $P > |z|$) in decreasing order (according to the values of *Coef.*). In 2010, CP10 was no longer a significant latent structure impacting SUS10, but STA10 became a substantial contributor to SUS10. By 2015, CP15 returned as a powerful influencer, but STA15 dropped out as a

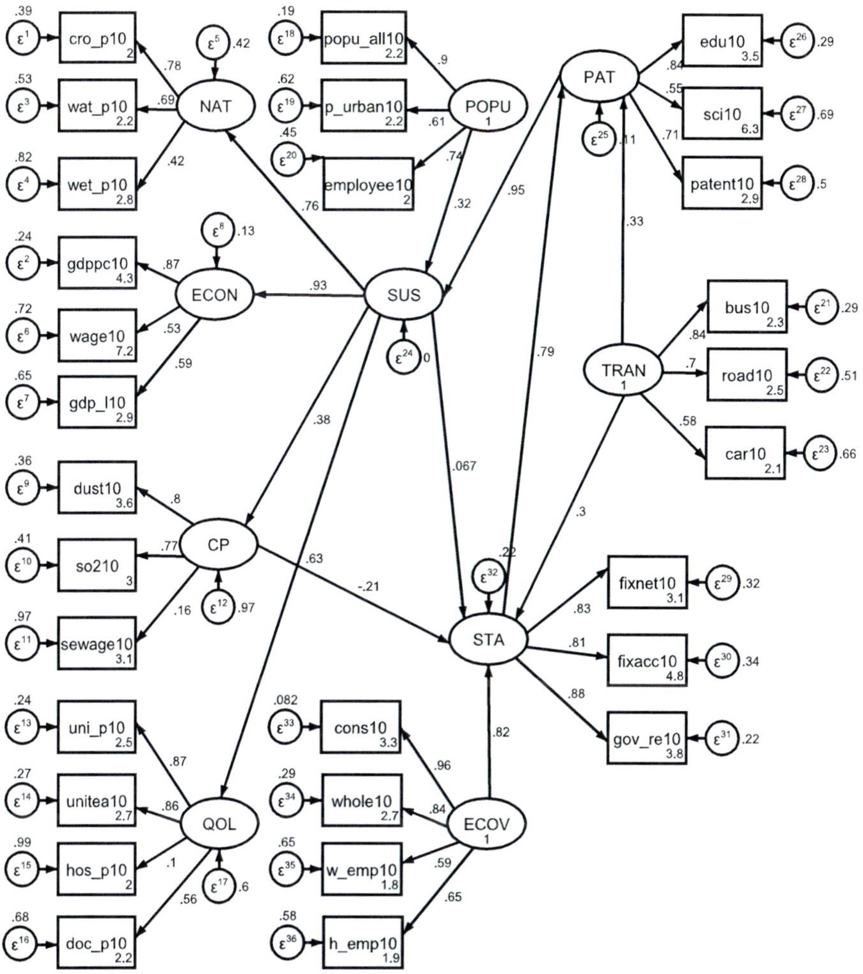

Fig. 5.6 The structural path diagram of urban sustainability status of Chinese prefecture cities in 2010

contributor to SUS2015. This time-sliced analysis of SUS indicated that the notion of urban sustainability was not a dominant objective in 2000 or at least was not reflected in the 2000 structural path diagram. By 2005, SUS became a known objective. Excellent economic performance facilitated clean production, natural resource preservation, and improvement of life quality. Afterward, the swing between state investment priorities and clean production priorities became a primary limiting factor influencing SUS.

The time-sliced SEM models were implemented in the statistical software package, Stata. The Stata codes running the SEM model, SUS estimation, and factor scores of the SUS's underlying causal constructs in 2000 were provided in Fig. 5.7

Table 5.4 The coefficients and associated statistics for the latent variables of SEM at four time points

Expl. L.[a]	Corr. L.[b]	Coef.	Std. Err.	z	$P > \|z\|$
Year 2000					
NAT00	SUS00	0.702	1.076	0.650	0.514
ECON00	SUS00	0.478	1.056	0.450	0.651
CP00	SUS00	0.234	0.203	1.150	0.250
STA00[c]	CP00	−0.033	0.993	−0.030	0.973
	SUS00[c]	−0.133	0.286	−0.470	0.642
	TRAN00	0.544	1.230	0.440	0.658
	ECOV00	0.889	0.408	2.180	0.029
QOL00	SUS00	0.376	4.959	0.080	0.940
SUS00[d]	PAT00	0.756	2.607	0.290	0.772
	POPU00	0.691	3.186	0.220	0.828
PAT00[e]	**STA00**	0.511	0.232	2.200	**0.028**
	TRAN00	0.404	1.496	0.270	0.787
Year 2005[f]					
NAT05	**SUS05**	0.641	0.234	2.740	**0.006**
ECON05	**SUS05**	0.769	0.037	21.010	**0.000**
CP05	**SUS05**	0.322	0.109	2.950	**0.003**
STA05	CP05	−0.029	0.073	−0.400	0.692
	SUS05	−0.094	0.878	−0.110	0.915
	TRAN05	0.201	0.550	0.370	0.714
	ECOV05	0.991	1.055	0.940	0.347
QOL05	**SUS05**	0.512	0.199	2.570	**0.010**
SUS05	PAT05	0.868	0.343	2.530	**0.011**
	POPU05	0.532	0.307	1.730	0.083
PAT05	STA05	0.783	1.754	0.450	0.655
	TRAN05	0.398	1.554	0.260	0.798
Year 2010					
NAT10	**SUS10**	0.761	0.386	1.970	**0.048**
ECON10	**SUS10**	0.934	0.424	2.200	**0.028**
CP10	SUS10	0.382	0.337	1.140	0.256
STA10	CP10	−0.205	0.130	−1.580	0.114
	SUS10	0.067	0.016	4.240	**0.000**
	TRAN10	0.295	0.200	1.470	0.141
	ECOV10	0.817	0.120	6.790	**0.000**
QOL10	**SUS10**	0.632	0.115	5.480	**0.000**
SUS10	PAT10	0.947	0.666	1.420	0.155
	POPU10	0.324	0.950	0.340	0.733
PAT10	STA10	0.794	0.547	1.450	0.146
	TRAN10	0.329	0.297	1.100	0.269
Year 2015					
NAT15	**SUS15**	0.785	0.077	10.200	**0.000**

(continued)

Table 5.4 (continued)

Expl. L.[a]	Corr. L.[b]	Coef.	Std. Err.	z	$P > |z|$
ECON15	**SUS15**	1.000	0.096	10.420	**0.000**
CP15	**SUS15**	0.509	0.111	4.570	**0.000**
STA15	CP15	−0.120	0.180	−0.670	0.505
	SUS15	−0.134	0.378	−0.350	0.723
	TRAN15	0.378	0.167	2.270	0.023
	ECOV15	0.913	0.219	4.170	0.000
QOL15	**SUS15**	0.510	0.119	4.290	**0.000**
SUS15	**PAT15**	0.860	0.105	8.220	**0.000**
	POPU15	0.459	0.172	2.680	**0.007**
PAT15	**STA15**	0.897	0.119	7.550	**0.000**
	TRAN15	0.086	0.145	0.590	0.552

[a]Expl. L.: An exploratory latent variable in the SEM path diagram
[b]Corr. L.: A corresponding latent variable in the SEM path diagram
[c]STA00 is an explanatory latent variable to four corresponding latent variables in the SEM path diagram. SUS00 is left adjusted to emphasize the chapter's focus of urban sustainability
[d]SUS00 in this latent structure is an explanatory latent variable and is rightly adjusted to indicate this latent variable is not the corresponding latent
[e]The black-highlighted font indicates this causal relationship is statistically significant
[f]The above notions are applied to the tables in the following sections

```
sem (NAT00 -> cro_p0 wat_p0 wet_p0) (ECON00 -> gdppc0 wage0 gdp_l0) (CP00 -> dust0
so20 sewage0 STA00) (QOL00 -> uni_p0 unitea0 hos_p0 doc_p0) (SUS00 -> NAT00 ECON00
CP00 QOL00 STA00) (POPU00 -> popu_all0 p_urban0 employee0 SUS00) (PAT00 -> edu0
sci0 patent0 SUS00) (STA00 -> fixnet0 fixacc0 gov_re0 PAT00) (ECOV00 -> cons0 whole0
w_emp0 h_emp0 STA00) (TRAN00 -> bus0 road0 car0 PAT00 STA00), covstruct(_lexogenous,
diagonal) standardized iterate(100) tolerance(1e-2) latent(NAT ECON CP QOL POPU TRAN
SUS PAT STA ECOV) nocapslatent

predict SUS, xblatent(SUS)
predict NAT, latent(NAT)
predict ECON, latent(ECON)
predict CP, latent(CP)
predict QOL, latent(QOL)
predict PAT, latent(PAT)
predict STA, latent(STA)
predict ECOV, latent(ECOV)
predict TRAN, latent(TRAN)
predict POPU, latent(POPU)
```

Fig. 5.7 Stata sample code for the 2000 SEM model

as an illustration. The Stata codes for 2005, 2010, and 2015 are similar except that the time suffix shall be replaced with the corresponding year suffix.

When a parenthesis in Fig. 5.7 consists of one latent variable (Capital Cases such as NAT00), one path (->), and several measured variables (lower cases such as cro_p0, wat_p0, and wet_p0), it comprises one measurement component of the SEM model. When a parenthesis consists of all latent variables, such as SUS00 -> NAT00

ECON00 CP00 QOL00 STA00, it depicts the primary construct model. When a parenthesis consists of one latent variable, one path, and a set of mixed measured and latent variables, such as CP00 -> dust0 so20 sewage0 STA00, the first latent variables and the measured variables compose a measurement component of the SEM model. The latent variables and the path are part of the structural model. For example, CP00 -> STA00 is part of the structural model, which indicates that CP00 responds to STA00 (or STA00 explains CP00).

Another critical analysis is to estimate SUS values and rank the SUS estimates against the factor scores of the primary constructs of the urban sustainability model of 278 Chinese prefecture cities. The Stata command, *predict*, is used to perform this task. The command, *predict*, in Stata SEM computes or creates new variables containing observation-by-observation values of estimated factor scores or predicted values of the latent and observed endogenous variables. The option, *xblatent*, is to make the linear predictions for a specified endogenous latent variable. The option, *latent*, is to estimate the factor scores of specified latent variables. In our SEM model, we want to generate the linear predicted (or estimated) values of SUS and the factor scores for other latent variables.

5.3.3 *Cross-lagged Causal Constructs of Urban Sustainability Status over 4 Years*

This chapter designed a cross-lagged panel model (across 278 cities and over four periods) to measure and examine the legacy effects and temporal dynamics of urban sustainability in Chinese prefecture cities. The cross-lagged causal structural was depicted in Fig. 5.8, which involved ten constructs and 29 observed variables, and looped through 4 years, being the most sophisticated cross-lagged structural model reported so far. However, because of the complexity of the cross-lagged model, no tools can plot out the derived structural path diagram with associated coefficients for multiple years. Instead, the coefficients and related statistics between these causal constructs were reported in Table 5.5. The coefficients and associated statistics between the causal constructs and their corresponding observed explanatory variables were reported in this chapter – Complemental Information.

The coefficients and statistics of the cross-lagged SEM model reflected temporal interactions and legacy effects between various causal structures over the four periods as a whole system. When including the cross-lagged autoregressive effects, we found that the urban sustainability in 2000 (SUS00) was mainly attributed to the natural resources (NAT00) (Table 5.5). In 2005, natural resources (NAT05), the previous sustainability status (SUS00), quality of life (QOL05), and state investment (STA05) were strongly affecting SUS05. NAT05 had the highest *Coef.* value 0.468. SUS00, QOL05, and STA05 had the *Coef.* values of 0.294, 0.054, and 0.024, respectively. By 2010, it was the first time that clean production (CP10) and economic performance (ECON10) exhibited significant contributions to SUS10.

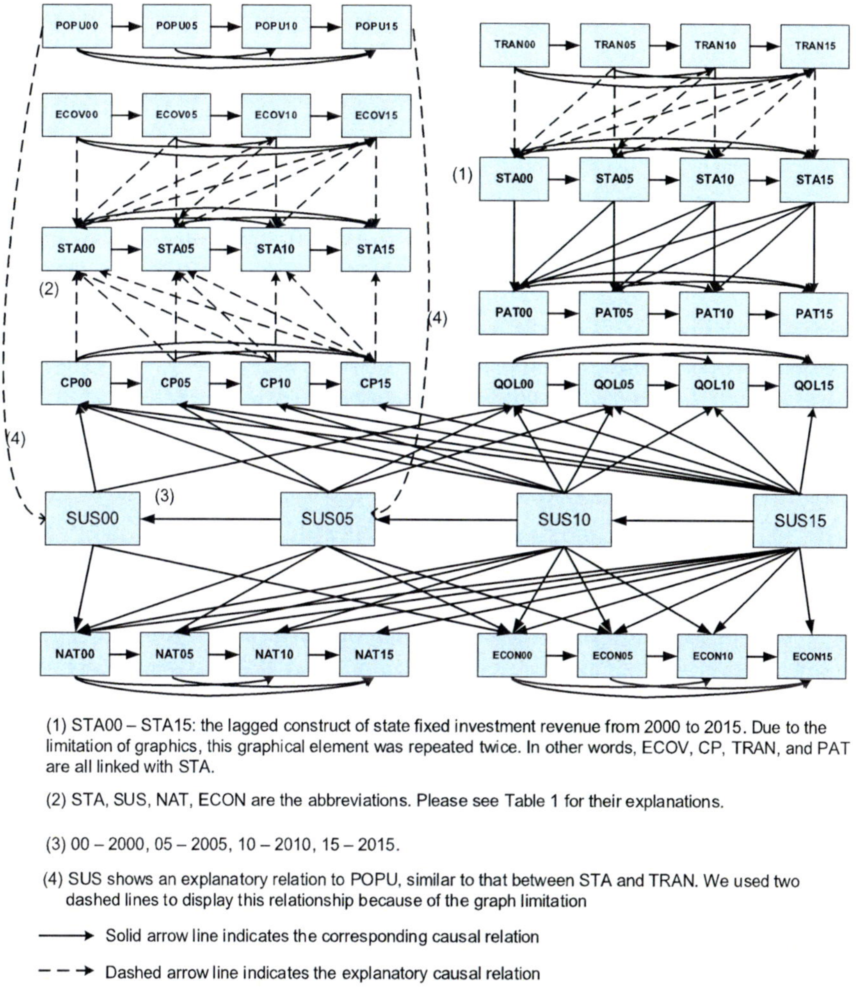

Fig. 5.8 The structural path diagram of the cross-lagged SEM with panel data of Chinese prefecture cities

According to the *Coef.* values, NAT10 (*Coef.* = 0.375), SUS05 (0.345), SUS00 (0.292), ECON10 (0.055), CP10 (0.039), QOL10 (0.039), and STA10 (0.035) significantly impacted SUS10 in decreasing order. In 2015, CP15 and QOL15 dropped out from the significant contributor list to urban sustainability (SUS15). The following constructs, NAT15 (*Coef.* = 0.464), SUS00 (0.261), SUS05 (0.234), SUS10 (0.21), ECON15 (0.022), and STA15 (0.022), significantly impacted SUS15.

Compared with the time-sliced SEM model, the cross-lagged SEM panel model revealed that SUS existed in 2000 but was only related to natural resources. The cross-lagged model traced the notion and presence of SUS back to 2000. However, this cross-lagged model did not find that CP05 significantly contributed to SUS05 in

Table 5.5 The coefficients and associated statistics for the latent variables of the cross-lagged SEM

Expl. L.[a]	Corr. L.[b]	Coef.	Std. Err.	z	P > |z|
Year 2000					
NAT00	**SUS00**	0.764	0.023	32.650	**0.000**
ECON00	SUS00	0.024	0.390	0.060	0.952
CP00	SUS00	0.021	0.210	0.100	0.922
STA00[c]	SUS00[c]	−0.032	0.129	−0.250	0.803
	CP00	−0.015	0.045	−0.330	0.743
	ECOV00	0.488	0.073	6.650	**0.000**
	TRAN00	0.632	0.020	31.040	**0.000**
QOL00	SUS00	0.004	0.240	0.020	0.986
SUS00[d]	**PAT00**[e]	0.058	0.004	13.080	**0.000**
	POPU00	0.771	0.039	19.730	**0.000**
PAT00	STA00	0.147	0.204	0.720	0.472
	TRAN00	0.066	0.128	0.510	0.607
Year 2005[f]					
NAT05	NAT00	0.187	0.174	1.080	0.282
	SUS05	0.468	0.026	18.120	**0.000**
ECON05	ECON00	0.101	0.168	0.600	0.550
	SUS05	0.060	0.172	0.350	0.727
CP05	CP00	0.104	0.071	1.470	0.143
	SUS05	0.029	0.146	0.200	0.841
STA05	STA00	0.145	0.226	0.640	*0.523*
	CP05	0.024	0.078	0.310	0.753
	SUS05	0.038	0.001	35.880	**0.000**
	ECOV05	0.084	0.161	0.520	0.601
	TRAN05	0.063	0.092	0.690	0.492
QOL05	QOL00	0.094	0.075	1.250	0.210
	SUS05	0.054	0.001	64.150	**0.000**
SUS00	**SUS05**	0.294	0.020	14.680	**0.000**
SUS05	POPU05	0.033	0.196	0.170	0.868
	PAT05	0.058	0.013	4.530	**0.000**
POPU05	POPU00	0.120	0.078	1.540	0.124
PAT05	PAT00	0.076	0.255	0.300	0.766
	STA05	0.074	0.017	4.490	**0.000**
	TRAN05	0.046	0.101	0.460	0.645
ECOV05	ECOV00	0.102	0.082	1.250	0.212
TRAN05	TRAN00	0.120	0.096	1.250	0.210
Year 2010					
NAT10	NAT00	0.188	0.677	0.280	0.781
	NAT05	0.123	0.408	0.300	0.763
	SUS10	0.375	0.046	8.170	**0.000**
ECON10	ECON00	0.081	0.012	6.720	0.000
	ECON05	0.100	0.104	0.960	0.336

(continued)

Table 5.5 (continued)

Expl. L.[a]	Corr. L.[b]	Coef.	Std. Err.	z	$P > \|z\|$
	SUS10	0.055	0.002	27.310	**0.000**
CP10	CP00	0.070	0.162	0.430	0.666
	CP05	0.080	0.173	0.460	0.644
	SUS10	0.039	0.001	30.040	**0.000**
STA10	STA00	0.119	0.230	0.520	*0.606*
	STA05	0.091	0.141	0.650	*0.515*
	CP10	0.007	0.077	0.090	0.930
	SUS10	0.035	0.001	25.780	**0.000**
	ECOV10	0.075	0.133	0.560	0.574
	TRAN10	0.058	0.090	0.650	0.518
QOL10	QOL00	0.083	0.124	0.670	0.503
	QOL05	0.099	0.125	0.790	0.430
	SUS10	0.039	0.001	29.850	**0.000**
SUS00	**SUS10**	0.292	0.029	10.130	**0.000**
SUS05	**SUS10**	0.345	0.012	28.180	**0.000**
SUS10	POPU10	0.021	0.213	0.100	0.922
	PAT10	0.114	0.214	0.530	0.594
POPU10	POPU05	0.117	0.073	1.600	0.109
	POPU00	0.114	0.007	15.210	**0.000**
PAT10	PAT00	0.081	0.118	0.690	0.492
	PAT05	0.076	0.102	0.740	0.458
	STA10	0.086	0.113	0.770	0.442
	TRAN10	0.064	0.087	0.740	0.459
ECOV10	ECOV05	0.107	0.202	0.530	0.596
	ECOV00	0.092	0.210	0.440	0.663
TRAN10	TRAN05	0.097	0.070	1.380	0.169
	TRAN00	0.096	0.001	129.570	**0.000**
Year 2015					
NAT15	NAT00	0.230	0.339	0.680	0.497
	NAT05	0.149	0.979	0.150	0.879
	NAT10	0.145	0.836	0.170	0.862
	SUS15	0.464	0.043	10.730	**0.000**
ECON15	ECON00	0.064	0.418	0.150	0.877
	ECON05	0.083	0.565	0.150	0.884
	ECON10	0.097	0.002	47.160	**0.000**
	SUS15	0.022	0.001	29.640	**0.000**
CP15	CP00	0.062	0.159	0.390	0.698
	CP05	0.065	0.173	0.370	0.708
	CP10	0.077	0.131	0.590	0.556
	SUS15	0.019	0.148	0.130	0.897
STA15	STA00	0.093	0.225	0.410	0.680
	STA05	0.074	0.173	0.430	0.668

(continued)

Table 5.5 (continued)

Expl. L.[a]	Corr. L.[b]	Coef.	Std. Err.	z	$P > \|z\|$
	STA10	0.089	0.196	0.450	0.651
	CP15	0.003	0.082	0.040	0.970
	SUS15	0.022	0.001	33.640	**0.000**
	ECOV15	0.075	0.188	0.400	0.688
	TRAN15	0.056	0.098	0.570	0.568
QOL15	QOL00	0.075	0.116	0.650	0.516
	QOL05	0.088	0.165	0.540	0.592
	QOL10	0.099	0.166	0.590	0.552
	SUS15	0.021	0.106	0.200	0.841
SUS00	**SUS15**	0.261	0.027	9.730	**0.000**
SUS05	**SUS15**	0.234	0.007	32.440	**0.000**
SUS10	**SUS15**	0.210	0.008	26.850	**0.000**
SUS15	POPU15	0.035	0.198	0.170	0.861
	PAT15	0.026	0.225	0.120	0.908
POPU15	POPU05	0.117	0.260	0.450	0.653
	POPU10	0.116	0.241	0.480	0.631
	POPU00	0.115	0.230	0.500	0.618
PAT15	PAT00	0.048	0.124	0.390	0.699
	PAT05	0.044	0.114	0.380	0.702
	PAT10	0.082	0.248	0.330	0.740
	STA15	0.068	0.219	0.310	0.758
	TRAN15	0.038	0.096	0.390	0.695
ECOV15	ECOV05	0.085	0.262	0.330	0.745
	ECOV10	0.094	0.186	0.500	0.614
	ECOV00	0.069	0.194	0.360	0.722
TRAN15	TRAN05	0.081	0.163	0.500	0.618
	TRAN10	0.097	0.130	0.740	0.458
	TRAN00	0.085	0.172	0.500	0.618

[a]Expl. L.: An exploratory latent variable in the SEM path diagram
[b]Corr. L.: A corresponding latent variable in the SEM path diagram
[c]STA00 is an explanatory latent variable to four corresponding latent variables in the SEM path diagram. SUS00 is left adjusted to emphasize the chapter's focus on urban sustainability
[d]SUS00 in this latent structure is an explanatory latent variable and is rightly adjusted to indicate this latent variable is not the corresponding latent
[e]The black-highlighted font indicates this causal relationship is statistically significant
[f]The above notions are applied to the tables in the following sections

2005. Instead, the cross-lagged model recognized CP10's significant contribution to SUS10 in 2010, contrasting with the time-sliced model's finding that CP10 did not significantly contribute to SUS10 in 2010. In addition, the cross-lagged model omitted CP15 as a significant contributor to SUS15 in 2015. In brief, the cross-lagged model signified that SUS's natural resource consumption and the historical states (i.e., the legacy effects) were the primary contributors to SUS. The cross-lagged model downgraded the influences of clean production and the quality of life on urban sustainability in China.

The outcomes of the cross-lagged SEM model were more meaningful. The cross-lagged model revealed a historical and dynamic picture of urban sustainability evolution in China. China's urban sustainability was exclusively attributed to natural resource consumption in 2000. In other words, China's urban development heavily relied on exploiting natural resources at that time. Even after 2000, the natural resource consumption construct remained the most significant contributor to urban sustainability. Second, from 2005 to 2015, the lagged effect of historical urban sustainability status showed the second most substantial impact on the current urban sustainability condition.

Third, economic performance only significantly impacted urban sustainability after 2010. Moreover, economic performance became the number one construct impacting urban sustainability in 2015, proving that economic growth was too hot around 2015. As a result, the constructs of clean production and quality of life were downgraded.

Fourth, the state investment was not positively leading to urban sustainability at the beginning of 2000 but gradually increased its impact on urban sustainability. However, state investment's contribution scale to urban sustainability was much less compared with the natural resource consumption and the legacy effects of historical sustainability conditions.

On the other hand, urban sustainability significantly progressed when China entered the year 2005. Quality of life dramatically improved and became a noticeable contributor to urban sustainability based on its *Coef.* value. The reliance on natural resource consumption was reduced (the *Coef.* value of NAT was reduced from 0.764 in 2000 to 0.468 in 2005 (Table 5.5)). By 2010, China's urban sustainability entered a new era. All five vital causal constructs, NAT, ECON, CP, QOL, and STA, helped improve urban sustainability in decreasing order. The contribution of NAT to SUS was further reduced from 0.468 to 0.375. CP (clean production) was the first time becoming a significant contributor to urban sustainability. The policy changes supported the noticeable sustainability advancement since 2005 during that period. In 2006, the Fourth Session of the Tenth National People's Congress passed the Outline of the Eleventh Five-Year Plan for National Economic and Social Development (Chen et al., 2017). The Outline requires that 1.8 billion mu of arable land be retained, which has a binding legal power and becomes an insurmountable red line. This policy reduced reliance on natural resource exploitation and stimulated clean production and sustainable economic development (Cao et al., 2009). Unfortunately, the trend of advancing sustainable urban development stopped between 2010 and 2015. The dependence of SUS on NAT increased from 0.375 in 2010 to 0.464 in 2015, returning to the level of 2005.

The cross-lagged SEM model with panel data is implemented in Stata 16. The dataset is provided in the book. The Stata codes of the cross-lagged SEM with panel data over the four periods (2000, 2005, 2010, and 2015) are given in Fig. 5.9.

```
sem 'define the latent structure model in 2000' 'define the latent structure model in 2005' 'define the latent structure model in 2010' 'define the latent structure model in 2015' 'define the cross lagged SEM model'
```

Where 'define the latent structure model in 2000' consists of the following codes:

```
(NAT00 -> cro_p0 wat_p0 wet_p0) (ECON00 -> gdppc0 wage0 gdp_I0) (CP00 -> dust0 so20 sewage0 STA00) (QOL00 -> uni_p0 unitea0 hos_p0 doc_p0) (SUS00 -> NAT00 ECON00 CP00 QOL00 STA00) (POPU00 -> popu_all0 p_urban0 employee0 SUS00) (PAT00 -> edu0 sci0 patent0 SUS00) (STA00 -> fixnet0 fixacc0 gov_re0 PAT00) (ECOV00 -> cons0 whole0 w_emp0 h_emp0 STA00) (TRAN00 -> bus0 road0 car0 PAT00 STA00)
```

The functions of 'define the latent structure model in 2005, 2010, and 2015' have similar codes to the function of 'define the latent structure model in 2000.' The only difference is to replace the year suffix with the corresponding year suffix. The code explanation is similar to Sample Code 1.

The function of 'define the cross-lagged SEM model' contains the following codes:

```
(NAT05 <- NAT00) (NAT10 <- NAT00 NAT05) (NAT15 <- NAT00 NAT05 NAT10) (POPU05 <- POPU00) (POPU10 <- POPU00 POPU05) (POPU15 <- POPU00 POPU05 POPU10) (TRAN05 <- TRAN00) (TRAN10 <- TRAN00 TRAN05) (TRAN15 <- TRAN00 TRAN05 TRAN10) (ECON05 <- ECON00) (ECON10 <- ECON00 ECON05) (ECON15 <- ECON00 ECON05 ECON10) (SUS05 <- SUS00) (SUS10 <- SUS00 SUS05) (SUS15 <- SUS00 SUS05 SUS10) (PAT05 <- PAT00) (PAT10 <- PAT00 PAT05) (PAT15 <- PAT00 PAT05 PAT10) (CP05 <- CP00) (CP10 <- CP00 CP05) (CP15 <- CP00 CP05 CP10) (STA05 <- STA00) (STA10 <- STA00 STA05) (STA15 <- STA00 STA05 STA10) (QOL05 <- QOL00) (QOL10 <- QOL00 QOL05) (QOL15 <- QOL00 QOL05 QOL10) (ECOV05 <- ECOV00) (ECOV10 <- ECOV00 ECOV05) (ECOV15 <- ECOV00 ECOV05 ECOV10), standardized iterate(100) nolog
```

Fig. 5.9 Sample code for implementing the cross-lagged SEM with panel data model

5.3.4 Urban Sustainability Status Rankings in 2000, 2005, 2010, and 2015

The factor scores of the causal constructs in the cross-lagged SEM with panel data were used to estimate the SUS rankings to reveal how they interacted with SUS (Schumacker, 2002; Joreskog et al., 2006; Zyphur et al., 2020). We estimated the SUS values at four time points, 2000, 2005, 2010, and 2015. We chose the top 30 cities at each year and reported their SUS estimates in Table 5.6. We discovered the following findings when comparing the top 30 cities with the highest SUS estimates in 2000, 2005, 2010, and 2015.

1. *Fifteen cities appeared in the top 30 cities in all 4 years*

   Fifteen cities appeared in the top 30 lists in all 4 years (the city name is marked green in Table 5.6). These large cities have a resident population of more than five million. In general, the larger the size of the city, the higher the level of economic development and social progress, and the better the total capacity for sustainable development (Sun et al., 2016). Among the 15 cities, Wuhan, Shenyang, Guangzhou, Nanjing, Hangzhou, Changsha, Jinan, Dalian, and Fuzhou are provincial capital cities. In addition, except for Wuhan and Changsha, the remaining 13 cities are located in eastern China. The process of China's reform and opening-up has advanced from East to West. Sustainable urban development status matched well with urban primacy, economic magnitude, and city administrative level. This match also reveals regional differences between the East and West (Peng et al., 2016).

Table 5.6 Top 30 cities with the highest sustainability estimates in 2000, 2005, 2010, and 2015

City2000[1]	City2005	City2010	City2015	City Name in Chinese 2015	Freq. in 4Y by 2015[2]	Sust. Rank in 2015[3]
Guangzhou	Shenzhen	Nanjing	Wuhan City	武汉市	4	1
Shenyang	Guangzhou	Guangzhou	Shenyang	沈阳市	4	2
Wuhan	Zhuhai	Hangzhou	Changchun	长春市	2	3
Qingdao	Foshan	Wuxi	Hefei	合肥市	3	4
Nanjing	Xiamen	Wuhan	Xi'an	西安市	2	5
Dalian	Hangzhou	Suzhou	Suzhou	苏州市	4	6
Jinan	Nanjing	Qingdao	Guangzhou	广州市	4	7
Hangzhou	Wuxi	Dalian	Nanjing	南京市	4	8
Ningbo	Suzhou	Shenyang	Zhenzhou	郑州市	3	9
Shenzhen	Dongying	Shenzhen	Qingdao	青岛市	4	10
Suzhou	Ningbo	Chengdu	Hangzhou	杭州市	4	11
Zhengzhou	Jinan	Hefei	Chengdu	成都市	3	12
Kunming	Shenyang	Ningbo	Changsha	长沙市	4	13
Xiamen	Qingdao	Jinan	Nanchang	南昌市	1	14
Chengdu	Wuhan	Changsha	Nantong	南通市	1	15
Changchun	Dalian	Yantai	Jinan	济南市	4	16
Zhuhai	Zhongshan	Xiamen	Xuzhou	徐州市	1	17
Xi'an	Fuzhou	Zhuhai	Wuxi	无锡市	4	18
Yantai	Kramay	Jiaxing	Dalian	大连市	4	19
Zibo	Zhengzhou	Yangzhou	Fuzhou	福州市	4	20
Harbin	Changsha	Weihai	Yantai	烟台市	4	21
Wuxi	Zibo	Taiyuan	Harbin	哈尔滨市	3	22
Wenzhou	Yantai	Foshan	Jining	济宁市	1	23
Fuzhou	Taiyuan	Dongying	Yangzhou	扬州市	2	24
Taiyuan	Kunming	Zibo	Weifang	潍坊市	1	25
Shijiazhuang	Dongguan	Zhenjiang	Ningbo	宁波市	4	26
Dongguan	Daqing	Zhongshan	Shengzhen	深圳市	4	27
Hefei	Jiaxing	Fuzhou	Kunming	昆明市	3	28
Changsha	Wenzhou	Shaoxing	Luoyang	洛阳市	1	29
Zhongshan	Harbin	Dongguan	Shijiazhuang	石家庄市	2	30

[a]**The top–bottom orders in City2000, City2005, City 2010, and City2015 reflect their ranking order, respectively**

[b]The frequency is the repeating occurrences of a city over 4 years, which is one of the top 30 cities that have highest urban sustainability estimates

[c]The ranking is based on the urban sustainability estimates in 2015

2. *Five cities made into the top 30 cities lists three times, while four cities got in the top lists two times out of 4 years*

Xi'an, Chengdu, and Kunming are located in southwestern China; Zhengzhou, Hefei, and Shijiazhuang in northcentral China; and Harbin and Changchun in northeastern China. Eight out of the nine cities (except Yangzhou) are provincial capital cities in inland China with more than five million populations. Yangzhou is the population and economic center in the northern Jiangsu Province and is located in northcentral China. The inland capital cities generally developed slower than their counterparts in east coastal China. They were important transferring sites of industrialization from the eastern region in the early stages of China's economic reforms, and they witnessed rigorous industrial upgrading in recent years. The periodical changes of urban sustainability status of these inland capital cities exhibited an inverted U-curve relationship, the environmental Kuznets curve (Grossman & Krueger, 1991). With industrialization, these cities first produced massive pollution emissions and consumed vast volumes of natural resources (Huang & Huang, 2015). In recent years, these cities were at the turning point of sustainable urban development, where the level of sustainable development was turning around from a decline. As a result, these inland capital cities would appear intermittently in the top 30 rankings.

In addition, the Northeast region has always been China's old industrial base. After China's economic reform and opening up in 1978, the southeast coastal cities observed rapid growth because of the port advantages. However, the Northeast region in the 1990s fell behind the competition and suffered rapid economic decline. Since 2003, China has embarked on a revitalization strategy in the Northeast, and the quality of economic development has been continuously improved, especially in Shenyang, Changchun, and other provincial capitals. The level of sustainable development in these cities has been rising, and got in the country's first echelon (Xiang, 2018).

3. *There were six "black horses" in the top 30 lists*

Six cities, Nanchang, Nantong, Xuzhou, Jining, Langfang, and Luoyang, were ranked in the top 30 sustainable cities only one time in 2015. Nanchang is the capital city of the inland province, Jiangxi, and has favorable political and financial advantages. Nantong is the central city of the northern wing of the Yangtze River Delta City Cluster. Xuzhou is the central city of the Huaihai Economic Zone. Luoyang is a subcentral city of the China Central Plains City Cluster. With the favorable support of China's Urban Cluster Policy, the sustainable development level of these three cities was continuously improving. In addition, Langfang was approved as a national sustainable development experimental zone in 2008 and passed accreditation from the National Sustainable Development Experimental Zone Committee in 2013. Jining is a relatively small Shandong Province city with rich natural resources. Jining is a vital coal energy base and agricultural production base. Its development and total economic output once exceeded Nanchang and other provincial capital cities. In 2007, the State Council of China issued Opinions on Promoting the Sustainable Development of Resource-based Cities (SCC, 2007). Jining has been a model city for implementing this policy, significantly improving its sustainable development capacity. Therefore, the good urban sustainability scoring of Jining has benefited from the financial incentives of this state policy.

5.3.5 Relationship Between the Rankings of SUS and Key Causal Constructs

We further compared the rankings of SUS estimates and the orders of the factor scores of these directly related causal constructs (Table 5.7). We computed three statistics to examine the SUS rankings in 2000, 2005, 2010, and 2015 and investigate the relationships between the SUS rankings and the factor scores of the underlying casual structures of SUS.

The first one is the frequencies of the cities ranked in the top 30 cities with the highest SUS estimates in the 4 years, 2000, 2005, 2010, and 2015. The frequencies are counted by using the Excel function SUM(--EXACT(data-range, a-city-name)). The argument of *data-range* defines the lists of city names on the excel worksheet. *A-city-name* is the city you are examining if it appears in the lists of the top 30 cities in 2000, 2005, 2010, and 2015. When it occurs, the EXACT function returns 1, and the SUM function summarizes its occurrences over the lists.

The second statistic is *NotMatched*, which is the number of the top 30 cities with the highest urban sustainability estimates that do not appear in the top 30 cities based on the factor scores of a latent variable. This number indicates the ranking difference between the urban sustainability estimates and the factor scores of this latent variable. The Excel function computing the number of *NotMatched* is:

$$\text{SUMPRODUCT}(--(\text{ISNUMBER}(\text{MATCH}(\text{data}-\text{range}-1,\text{data}-\text{range}-2,0))))$$

where data-range-1 is the ranking order of the top 30 cities in SUS and data-range-2 is the ranking order of these top 30 SUS cities in a latent variable.

The third one is *unique*, the number of the top 30 cities based on the factor scores of a latent variable that do not appear in the lists of the top 30 cities of the urban sustainability and the remaining latent variables. The *unique* statistic indicates the heterogeneity of this latent variable compared to the SUS and other latent variables. The cells with unique values in Excel are highlighted through the *Conditional Formatting* command in the *Style* group on the *Home* tab.

The number of *not-matched* is the number of the top 30 cities with the highest urban sustainability estimates that were not within the top 30 cities based on the factor scores of these causal constructs. The *not-matched* number indicated the ranking difference between the urban sustainability estimates and the factor scores of these causal constructs that underlined urban sustainability. The number of *unique* suggested that a city only appeared once in the top 30 factor scores of one causal construct but did not appear in the top 30 cities of the remaining causal constructs. This *unique* number indicated heterogeneity of this latent variable compared to the others.

These statistics revealed two critical findings. First, there existed high numbers of *not-matched* and *unique* among the causal constructs, CP (clean production), STA (state investment), and NAT (natural resource consumption). Thus, CP, STA,

Table 5.7 Comparison of the rankings of the sustainability estimates and the factor scores of the critical latent variables

City2015	SUS15_R[a]	NAT15_R[b]	ECON15_R	CP15_R	QOL15_R	STA15_R
Wuhan City	1	1	1	1	2	18
Shenyang	2	2	2	48	3	15
Changchun	3	3	3	49	8	16
Hefei	4	7	4	50	1	19
Xi'an	5	16	5	2	5	33
Suzhou	6	18	7	24	17	28
Guangzhou	7	20	8	3	21	36
Nanjing	8	22	10	73	23	23
Zhenzhou	9	27	11	52	19	24
Qingdao	10	26	9	4	18	35
Hangzhou	11	44	12	27	10	41
Chengdu	12	55	18	6	4	43
Changsha	13	68	20	7	16	45
Nanchang	14	4	6	5	15	21
Nantong	15	13	15	9	92	57
Jinan	16	30	16	91	11	71
Xuzhou	17	24	22	64	133	56
Wuxi	18	21	13	55	20	20
Dalian	19	32	17	41	12	17
Fuzhou	20	99	30	63	29	47
Yantai	21	87	29	12	64	51
Harbin	22	6	19	53	13	74
Jining	23	28	25	137	132	53
Yangzhou	24	5	14	10	69	50
Weifang	25	49	26	90	80	63
Ningbo	26	79	31	11	30	40
Shengzhen	27	33	21	8	22	13
Kunming	28	91	37	115	6	30
Luoyang	29	84	47	132	135	55
Shijiazhuang	30	39	27	136	27	61
Subtotal		***2015***				
NotMatched[c]	56	12	3	16	7	18
Unique[d]	33	8	2	7	5	11
2010						
NotMatched	39	10	0	12	8	9
Unique	25	7	0	10	4	4
2005						
NotMatched	52	12	3	18	9	10
Unique	27	7	1	9	6	4
2000						
NotMatched	68	12	12	19	12	13
Unique	33	7	4	12	7	3

(continued)

Table 5.7 (continued)

[a]SUS15_R: The 30 cities in 2015 that had highest estimates of urban sustainability according to the SEM model (Fig. 5.3)
[b]NAT15_R (ECON15_R, CP15_R, QOL15_R, STA15_R): the rankings of the top 30 cities with the highest urban sustainability estimates in these latent variables, NAT15, ECON15, CP15, QOL15, and STA15, respectively. These rankings were based on the factor scores of these latent variables
[c]NotMatched: The number of the top 30 cities with the highest urban sustainability estimates were not within the top 30 cities based on the factor scores of a latent variable. This number indicated the ranking difference between the urban sustainability estimates and the factor scores of this latent variable
[d]Unique: This was the number of the top 30 cities based on the factor scores of a latent variable that did not appear in the lists of the top 30 cities of the urban sustainability and the remaining latent variables. This unique number indicated heterogeneity of this latent variable in comparison with the others. The cities that were unique in comparison with the rankings of the urban sustainability and other latent variables in 2015 are marked in light pink color in this table

and NAT were the primary causal constructs that determined the dynamics of urban sustainability. Second, the number of unique and not-matched cities decreased gradually from 2000 to 2010 but increased obviously after 2010 (Table 5.7). This finding confirmed that economic growth had been overheated since 2010.

5.3.6 Revisit the Rankings of SUS and Causal Constructs from the Perspective of Multi-objective Optimization (MOO) Approach

The rankings of SUS are estimated through the linear predictions of the super structure (i.e., the latent structure) of the cross-lagged SEM model. The higher a city's SUS ranking, the larger the factor scores of the city's underlying latent variables (Table 5.8). The SUS ranking of +1 indicates that this city is rated Top 1 in its sustainability level, while the SUS ranking of −1 means it is ranked Bottom 1. The factor scores are converted to the Likert scale 1–10. Ten is the highest score while 1 is the lowest score. Table 5.8 reveals that the top-ranking cities have the highest factor scores. The factor scores of the top 10 SUS ranked cities are almost identical 10.0. However, the factors scores of the bottom 10 SUS ranked cities are most at 1.0.

However, if we adopt multi-objective optimization approaches to assess relationships between the SUS rankings and the factor scores of the underlying latent variables, the outcomes will differ. For instance, we assume two scenarios to construct the Pareto fronts. The first one is to maximize all ten constructs: the sustainable urban status (SUS), cleaner production (CP), economic performance (ECON), economic vitality (ECOV), natural resources consumption (NAT), patents and R & D status (PAT), population resources (POPU), quality of life (QOL), state investment (STA), and transportation development (TRAN). This assumption is analogous to the cross-lagged SEM model. The sole difference is that the SEM model uses the

Table 5.8 The top and bottom SUS-ranked 30 cities and the associated factor (latent variable) scores

City	Chinese name	SUS ranking	SUS	CP	ECON	QOL	PAT	STA	ECOV	NAT	TRAN	POPU
Wuhan	武汉市	1	10	10	10	10	10	10	10	10	10	10
Shenyang	沈阳市	2	10	9	10	10	10	10	10	10	10	10
Changchun	长春市	3	10	9	10	10	10	10	10	10	10	10
Hefei	合肥市	4	10	9	10	10	10	10	10	10	10	10
Xi'an	西安市	5	10	10	10	10	10	9	10	10	9	10
Suzhou	苏州市	6	10	10	10	10	10	9	10	10	9	10
Guangzhou	广州市	7	10	10	10	10	9	9	10	10	9	10
Nanjing	南京市	8	10	8	10	10	10	10	9	10	10	10
Zhengzhou	郑州市	9	10	9	10	10	10	10	9	10	10	10
Qingdao	青岛市	10	10	10	10	10	10	9	10	10	10	10
Hangzhou	杭州市	11	10	10	10	10	9	9	10	9	9	10
Chengdu	成都市	12	10	10	10	10	9	9	10	9	9	10
Changsha	长沙市	13	10	10	10	10	9	9	10	8	9	10
Nanchang	南昌市	14	10	10	10	10	10	10	10	10	8	9
Nantong	南通市	15	10	10	10	7	8	8	10	10	7	10
Jinan	济南市	16	10	7	10	10	8	8	8	9	10	10
Xuzhou	徐州市	17	10	8	10	6	8	8	9	10	7	10
Wuxi	无锡市	18	10	9	10	10	10	10	10	10	9	9
Dalian	大连市	19	10	9	10	10	10	10	9	9	10	9
Fuzhou	福州市	20	10	8	9	9	9	9	9	7	9	9
Yantai	烟台市	21	10	10	9	8	9	9	10	7	9	9
Harbin	哈尔滨市	22	10	9	10	10	8	8	9	10	8	10
Jining	济宁市	23	10	6	10	6	8	9	10	9	7	10
Yangzhou	扬州市	24	10	10	10	8	8	9	10	10	6	9
Weifang	潍坊市	25	10	7	10	8	8	8	10	9	5	10

Ningbo	宁波市	26	10	10	9	9	9	9	9	8	9	9
Shenzhen	深圳市	27	10	10	10	10	10	10	10	9	10	8
Kunming	昆明市	28	9	6	9	10	10	9	9	7	10	9
Luoyang	洛阳市	29	9	6	9	6	9	9	9	7	8	9
Shijiazhuang	石家庄市	30	9	6	10	10	8	8	9	9	8	10
Hegang	鹤岗市	−30	2	1	2	4	2	2	2	7	4	2
Ulanqab	乌兰察布	−29	2	2	1	7	4	3	1	2	10	1
Shuangyashan	双鸭山市	−28	1	1	2	9	3	3	1	2	10	1
Fuxin	阜新市	−27	1	1	2	7	3	3	3	5	5	3
Tianshui	天水市	−26	1	7	2	6	1	1	3	2	2	4
Guangyuan	广元市	−25	1	5	1	3	2	2	2	2	2	3
Baiyin	白银市	−24	1	1	1	2	3	3	2	1	5	1
Chaoyang	朝阳市	−23	1	1	1	2	2	2	3	3	3	4
Pu'er	普洱市	−22	1	5	1	3	4	4	3	1	3	1
Wuwei	武威市	−21	1	2	1	6	2	3	3	1	2	1
Black River	黑河市	−20	1	2	1	3	2	2	1	1	4	2
Baoshan	保山市	−19	1	3	1	3	2	2	3	1	1	3
Ankang	安康市	−18	1	6	1	5	1	1	1	1	1	3
Suihua	绥化市	−17	1	7	1	1	1	1	1	3	1	6
Guang'an	广安市	−16	1	5	1	1	1	1	2	4	1	5
City	毕节市	−15	1	1	1	1	1	1	1	1	1	8
Laibin	来宾市	−14	1	2	1	3	1	1	1	3	2	3
Qitaihe	七台河市	−13	1	1	1	5	2	2	2	4	4	1
Guigang	贵港市	−12	1	3	1	1	1	1	1	4	1	6
Lincang	临沧市	−11	1	2	1	2	3	3	3	1	1	1

(continued)

Table 5.8 (continued)

City	Chinese name	SUS ranking	SUS	CP	ECON	QOL	PAT	STA	ECOV	NAT	TRAN	POPU
Hechi	河池市	−10	1	1	1	2	1	1	1	1	3	3
Shangluo	商洛市	−9	1	5	1	4	1	1	1	1	1	3
Zhangjiajie	张家界市	−8	1	2	1	3	2	2	2	1	5	1
Pingliang	平凉市	−7	1	1	1	2	1	1	1	1	2	2
Yichun	伊春市	−6	1	1	1	3	2	2	1	1	5	1
Bazhong	巴中市	−5	1	7	1	1	1	1	1	1	1	5
Guyuan	固原市	−4	1	1	1	3	1	1	1	1	4	1
Zhaotong	昭通市	−3	1	3	1	1	1	1	1	1	1	5
Dingxi	定西市	−2	1	2	1	1	1	1	1	1	1	1
Longnan	陇南市	−1	1	2	1	1	1	1	1	1	1	2

linear relationships between SUS and other construct's factor scores to predict SUS. The latter adopts the multi-objective maximization to reveal which cities often fall within the Pareto front.

The second scenario is that the multi-objectives consist of maximizing the six constructs: SUS, CP, ECON, QOL, PAT, and STA while minimizing the remaining constructs: ECOV, NAT, TRAN, and POPU. The rationale for this scenario is that sustainable development does not seek overall development but advocates sustainable development. Reducing consumer consumption (ECOV) and natural resources consumption (NAT), promoting healthy travel (cutting TRAN), and avoiding overpopulation (POPU) are essential means toward sustainability (Song et al., 2021, 2022). There are 4 years of data (2000, 2005, 2010, and 2015). We applied the NSGA-III algorithm to construct annual Pareto fronts under two scenarios in 4 years, 2000, 2005, 2010, and 2015. The results are reported in Tables 5.9 and 5.10.

The outcomes of the maximization scenario have some similarities to the cross-lagged SEM model results (Table 5.9). Notably, the cities that appeared twice and more in the maximization Pareto fronts were often included in the top 30 SUS cities list. However, Tangshan, Foshan, Karamay, Weihai, and Huizhou in the Pareto fronts were not ranked very high by the CLPM model. Moreover, among the cities only once falling within the Pareto fronts, many of them were in the top 30 SUS ranked cities. The difference is that the cities in the top 30 SUS ranked list usually had overall high factor scores in all constructs. However, the cities falling in the Pareto fronts did not necessarily have high factor scores for all constructs. As we know, the Pareto fronts were generated by the multi-objective optimization algorithm, which only requires that some constructs are better than others.

It is surprising to find out that the mixed maximization–minimization scenario produced very different lists for the SEM-ranked top SUS cities and the cities falling in the Pareto fronts (Table 5.10). First, only a few cities that were included in the top 30 SUS cities, such as Zhengzhen, Hefei, Dalian, Guangzhou, Wuhan, Nanjing, and Hangzhou, appeared in the mixed scenario Pareto fronts. Second, most cities in the Pareto fronts did not show up in the top 30 SUS list. Many of them displayed very low levels of construct factor scores and, particularly, the lowest scores in ECOV, NAT, TRAN, and POPU. As we explained previously, these factor scores were minimized in the multi-objective optimization. For instance, the cities of Ya'an, Dingxi, Longnan, and Lincang disclosed the lowest scores in these constructs.

5.4 Discussion and Conclusion

In summary, SUS was highly corresponding to NAT, ECON, CP, QOL, and STA, although the strength and significance of their causal relationships changed overtimes. SUS and NAT displayed the most coherent relationship, which indicates that sustainable urban development in Chinese prefecture cities heavily relies on high natural resource consumptions (Xiao et al., 2018). SUS and QOL exhibited the second coherent relationship. However, SUS and CP showed the most sensitive

Table 5.9 The list of cities falling in the Pareto fronts in 4 years and the associated levels of SUS and causal constructs under the scenario of maximization

City	Chinese name	Years in Pareto	SUS	CP	ECON	QOL	PAT	STA	ECOV	NAT	TRAN	POPU
Wuhan	武汉市	4	10	10	10	10	10	10	10	10	10	10
Shenyang	沈阳市	4	10	9	10	10	10	10	10	10	10	10
Shenzhen	深圳市	4	10	10	10	10	10	10	10	9	10	8
Tangshan	唐山市	3	8	4	8	6	7	7	8	10	6	10
Guangzhou	广州市	3	10	10	10	10	9	9	10	10	9	10
Changchun	长春市	3	10	9	10	10	10	10	10	10	10	10
Qingdao	青岛市	3	10	10	10	10	10	9	10	10	10	10
Foshan	佛山市	2	9	10	9	6	9	9	10	9	7	8
Karamay	克拉玛依市	2	7	6	7	8	10	9	7	7	10	2
Nantong	南通市	2	10	10	10	7	8	8	10	10	7	10
Xiamen	厦门市	2	9	9	9	9	10	10	9	9	10	7
Hefei	合肥市	2	10	9	10	10	10	10	10	10	10	10
Dalian	大连市	2	10	9	10	10	10	10	9	9	10	9
Weihai	威海市	2	9	10	9	9	10	10	9	9	10	6
Huizhou	惠州市	2	9	10	9	7	9	9	9	7	9	7
Wuxi	无锡市	2	10	9	10	10	10	10	10	10	9	9
Hangzhou	杭州市	2	10	10	10	10	9	9	10	9	9	10
Suzhou	苏州市	2	10	10	10	10	10	9	10	10	9	10
Xi’an	西安市	2	10	10	10	10	10	9	10	10	9	10
Dongguan	东莞市	1	9	9	9	10	10	10	10	9	10	6
Dongying	东营市	1	8	10	9	8	10	10	8	10	10	4
Zhongshan	中山市	1	8	10	9	9	9	10	10	9	7	5
Urumqi	乌鲁木齐市	1	9	5	10	10	10	10	9	10	9	7
Baoding	保定市	1	4	7	5	8	3	3	4	6	8	10
Nanjing	南京市	1	10	8	10	10	10	10	9	10	10	10

Nanchang	南昌市	1	10	10	10	10	10	10	10	10	8	9
Hohhot	呼和浩特市	1	9	5	9	10	10	10	8	9	10	5
Daqing	大庆市	1	6	10	7	9	7	6	7	9	9	6
Taiyuan	太原市	1	9	6	10	10	10	10	9	9	10	8
Ningde	宁德市	1	4	8	5	2	5	5	5	3	3	4
Ningbo	宁波市	1	10	10	9	9	9	9	9	8	9	9
Xuzhou	徐州市	1	10	8	10	6	8	8	9	10	7	10
Chengdu	成都市	1	10	10	10	10	9	9	10	9	9	10
Jieyang	揭阳市	1	3	9	3	1	2	2	6	4	1	9
Shanwei	汕尾市	1	2	7	2	1	2	2	1	4	3	4
Jinan	济南市	1	10	7	10	10	8	8	8	9	10	10
Jining	济宁市	1	10	6	10	6	8	9	10	9	7	10
Haikou	海口市	1	8	10	8	10	8	8	8	8	9	5
Huai'an	淮安市	1	9	6	9	7	8	8	10	10	3	9
Zhangzhou	漳州市	1	8	8	8	7	7	8	8	4	6	6
Weifang	潍坊市	1	10	7	10	8	8	8	10	9	5	10
Chaozhou	潮州市	1	6	9	6	2	4	4	5	8	2	4
Yantai	烟台市	1	10	10	9	8	9	9	10	7	9	9
Zhuhai	珠海市	1	9	9	9	9	10	10	9	9	10	5
Yancheng	盐城市	1	9	8	9	4	8	8	10	9	3	9
Zhoushan	舟山市	1	8	10	8	9	10	10	10	7	7	3
Wuhu	芜湖市	1	9	8	9	9	9	10	9	10	7	6
Zhengzhou	郑州市	1	10	9	10	10	10	10	9	10	10	10
Yinchuan	银川市	1	8	6	9	9	10	10	9	9	10	5
Zhenjiang	镇江市	1	9	10	9	8	10	10	8	10	10	6
Changsha	长沙市	1	10	10	10	10	9	9	10	8	9	10

Table 5.10 The list of cities falling in the Pareto fronts in 4 years and the associated levels of SUS and causal constructs under the scenario of mixed maximization and minimization (the first six columns maximization and the last four columns minimization)

City	Chinese name	Years in Pareto	SUS	CP	ECON	QOL	PAT	STA	ECOV	NAT	TRAN	POPU
Shenzhen	深圳市	4	10	10	10	10	10	10	10	9	10	8
Ya'an	雅安市	4	4	6	3	8	5	5	5	2	1	1
Lijiang	丽江市	3	3	3	3	6	6	6	3	2	8	1
Wuzhong	吴忠市	3	3	1	2	3	5	5	1	3	9	1
Guyuan	固原市	3	1	1	1	3	1	1	1	1	4	1
Dingxi	定西市	3	1	2	1	1	1	1	1	1	1	1
Bazhong	巴中市	3	1	7	1	1	1	1	1	1	1	5
Zhangjiajie	张家界市	3	1	2	1	3	2	2	2	1	5	1
Zhangye	张掖市	3	2	1	2	5	4	4	2	3	3	1
Zhaotong	昭通市	3	1	3	1	1	1	1	1	1	1	5
Maoming	茂名市	3	3	9	4	1	1	2	7	5	1	9
Guigang	贵港市	3	1	3	1	1	1	1	1	4	1	6
Longnan	陇南市	3	1	2	1	1	1	1	1	1	1	2
Huangshan	黄山市	3	4	8	4	7	6	6	6	2	3	1
Lincang	临沧市	2	1	2	1	2	3	3	3	1	1	1
Baoshan	保山市	2	1	3	1	3	2	2	3	1	1	3
Xiamen	厦门市	2	9	9	9	9	10	10	9	9	10	7
Hefei	合肥市	2	10	9	10	10	10	10	10	10	10	10
Shangluo	商洛市	2	1	5	1	4	1	1	1	1	1	3
Dalian	大连市	2	10	9	10	10	10	10	9	9	10	9
Guangzhou	广州市	2	10	10	10	10	9	9	10	10	9	10
Wuwei	武威市	2	1	2	1	6	2	3	3	1	2	1
Wuhan	武汉市	2	10	10	10	10	10	10	10	10	10	10
Shanwei	汕尾市	2	2	7	2	1	2	2	1	4	3	4
Chizhou	池州市	2	6	4	6	6	7	7	5	7	7	1

Silver	白银市	2	1	1	1	2	3	3	2	1	5	1
Jinchang	金昌市	2	4	1	4	8	7	7	6	2	6	1
Black River	黑河市	2	1	2	1	3	2	2	1	1	4	2
Dongguan	东莞市	1	9	9	9	10	10	10	10	9	10	6
Dongying	东营市	1	8	10	9	8	10	10	8	10	10	4
Linfen	临汾市	1	3	2	3	7	4	4	4	2	3	6
Yeosu	丽水市	1	5	4	4	5	6	6	7	2	5	2
Urumqi	乌鲁木齐市	1	9	5	10	10	10	10	9	10	9	7
Yunfu	云浮市	1	3	5	2	1	3	4	4	2	2	2
Yichun	伊春市	1	1	1	1	3	2	2	1	1	5	1
Liupanshui	六盘水市	1	5	1	4	2	7	7	4	2	6	2
Nanjing	南京市	1	10	8	10	10	10	10	9	10	10	10
Ji'an	吉安市	1	5	4	4	2	4	4	2	3	3	6
Hohhot	呼和浩特市	1	9	5	9	10	10	10	8	9	10	5
Siping	四平市	1	2	4	3	7	4	4	5	5	4	4
Taiyuan	太原市	1	9	6	10	10	10	10	9	9	10	8
Weihai	威海市	1	9	10	9	9	10	10	9	9	10	6
Ankang	安康市	1	1	6	1	5	1	1	1	1	1	3
Anshun	安顺市	1	2	2	2	4	2	2	1	2	2	3
Yichun	宜春市	1	7	3	6	4	5	5	5	5	2	7
Baoji	宝鸡市	1	4	7	5	4	2	1	1	3	7	6
Chongzuo	崇左市	1	2	6	2	7	3	3	3	1	1	2
Bayannaur	巴彦淖尔市	1	3	2	3	6	6	6	4	3	3	1
Xinzhou	忻州市	1	2	1	1	7	4	4	4	1	2	3
Chengdu	成都市	1	10	10	10	10	9	9	10	9	9	10

(continued)

Table 5.10 (continued)

City	Chinese name	Years in Pareto	SUS	CP	ECON	QOL	PAT	STA	ECOV	NAT	TRAN	POPU
Fuzhou	抚州市	1	6	6	5	4	3	4	3	5	2	6
Wuxi	无锡市	1	10	9	10	10	10	10	10	10	9	9
Laibin	来宾市	1	1	2	1	3	1	1	1	3	2	3
Hangzhou	杭州市	1	10	10	10	10	9	9	10	9	9	10
Songwon	松原市	1	3	7	4	2	5	5	5	8	7	4
Wuzhou	梧州市	1	3	8	3	2	3	3	2	2	4	4
Yulin	榆林市	1	5	2	5	2	6	6	7	3	4	5
City	毕节市	1	1	1	1	1	1	1	1	1	1	8
Yongzhou	永州市	1	2	6	2	2	1	1	2	2	2	7
Hanzhong	汉中市	1	2	1	2	5	2	2	4	1	1	4
Shenyang	沈阳市	1	10	9	10	10	10	10	10	10	10	10
Heyuan	河源市	1	4	8	3	1	4	3	3	3	5	3
Jinan	济南市	1	10	7	10	10	8	8	8	9	10	10
Haikou	海口市	1	8	10	8	10	8	8	8	8	9	5
Qingyuan	清远市	1	4	3	3	1	3	3	4	3	2	5
Yulin	玉林市	1	2	7	2	1	1	1	2	2	1	7
White	白城市	1	3	5	3	7	5	5	5	6	4	2
Hakusan	白山市	1	4	7	3	4	6	6	4	1	5	2
Zhoushan	舟山市	1	8	10	8	9	10	10	10	7	7	3
Ziyang	资阳市	1	2	8	3	1	1	1	2	4	1	7
Liaoyuan	辽源市	1	3	4	4	4	5	5	4	6	7	1
Dazhou	达州市	1	2	4	2	1	1	1	2	2	1	8
Tongchuan	铜川市	1	3	3	4	6	5	5	4	6	2	2
Yinchuan	银川市	1	8	6	9	9	10	10	9	9	10	5
Zhenjiang	镇江市	1	9	10	9	8	10	10	8	10	10	6
Qingdao	青岛市	1	10	10	10	10	10	9	10	10	10	10
Yingtan	鹰潭市	1	6	6	7	5	8	8	5	6	6	1

causal relations. SUS and CP only displayed a solid and positive relationship in one of the 4 years, 2010. The explanation was that the sustainable urban development in China was still in its early stage. As a result, the priority of clean production was still lower than the priority of economic development (Cao et al., 2009).

Furthermore, the influence of state investment on urban sustainability fluctuated and was not stable (Zhong et al., 2020). The changing nature of state funding concerning urban sustainability further confirmed that the priority of China's state investment was placed on economic development. Incredibly, there was a reversion from sustainable urban growth between 2010 and 2015. From the perspective of individual city sustainability estimation, the coastal large cities were doing much better than their inland counterparts.

Nevertheless, these findings confirm that a total system approach based on the society and nature coevolution theory is vital for examining urban sustainability. The urban sustainability measurements proposed in this chapter describe a complex and an integrated urban system consisting of multiple and interconnected elements involving all sectors of human society and the environment. Some features are tangible and can be measured, while others are intangible and cannot be observed. The adoption of causal constructs through latent variables in SEM is a workable solution to represent the interactions between these elements. In other words, the concept and method of the cross-lagged panel model (CLPM) are powerful for revealing the underlying causal structures of an urban system.

Moreover, the findings confirm that urban sustainability is an evolving concept involving history, representing the current, and predicting the future. It is critical to consider lagged and cross-lagged effects of an urban system and its substructures when examining its sustainability. The cross-lagged SEM with panel data developed in this chapter provides an effective means for examining urban sustainability, its underlying constructs, and their joint lagged effects. This chapter gives a comprehensive illustration of this modeling framework.

Furthermore, the measurement of urban sustainability is not a single composite score but involves a large set of trade-offs between sub-aims of urban sustainability. Many methods can assess the trade-offs between urban sustainability levels and their underlying constructs. This chapter has made an initial attempt to investigate such trade-offs. The linear prediction associated with CLPM and the maximization method from the multi-objective optimization (MOO) assess the additive effects of the underlying constructs. As a result, the outcomes from both algorithms are close though they disclose some variations.

However, if MOO includes mixed maximization and minimization functions, it generates dramatically different results from the SEM and the all-maximized MOO functions. The mixed MOO functions approach enables researchers and modelers to examine various forms of trade-offs. There are ten constructs in this chapter, which can produce many forms of mixed MOO functions. Therefore, the mixed MOO approach can handle dynamic trade-offs between the SUS levels and their underlying constructs. For instance, when minimizing the factor scores of ECOV, NAT, TRAN, and POPU, most of the top 30 SUS cities based on the cross-lagged SEM dropped out from the top list. As we explained previously, these factor scores were minimized in the multi-objective optimization. For instance, many SUS-ranked low cities, such as Ya'an, Dingxi, Longnan, and Lincang, fell into the mixed MOO Pareto fronts. These cities have low factor scores in these constructs, which are assumed to achieve sustainability if low carbon urban growth is one critical goal (Baeumler et al., 2012). This different view of sustainability indicates that a viable or effective urban sustainability assessment framework shall support sustainability evaluation under various notions (Gao & Zhang, 2020). For instance, it has been proposed that one viable solution to hot urban development in China will be low carbon urban growth (Li & Lin, 2022). The illustration of the mixed MOO functions in this chapter confirms that empirical evaluation of urban sustainability from underlying causal structures and legacy effects is a promising systematic approach to assessing urban sustainability.

Therefore, when measuring urban sustainability, we need to follow the society–nature coevolution theory, untangle underlying causal structures of an urban system, and quantify sustainability as dynamic and evolving quantities. The rules developed in this chapter are conceptually suitable for measuring urban sustainability in other regions. In the near future, it warrants a thorough exploration of urban sustainable causal structures and a comprehensive validation of the models and the model results.

There are some limitations in the current chapter. The data items pertaining China's urban sustainability are collected mainly from China's statistical yearbooks. We are severely limited by the availability of measured data items when we explore the causal constructs that underly the urban sustainability of the Chinese prefecture cities. Second, there should be many ways to define a causal construct structure that determines urban sustainability if more data items are available. Therefore, we should validate this new way of measuring urban sustainability in other countries. Third, only one mixed set of multi-objective optimization functions was experimented in this chapter. More experiments on mixed MOO functions are needed to assess urban sustainability. Fourth, the time-series of the case study in this chapter is very short. The findings of the case study of Chinese 278 prefecture cities based on this short time duration may not correctly reflect the actual historical facts of urban sustainable development in China. Fifth, although we attempt to develop a comprehensive method to measure the urban sustainability of Chinese prefecture cities, it is difficult to validate the findings of the case study because we are developing a new approach but also are limited by the lack of historical studies of Chinese city sustainability in current literature for validation. Finally, a good design or formalization of a historical (or cross-lagged) urban structural equation model depends on many factors, including the modelers' knowledge and experience, the data completeness, the availability of supporting references. These limitations warrant more explorations and validations of this newly developed cross-lagged structural equation model with panel data in near future.

References

Ali-Toudert, F., & Ji, L. (2017). Modeling and measuring urban sustainability in multicriteria based systems – A challenging issue. *Ecological Indicators, 73*(2017), 597–611.

Ali-Toudert, F., Ji, L., Fährmann, L., & Czempik, S. (2019). Comprehensive assessment method for sustainable urban development (CAMSUD)—A new multi-criteria system for planning, evaluation and decision-making. *Progress in Planning, 140*, 100430.

Allison, P. D., Williams, R., & Moral-Benito, E. (2017). *Maximum likelihood for cross-lagged panel models with fixed effects* (p. 3). Sociological Research for a Dynamic World. https://doi.org/10.1177/2378023117710578

Austen, E., & Griffiths, S. (2022). Weight stigma predicts reduced psychological well-being and weight gain among sexual minority men: A 12-month longitudinal cohort study using random intercept cross-lagged panel models. *Body Image, 2022*, 19–29. https://doi.org/10.1016/j.bodyim.2021.10.006

Baeumler, A., Ijjasz-Vasquez, E., & Mehndiratta, S. (2012). *Sustainable low-carbon city development in China*. World Bank.

Bai, X., Surveyer, A., Elmqvist, T., Gatzweiler, F. W., Güneralp, B., Parnell, S., Prieur-Richard, A.-H., Shrivastava, P., Siri, J. G., Stafford-Smith, M., Toussaint, J.-P., & Webb, R. (2016). Defining and advancing a systems approach for sustainable cities. *Current Opinion in Environmental Sustainability, 23*, 69–78. https://doi.org/10.1016/j.cosust.2016.11.010

Baltagi, B. H., Song, S. H., Jung, B. C., & Koh, W. (2007). Testing for serial correlation, spatial autocorrelation and random effects using panel data. *Journal of Econometrics, 140*, 5–51.

Batty, M. (2013). *The new science of cities*. MIT Press.

Batty, M. (2018). *Inventing future cities*. MIT Press.

Benson, E. M., & Bereitschaft, B. (2020). Are LEED-ND developments catalysts of neighborhood gentrification? *International Journal of Urban Sustainable Development, 12*(1), 73–88. https://doi.org/10.1080/19463138.2019.1658588

Brandon, P. S., & Lombardi, P. (2010). *Evaluating sustainable development in the built environment* (2nd ed.). Chichester, West Sussex.

Brown, D. G., Agrawal, A., Sass, D. A., Wang, J., Hua, J., & Xie, Y. (2013). Responses to climate and economic risks and opportunities across national and ecological boundaries: Changing household strategies on the Mongolian Plateau. *Environmental Research Letters, 8*, 045011.

Browne, M. W., & Cudeck, R. (1993). Alternative ways of assessing model fit. In K. A. Bollen & J. S. Long (Eds.), *Reprinted in testing structural equation models* (pp. 136–162). SAGE.

Burić, I., Zuffianò, A., & López-Pérez, B. (2022). Longitudinal relationship between teacher self-efficacy and work engagement: Testing the random-intercept cross-lagged panel model. *Contemporary Educational Psychology, 70*, 102092. https://doi.org/10.1016/j.cedpsych.2022.102092

Byrne, B. M. (2016). *Structural equation modeling with Amos: Basic concepts, applications, and programming* (3rd ed.). Routledge.

Cao, J., Garbaccio, R., & Ho, M. (2009). China's 11th five-year plan and the environment: Reducing SO2 emissions. *Review of Environmental Economics and Policy*, 1–20. https://doi.org/10.1093/reep/rep006

Chelleri, L., Waters, J. J., Olazabal, M., & Minucci, G. (2015). Resilience trade-offs: Addressing multiple scales and temporal aspects of urban resilience. *Environment and Urbanization*, (January). https://doi.org/10.1177/0956247814550780

Chen, D., Li, O. Z., & Xin, F. (2017). Five-year plans, China finance and their consequences. *China Journal of Accounting Research, 10*(3), 189–230. https://doi.org/10.1016/j.cjar.2017.06.001

Cole, R. J., & Valdebenito, M. J. (2013). The importation of building environmental certification systems: International usages of BREEAM and LEED. *Building Research & Information, 41*(6), 662–676. https://doi.org/10.1080/09613218.2013.802115

Corne, D. W., Jerram, N. R., Knowles, J. D., & Oates, M. J. (2001, July). PESA-II: Region-based selection in evolutionary multiobjective optimization. In *Proceedings of the 3rd Annual conference on genetic and evolutionary computation, San Francisco, CA, USA* (pp. 283–290).

Cutaia, F. (2016). The use of landscape indicators in environmental assessment BT. In F. Cutaia (Ed.), *Strategic environmental assessment: Integrating landscape and urban planning* (pp. 29–43). Springer International Publishing. https://doi.org/10.1007/978-3-319-42132-2_4

Dasgupta, P. (2009). The welfare economic theory of green national accounts. *Environmental and Resource Economics, 42*, 3–38.

Dawson, T. P., Rounsevell, M. D. A., Kluvánková-Oravská, T., Chobotová, V., & Stirling, A. (2010). Dynamic properties of complex adaptive systems: Implications for the sustainability of service provision. *Biodiversity and Conservation, 19*(10), 2843–2853. https://doi.org/10.1007/s10531-010-9892-z

De Blander, R. (2020). Iterative estimation correcting for error auto-correlation in short panels, applied to lagged dependent variable models. *Econometrics and Statistics*. https://doi.org/10.1016/j.ecosta.2020.02.001

Deb, K. (2011). *Multi-objective optimization using evolutionary algorithms*. Wiley, New York, NY, USA.

Deb, K., & Jain, H. (2014). An evolutionary many-objective optimization algorithm using reference-point-based nondominated sorting approach, Part I: Solving problems with box constraints. *IEEE Transactions on Evolutionary Computation, 2014*(18), 577–601.

Deb, K., Pratap, A., Agarwal, S., & Meyarivan, T. (2022). A fast and elitist multi-objective genetic algorithm: NSGA-II. *IEEE Transactions on Evolutionary Computation, 2002*(6), 182–197.

Demeke, A. B., Keil, A., & Zeller, M. (2011). Using panel data to estimate the effect of rainfall shocks on smallholders food security and vulnerability in rural Ethiopia. *Climatic Change*, 108, 185–206.

Elhorst, P. (2010). Applied spatial econometrics: Raising the bar. *Spatial Economic Analysis, 5*, 9–28.

Esso, L. J. (2010). Threshold co-integration and causality relationship between energy use and growth in seven African countries. *Energy Economics, 32*, 1383–1391.

Fernandes, A. A. R., Hutahayan, B., Solimun, Arisoesilaningsih, E., Yanti, I., Astuti, A. B., Nurjannah, & Amaliana, L. (2019). Comparison of curve estimation of the smoothing spline nonparametric function path based on PLS and PWLS in various levels of heteroscedasticity. IOP conference series. *Materials Science and Engineering; Bristol, 546*(5). https://doi.org/10.1088/1757-899X/546/5/052024

Filho, A. M., da Silva, M. F., & Zebende, G. F. (2014). Autocorrelation and cross-correlation in time series of homicide and attempted homicide. *Physica A: Statistical Mechanics and Its Applications, 400*, 12–19. https://doi.org/10.1016/j.physa.2014.01.015

Fonseca, C. M., & Fleming, P. J. (1993, May 28). Multiobjective genetic algorithms. In *Proceedings of the IEE Colloquium on Genetic Algorithms for Control Systems Engineering, London, UK* (pp. 6/1–6/5).

Fu, B., Li, Y., Wang, Y., Zhang, B., Yin, S., Zhu, H., & Xing, Z. (2016). Evaluation of ecosystem service value of riparian zone using land use data from 1986 to 2012. *Ecological Indicators, 69*, 873–881. https://doi.org/10.1016/j.ecolind.2016.05.048

Gao, S., & Zhang, H. (2020). Urban planning for low-carbon sustainable development. *Sustainable Computing: Informatics and Systems, 28*, 100398.

Geoffrion, A. M. (1968). Proper efficiency and the theory of vector maximization. *Journal of Mathematical Analysis and Applications, 22*, 618–630.

Gibson, R. B., Hassan, S., Holtz, S., Tansey, J., & Whitelaw, G. (2005). *Sustainability assessment: Criteria and processes*. Earthscan.

Goodchild, M. F. (2016). GIS in the era of big data, *Cybergeo: European Journal of Geography* [En ligne], Les 20 ans de Cybergeo, mis en ligne le 25 avril 2016, consulté le 30 octobre 2022. URL: http://journals.openedition.org/cybergeo/27647

Grace, J. B. (2006). *Structural equation modeling and natural systems*. Cambridge University Press.

Grossman, G. M., & Krueger, A. B. (1991). *Environmental impact of a North American free trade agreement* (Working paper 3914). National Bureau of Economic Research.

Halkos, G., Leonti, A., & Sardianou, E. (2021). Activities, motivations and satisfaction of urban parks visitors: A structural equation modeling analysis. *Economic Analysis and Policy, 70*, 502–513. https://doi.org/10.1016/j.eap.2021.04.005

Hamaker, E. L., Kuiper, R. M., & Grasman, R. P. P. P. (2015). A critique of the cross-lagged panel model. *Psychological Methods, 20*(1), 102–116. https://doi.org/10.1037/a0038889

Hampton, S. E., Strasser, C., Tewksbury, J. J., Gram, W. K., Budden, A. E., Batcheller, A. L., Duke, C. S., & Porter, J. H. (2013). Big data and the future of ecology. *Frontiers in Ecology and the Environment, 11*, 156–162.

Han, Y., Gong, D., Jin, Y., & Pan, Q. (2017). Evolutionary multi-objective blocking lot-streaming flow shop scheduling with machine breakdowns. *IEEE Transactions on Cybernetics, 49*, 184–1967.

Horn, J., Nafpliotis, N., & Goldberg, D. E.(1994, June 27–29). A Niched Pareto genetic algorithm for multiobjective optimization. In *Proceedings of the First IEEE Conference on Evolutionary Computation, Orlando, FL, USA*.

Huang, X., & Huang, M. (2015). Relationship between sustainable urban development and economic growth based on energy analysis: A case study of Quanzhou City. *Progress in Geography, 34*(1), 38–47.

Hull, V., Tuanniu, M. N., & Liu, J. G. (2015). Synthesis of human-nature feedbacks. *Ecology and Society, 20*(3), Artn 17. https://doi.org/10.5751/Es-07404-200317

Hwang, C.-L., & Masud, A. S. M. (1979). *Multiple objective decision making, methods and applications: A state-of-the-art survey*. Springer-Verlag. ISBN 978-0-387-09111-2. Retrieved 29 May 2012.

Joreskog, K. G., Sorbom, D., & Wallentin, F. Y. (2006). *Latent variable scores and observational residuals*. Retrieved from https://citeseerx.ist.psu.edu/document?repid=rep1&type=pdf&doi=1deaafb53066650800697efd5fc8c261103350f6b

Kaur, H., & Garg, P. (2019). Urban sustainability assessment tools: A review. *Journal of Cleaner Production, 210*, 146–158. https://doi.org/10.1016/j.jclepro.2018.11.009

Kelly, S. (2011). Do homes that are more energy efficient consume less energy?: A structural equation model for England's residential sector. EPRG Working Paper 1117, Cambridge Working Paper in Economics 1139. https://ideas.repec.org/a/eee/energy/v36y2011i9p5610-5620.html.

Kline, R. B. (2011). *Principles and practice of structural equation modeling*. Guilford Press.

Kojima, R., Shinohara, R., Akiyama, Y., Yokomichi, H., & Yamagata, Z. (2021). Temporal directional relationship between problematic internet use and depressive symptoms among Japanese adolescents: A random intercept, cross-lagged panel model. *Addictive Behaviors, 120*, 106989. https://doi.org/10.1016/j.addbeh.2021.106989

Kubiszewski, I., Costanza, R., Franco, C., Lawn, P., Talberth, J., Jackson, T., & Aylmer, C. (2013). Beyond GDP: Measuring and achieving global genuine progress. *Ecological Economics, 93*, 57–68. https://doi.org/10.1016/j.ecolecon.2013.04.019

Li, Y., & Lin, G. C. S. (2022). The making of low-carbon urbanism: Climate change, discursive strategy, and rhetorical decarbonization in Chinese cities. *Environment and Planning C: Politics and Space, 40*(6), 1326–1345.

Li, S., Xie, Y., Brown, D. G., & Bai, Y. (2013). Spatial variability of the adaptation of grassland vegetation to climatic change in Inner Mongolia of China. *Applied Geography, 43*, 1–12.

Liang, J., Xie, Y., Sha, Z., & Zhou, A. (2020). Modeling urban growth sustainability in the cloud by augmenting Google Earth Engine (GEE). *Computers, Environment and Urban Systems, 84*, 101542.

Liu, Y., & Xie, Y. (2013a). Measuring the dragging effect of natural resources on economic growth: Evidence from a space–time panel filter modeling in China. *Annals of the Association of American Geographers, 103*(6), 1539–1551.

Liu, Y., & Xie, Y. (2013b). Asymmetric adjustment of the dynamic relationship between energy intensity and urbanization in China. *Energy Economics, 36*, 43–54.

Liu, J., Mooney, H., Hull, V., Davis, S. J., Gaskell, J., Hertel, T., Lubchenco, J., Seto, K. C., Gleick, P., Kremen, C., & Li, S. (2015). Systems integration for global sustainability. *Science, 347*, 1258832.

Longo, S., Montana, S. L. F., & Sanseverino, E. R. (2019). A review on optimization and cost-optimal methodologies in low energy buildings design and environmental considerations. *Sustainable Cities and Society, 2019*(45), 87–104.

Maes, C., & Vandenbosch, L. (2022). Adolescent girls' Instagram and TikTok use: Examining relations with body image-related constructs over time using random intercept cross-lagged panel models. *Body Image, 41*, 453–459. https://doi.org/10.1016/j.bodyim.2022.04.015

McPhearson, T., Pickett, S. T. A., Grimm, N. B., Niemelä, J., Alberti, M., Elmqvist, T., Weber, C., Haase, D., Breuste, J., & Qureshi, S. (2016). Advancing urban ecology toward a science of cities. *BioScience, 66*, 198–212. https://doi.org/10.1093/biosci/biw002

Merino-Saum, A., Halla, P., Superti, V., Boesch, A., & Binder, C. (2020). Indicators for urban sustainability: Key lessons from a systematic analysis of 67 measurement initiatives. *Ecological Indicators, 119*, 106879. https://doi.org/10.1016/j.ecolind.2020.106879

Miller-Rushing, A. J., Inouye, D. W., & Primack, R. B. (2008). How well do first flowering dates measure plant responses to climate change? The effects of population size and sampling frequency. *Journal of Ecology, 96*(6), 1289–1296.

Mori, K., & Christodoulou, A. (2012). Review of sustainability indices and indicators: Towards a new City Sustainability Index (CSI). *Environmental Impact Assessment Review, 32*(1), 94–106.

Mund, M., & Nestler, S. (2019). Beyond the Cross-Lagged Panel Model: Next-generation statistical tools for analyzing interdependencies across the life course. *Advances in Life Course Research, 41*, 100249. https://doi.org/10.1016/j.alcr.2018.10.002

Murakami, M., Takebayashi, Y., Harigane, M., Mizuki, R., Suzuki, Y., Ohira, T., Maeda, M., & Yasumura, S. (2020). Analysis of direction of association between radiation risk perception and relocation using a random-intercept and cross lagged panel model: The Fukushima Health Management Survey. *SSM – Population Health, 12*, 100706. https://doi.org/10.1016/j.ssmph.2020.100706

Napper, L. E., Kenney, S. R., Lac, A., Lewis, L. J., & LaBrie, J. W. (2014). A cross-lagged panel model examining protective behavioral strategies: Are types of strategies differentially related to alcohol use and consequences? *Addictive Behaviors, 39*(2), 480–486. https://doi.org/10.1016/j.addbeh.2013.10.020

Ness, B., Anderberg, S., & Olsson, L. (2010). Structuring problems in sustainability science: The multi-level DPSIR framework. *Geoforum, 41*(3), 479–488.

Oud, J. H. L., Folmer, H., Patuelli, R., & Nijkamp, P. (2012). Continuous-time modeling with spatial dependence. *Geographical Analysis, 44*(1), 29–46.

Parent, O., & LeSage, J. (2011). A space–time filter for panel data models containing random effects. *Computational Statistics and Data Analysis, 55*, 475–490.

Peng, C., Chen, Z., Wu, H., Sun, X., & Yao, N. (2016). Spatial-temporal differentiation of urban sustainable development in China based on ESDA. *China Population, Resources and Environment, 26*(2), 144–151. (in Chinese).

Petrie, M., Brunsell, S., & Nippert, J. (2012). Climate change alters growing season flux dynamics in mesic grasslands. *Theory of Applied Climatology, 107*, 427–440.

Pupphachai, U., & Zuidema, C. (2017). Sustainability indicators: A tool to generate learning and adaptation in sustainable urban development. *Ecological Indicators, 72*, 784–793.

Reed, W. R., & Zhu, M. (2017). On estimating long-run effects in models with lagged dependent variables. *Economic Modelling, 64*, 302–311. https://doi.org/10.1016/j.econmod.2017.04.006

Rogosa, D. R. (1980). A critique of cross-lagged correlation. *Psychological Bulletin, 88*, 245–258.

Roman, P., & Thiry, G. (2016). The inclusive wealth index. A critical appraisal. *Ecological Economics, 124*, 185–192.

Sabol, B. M., Kernsmith, P. D., Hicks, M. R., & Smith-Darden, J. P. (2020). Attitudes about aggression and perpetration of Adolescent Dating Aggression: A cross-lagged panel model. *Journal of Adolescence, 83*, 100–111. https://doi.org/10.1016/j.adolescence.2020.07.005

Schechter C. (2019). *CFA resulted "Warning: convergence not achieved." StataList: The Stata Forum*. https://www.statalist.org/forums/forum/general-stata-discussion/general/1497209-cfa-resulted-warning-convergence-not-achieved-in. Accessed on 9 Sept 2022.

Schumacker, R. E. (2002). Latent variable interaction modeling. *Structural Equation Modeling: A Multidisciplinary Journal, 9*(1), 40–54. https://doi.org/10.1207/s15328007sem0901_3

Scrieciu, S. S. (2007). Can economic causes of tropical deforestation be identified at a global level? *Ecological Economics, 62*(3–4), 603–612.

Sebastien, L., Bauler, T., & Lehtonen, M. (2014). Can indicators bridge the gap between science and policy? An exploration into the (Non)use and (Non)influence of indicators in EU and UK policy making. *Nature Culture, 9*(3), 316–343.

Sharifi, A. (2021). Urban sustainability assessment: An overview and bibliometric analysis. *Ecological Indicators, 121*, 107102. https://doi.org/10.1016/j.ecolind.2020.107102

Sharifi, A., & Murayama, A. (2013). A critical review of seven selected neighborhood sustainability assessment tools. *Environmental Impact Assessment Review, 38*(2013), 73–87.

Shumway, R. H., & Stoffer, D. S. (2017). *Time series analysis and its applications: With R examples*. Springer.

Simkin, V., Hodsoll, J., & Veale, D. (2022). The relationship between symptoms of obsessive compulsive disorder and depression during therapy: A random intercept cross-lagged panel model. *Journal of Behavior Therapy and Experimental Psychiatry, 76*, 101748. https://doi.org/10.1016/j.jbtep.2022.101748

Song, Y., Liu, D., & Wang, Q. (2021). Identifying characteristic changes in club convergence of China's urban pollution emission: A spatial-temporal feature analysis. *Energy Economics, 98*, 105243.

Song, Y., Yeung, G., Zhu, D., Xu, Y., & Zhang, L. (2022). Efficiency of urban land use in China's resource-based cities, 2000–2018. *Land Use Policy, 115*, 106009.

Sorjonen, K., Nilsonne, G., Melin, B., & Ingre, M. (2023). Uncertain inference in random intercept cross-lagged panel models: An example involving need for cognition and anxiety and depression symptoms. *Personality and Individual Differences, 201*, 111925. https://doi.org/10.1016/j.paid.2022.111925

Spindler, E. A. (2011). *Geschichte der Nachhaltigkeit: Vom Werden und Wirkeneines beliebten Begriffes* [Online]. https://www.nachhaltigkeit.info/media/1326279587phpeJPyvC.pdf

Srinivas, N., & Deb, K. (1994). Multi-objective optimization using nondominated sorting in genetic algorithms. *Evolutionary Computation, 1994*(2), 221–248.

SCC – State Council of China. (2007). *The Central People's Government of the People's Republic of China: Opinions on promoting the sustainable development of resources-based cities of the State Council*. http://www.gov.cn/zwgk/2007-12/24/content_841978.htm

Sun, X., Liu, X., Li, F., & Tao, Y. (2016). Comprehensive evaluation of sustainable development for different scale cities in China. *Acta Ecologica Sinica, 36*(17), 5590–5600. (in Chinese).

Szell, M. (2014). Connecting paradigms. *Science, 343*, 970.

Tan, Y., Jiao, L., Shuai, C., & Shen, L. (2018). A system dynamics model for simulating urban sustainability performance: A China case study. *Journal of Cleaner Production, 199*, 1107–1115.

Thrift, N. (2013). The cover commentary on the book. In Batty (Ed.), *The new science of cities*. The MIT Press.

UN (United Nations Department of Economic and Social Affairs, Population Division). (2022). *World population prospects 2022: Summary of results*. UN DESA/POP/2022/TR/NO. 3.

UN-HABITAT. (2015). *The city prosperity initiative (CPI) global city report – 2015*. https://smart-net.niua.org/content/ba3a1dcb-3012-44d6-87b5-fbaa28318de7

Verma, P., & Raghubanshi, A. S. (2018). Urban sustainability indicators: Challenges and opportunities. *Ecological Indicators, 93*, 282–291. https://doi.org/10.1016/j.ecolind.2018.05.007

von Edmund, A. S. (2012). *Geschichte der Nachhaltigkeit: Vom Werden und Wirkeneines Beliebten Begriffes* [Online]. Available online: https://www.nachhaltigkeit.info/media/1326279587phpeJPyvC.pdf. Accessed on 15 July 2022.

Williams, R., Allison, P. D., & Moral-Benito, E. (2018). Linear dynamic panel-data estimation using maximum likelihood and structural equation modeling. *The Stata Journal, 18*(2), 293–326. https://doi.org/10.1177/1536867X1801800201

Wong, C. (2015). A framework for 'City Prosperity Index': Linking indicators, analysis and policy. *Habitat International, 45*(1), 3–9.

World Bank Urban Development. (2020). Available online: https://www.worldbank.org/en/topic/urbandevelopment/overview#1. Accessed on 29 Dec 2021.

Xia, Y., & Yang, Y. (2019). RMSEA, CFI, and TLI in structural equation modeling with ordered categorical data: The story they tell depends on the estimation methods. *Behavior Research Methods, 51*, 409–428. https://doi.org/10.3758/s13428-018-1055-2

Xiang, N. (2018). Research on the classified evaluation of city sustainable development in China. *Science & Technology Progress and Policy., 35*(10), 121–129. https://doi.org/10.6049/kjjbydc.2017050763. (in Chinese).

Xiao, Y., Song, Y., & Wu, X. (2018). How far has China's urbanization gone? *Sustainability, 10*(8), 2953. https://doi.org/10.3390/su10082953

Xie, Y., Crary, D., Bai, Y., Cui, X., & Zhang, A. (2019). Modeling grassland ecosystem responses to coupled climate and socioeconomic influences from multi-spatial-and-temporal scales. *Journal of Environmental Informatics, 33*(1), 37–46.

Xie, Y., Fan, S., & Zhou, C. (2021). Examining ecosystem deterioration using a total socioenvironmental system approach. *Science of the Total Environment*, 784, 147171. https://authors.elsevier.com/a/1cyVJB8ccquVR

Xie, Y., Liu, C., Chang, S., & Jiang, B. (2022). Urban sustainability: Integrating socioeconomic and environmental data for multi-objective assessment. *Sustainability, 14*(15), 9142. https://doi.org/10.3390/su14159142

Xue, M., & Luo, Y. (2015). Dynamic variations in ecosystem service value and sustainability of urban system: A case study for Tianjin city, China. *Cities, 46*, 85–93. https://doi.org/10.1016/j.cities.2015.05.007

Zadeh, L. A. (1963). Optimality and non-scalar-valued performance criteria. *IEEE Transactions on Automatic Control, 8*, 59–60.

Zhang, X., Xie, Y., & Li, L. (2020). Four decades of urban and regional development and planning in China. In R. Thakur, A. Dutt, S. Thakur, & G. Pomeroy (Eds.), *Urban and regional planning and development*. Springer.

Zhang, H., Miller-Cotto, D., & Jordan, N. C. (2023). Estimating the co-development of executive functions and math achievement throughout the elementary grades using a cross-lagged panel model with fixed effects. *Contemporary Educational Psychology, 72*, 102126. https://doi.org/10.1016/j.cedpsych.2022.102126

Zhong, L., Li, X., Law, R., & Sun, S. (2020). Developing sustainable urbanization index: Case of China. *Sustainability., 12*(11), 4585. https://doi.org/10.3390/su12114585

Zitzler, E., Laumanns, M., & Thiele, L. (2001). SPEA2: Improving the strength pareto evolutionary algorithm. TIK-Report 2001, 103. https://doi.org/10.3929/ethz-a-004284029

Zyphur, M. J., Allison, P. D., Tay, L., Voelkle, M. C., Preacher, K. J., Zhang, Z., Hamaker, E. L., Shamsollahi, A., Pierides, D. C., Koval, P., & Diener, E. (2020). From data to causes I: Building a general cross-lagged panel model (GCLM). *Organizational Research Methods, 23*, 651–687.

Chapter 6
Constructing Spatially Compatible TSES Dataset: Examining Urban Declining in Southeast Michigan as the Case Study

This chapter develops a set of integrated areal interpolation techniques across socioeconomic and environmental spaces over land use and land cover patches. Urban systems involve many socioeconomic and environmental subsystems and their interactions and feedback. Socioeconomic and environmental subsystems have varied spatial scales, temporal rhythms, and distinct data characteristics. Studies of variations of ecological phenomena at multiple scales in space are well-informed traditions in environmental and geographical sciences (NRC, 2010). Empirical studies have demonstrated that spatial heterogeneity significantly impacts the diversity and resilience of ecosystems (Parrott & Meyer, 2012; Virah-Sawmy et al., 2009). The concepts of spatial heterogeneity and scale effects are true to socioeconomic systems. At the global scale, the land and natural resources that produce goods and provide ecosystem services in one place are not only interacting with its local socioeconomic conditions. Still, they are affecting and also affected by socio-environmental processes in distant places through telecoupling (Liu et al., 2018) or teleconnection (Seto et al., 2012). At the regional and local scales, productivity in a region is affected by human capital, R&D capability, and public infrastructure from neighboring areas (Bronzini & Piselli, 2009). Urban decline in city centers correlates with suburban sprawl, industrial reconstructing, demographic change, and socioeconomic inequality (Xie et al., 2018). There are many methods developed to investigate geographic spillover effects, such as the Granger causality test (Profillidis & Botzoris, 2019), spatial autoregressive model (Anselin & Smirnov, 1996), spatial error model (De Blander, 2020), spatial lag of X model (Shumway & Stoffer, 2017), and spatial Durbin model (Mur & Angulo, 2006).

More importantly, the data describing socioeconomic and environmental systems show distinct characteristics. Economic, demographic, and social statistics are collected over a nation, state, county, or an administrative enumeration area such as

---

Supplementary Information The online version contains supplementary material available at https://doi.org/10.1007/978-3-031-39594-9_6.

Y. Xie, *Total Socioenvironmental Systems*,
https://doi.org/10.1007/978-3-031-39594-9_6

census tracts in the United States. These data are often enumerated yearly or during a fixed time duration of 5 or 10 years. Surveys on specific events or policies are typical for collecting individual-level socioeconomic or policy information (Brown et al., 2013). On the other hand, environmental data are usually obtained through field samplings, such as air, plant, soil, and water samples. For a long time, airborne and spaceborne remote-sensed data have been necessary to obtain ecological and environmental information (Xie et al., 2008). In recent years, drone-based remote sensing has become an emergent and effective technique for extracting biophysical data (Sarigai et al., 2021; Sun et al., 2021; Xie et al., 2008). However, as described above, a spatial mismatch exists between socioeconomic and environmental datasets. Critical boundaries of environmental systems or ecosystems like vegetation zones or plant communities are influenced by precipitation and temperature gradients at macro-scales and topographical compositions at local scales (Sha et al., 2020; Xie et al., 2019).

On the contrary, socioeconomic systems boundaries are human determined, such as administrative, military, or political boarding lines. These two systems of boundaries are not conforming with each other. Subsequently, the attributes of a vegetation zone, such as temperature, precipitation, plant species, cannot easily be assembled with the attributes of a county or an administrative area, including population, demographics, GDP, and other socioeconomic indices. Even within an administrative area, the distributions of socioeconomic activities are heterogeneous. Residents and their economic activities may concentrate in a single city or a few large urban centers. Given this spatial mismatch between environmental and socioeconomic data sources, it is imperative to preprocess datasets of different spatial areas and inconsistent scales into a unified database with a common boundary.

There is a critical need to match environmental data collected over sapling sites or natural patches with socioeconomic data tabulated by administrative areas when investigating interactions and feedback embedded in sustainable urban system (SUS) studies. Three groups of techniques are reported in current literature to match data between different spatial boundaries. The first group is called areal interpolation (AI) and originates from geographical information science (GIS) (Burrough, 1986). AI is used to transform available data at a larger spatial unit to a finer spatial unit. In other words, AI transfers available attribute data from a larger spatial unit such as a county (source layer in the terminology of GIS) to a smaller spatial unit such as a township (target layer), where attribute data do not exist but are required to feed spatial analysis. AI's critical process is finding an intermediate layer or ancillary information to facilitate the attribute transfer from the source layer to the target layer. For example, four types of intermediate layers have been processed in GIS by overlaying the census tract population layer (the source) with the US street layer (the target layer) to transfer the population from the source layer to the target layer (Xie, 1995). These four intermediate layers are (a) the percentages of overlapped source and target layers; (b) the population densities by streets; (c) the hierarchically weighted population densities by streets; and (d) the house bearings (the house counts) by streets.

Land uses classified from remote sensing are overlayed with the Census enumeration districts to create an intermediate layer. The areas of various land use on the intermediate layer are treated as the explanatory variables in a regression model to validate the population counts estimated from census surveys (Yuan et al., 1997). The areal interpolation through GIS overlaying function is a classic GIS application.

The second group is called the computational geometry approach, including various techniques. One example is the so-called natural city. Traditionally, a city is delineated by administrative needs or statistical enumeration purposes. One alternative to defining cities without referencing economic models or political boundaries is the City Clustering Algorithm (CCA) (Rozenfeld Hernán et al., 2008). The CCA algorithm partitions a city territory into a fine square grid and assigns the number of inhabitants into each cell. Afterward, the adjacent cells with a nonzero population are merged together iteratively. Each of those grouped clusters represents a natural city. Later the concept of CCA was extended to use street nodes to delineate the boundaries of natural cities (Jiang & Jia, 2011; Jiang & Liu, 2012; Jiang et al., 2015).

Another example is the hierarchical networked urban landscape structure (Xie & Ma, 2015). Impervious land surfaces are classified from remote sensing images over a large region. The natural breaks optimization algorithm is implemented to decide an optimal number of urban land hierarchies. The discrete land patches are assigned into ordered subgroups according to a selected geometric or functional attribute. The patches at a hierarchical level are spatially allocated to the Voronoi cells at the same hierarchical level according to the spatial relationship between the patch centroids and the Voronoi cells. The hierarchical structure is constructed by iterating spatial partition and allocation operations at each hierarchical level. Taking the city of Beijing as an example, the Beijing metropolitan region consists of one central city (Beijing Inner City), six major suburban natural cities, 14 secondary suburban cities, 60 small natural cities, and 1668 natural urban neighborhoods (Xie & Ma, 2015).

Many other types of ancillary data are used to construct natural cities. For example, night-time satellite images (Bergs, 2018), percolation theory (Montero et al., 2021), fractal theory (Tannier et al., 2011), spatial interactions in social media (Ren et al., 2019), density thresholds (Arcaute et al., 2015), and Thiessen polygons partitions (Liu et al., 2019) have been adopted to delineate natural cities.

The third group is small-area population estimation with remote sensing images (Harvey, 2002). A mixture of mean reflectance and pixel counts in different residential land-use classes was used to estimate the population and dwelling unit numbers in 44 tertiary planning units in Kowloon, Hong Kong (Lo, 1995). Image tone and texture characteristics, urban pixel density, entropy, homogeneity, and distance from city centers were used to estimate dwelling densities using regression analysis in the 47 suburbs of Harare, Zimbabwe (Webster, 1996). Various combinations of Landsat six bands along with census tract population densities, percentage of the occupied area, and percentage of slums were used to build seven types of regression models to estimate the local population (Campos et al., 2020).

However, these three types of spatial data interpolation, configuration, and estimation face several common challenges. First, all these methods make assumptions and generate errors, and their accuracies vary and depend on their specific locations and the initial data characteristics. Second, the remote sensing-based methods' accuracy and estimation performance rely heavily on the availability and quality of training samples. In addition, some types of image classification and image preprocessing are involved, which are prone to many errors. Third, the computational geometry approach requires data-intensive geospatial computing. It also relies on assumptions or algorithms that approximate urban morphology or cluster bottom-up blocks or detect delineation of clustered bottom blocks. These algorithms are case-specific, prone to human interpretations, and lack transparent and compatible procedures.

Finally, the most serious challenge is that most of these case studies focus on local population estimation or validation but do not match two types of distinct boundaries of socioeconomic data and physio-geographic data (Martin & Bracken, 1993). This chapter makes the first attempt to develop a comprehensive and systematic set of spatial data interpolation tools to match these two types of boundaries. The chapter proposes that socioeconomic data are commonly enumerated over administrative areas but are seldom collected simultaneously with environmental samples. In other word, no comprehensive or complete socioeconomic statistic data accompany environmental data. A critical task of constructing a unified socioenvironmental database is to disaggregate the socioeconomic statistical data from large administrative areas to small environmental area units. The chapter further recommends that small but congruent natural land use and land cover patches are the basic environmental area unit. These smallest natural land use and land cover patches are basic natural patches (NPs) for spatial interpolation. The key breakthrough of this chapter is to use NPs as the basic spatial unit for constructing unified environmental and socioeconomic database. The socioeconomic data over large areas are disaggregated, interpolated, and allocated to NPs, in which no socioeconomic data are initially collected.

This systematic areal interpolation is called the natural patch (NP)-based socioeconomic data disaggregation and doubled-ratio reallocation. In other words, this paradigm consists of two procedures. The first one is the source–target–zone intersected subtotal adjustment, which disaggregates socioeconomic data from the source zones to target zones, adjusts the disaggregated values to local constraints, and improves the interpolation accuracies (Fig. 6.1). The second one is called the double-ratio reallocation procedure, which reallocates the disaggregated values to various natural patches. The disaggregation and reallocation processes involve many processes, concepts, and methods that do not exist at present. This chapter will first describe these processes through several pilot projects, and then synthesize them into several areal interpolation algorithms based on these illustrations.

The foundation of this newly proposed NP-based areal interpolation is the natural patch derived from remote sensing. The concept of NP comes from remote sensing technologies and the characteristics of ecological and environmental data (Xie et al., 2019). This concept has been used in land-use classification-based areal interpolation (Yuan et al., 1997) and small area population estimation with remote sensing images (Campos et al., 2020; Harvey, 2002; Lo, 1995; Webster, 1996). NP-based interpolation builds on NP, integrates with the techniques of natural cities (Jiang &

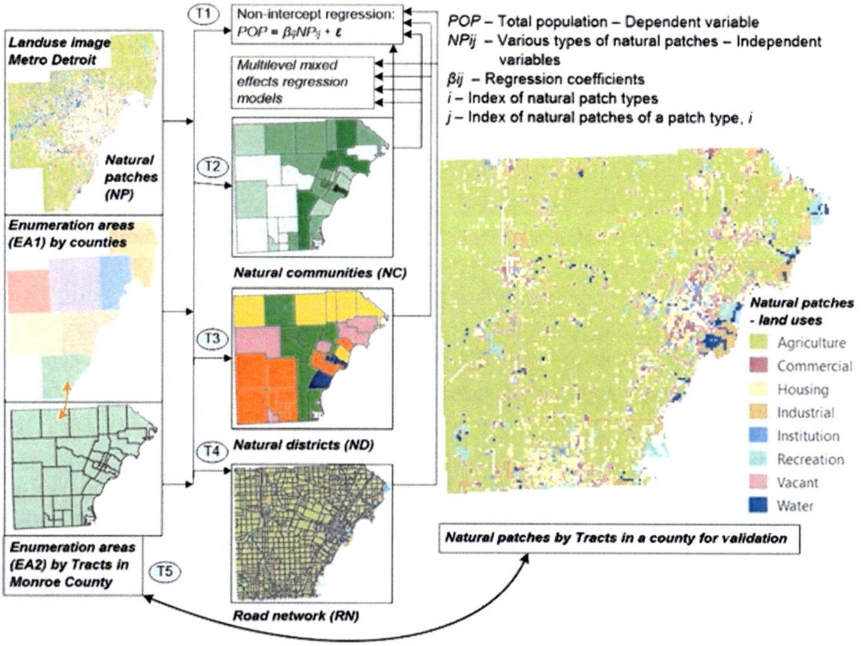

Fig. 6.1 The design diagram of areal interpolation

Jia, 2011) and hierarchical urban networks (Xie & Ma, 2015), and encompasses big data analytics. NP-based interpolation innovatively synthesizes current spatial interpolation methods and big data analytics to match two distinct socioeconomic and environmental data boundaries. We use the variable total population (POP) in Southeast Michigan (Metropolitan Detroit) as the illustration to explain the NP-based interpolation framework (Fig. 6.1).

POP is enumerated over the county, micro civil division (MCD), census tracts (CT), and block group (BG) in the United States. For the consideration of validation, we design two experiments. In the first one, we select MCD as the original source layer, CT as the target layer, and NP as the intermediate or training layer. This is the primary experiment, through which we will explain the concepts and techniques of the NP-based areal interpolation and the double-ratio reallocation. The second experiment treats CT as the source layer, BG as the target layer, and NP as the intermediate or training layer. We use this one to approve the concept that the spatial scale determines the interpolation accuracy. When we analyze the impact of POP on urban ecosystem service values, it is preferred to focus the attention on the natural patches (various types of land uses). However, POP and other socioeconomic data are generally unavailable at a finer geographical unit or by land uses (natural patches – NPs). In our case studies, the data of POP are available at the CT and BG geographies because we purposefully plan to use the available data to validate the interpolation results. However, we will remind readers that the attribute data at the target layer usually does not exist, which must be derived from some types of areal interpolation techniques or survey methods.

- Technique One (T1) – *Natural Patches-Based Areal Interpolation*: An intercept-suppressed (non-intercept) regression analysis (Yuan et al., 1997) will be carried out over the natural land-use patches within the source layer. For example,

$$\mathrm{POP}_i = \beta_{ij}\mathrm{NP}_{ij} + \varepsilon,$$

- where NP_{ij} are the independent variables (i.e., various land-use types) at *i*th MCD and *j*th natural patch; β_{ij} are the coefficients of the independent variables; and ε is the error term. POP_i is the population count at an MCD. Two new procedures, the source–target–zone intersected subtotal adjustment and the double-ratio reallocation procedure, are further applied to disaggregate and resemble socioeconomic statistics from the source zones to natural land-use patches.
- Technique Two (T2) – *Natural Communities-Based Areal Interpolation*: A multivariate geographic k-medoids clustering analysis (Ippoliti et al., 2019) will be performed to cluster the natural patches into natural communities by considering several environmental factors, such as a digital elevation model (DEM), slope, precipitation, temperature, and distance from major transportation line. A discriminant analysis can also be applied to determine how many clusters will minimize within-group and between-group differences (McLachlan, 2004). As a result, MCDs will be clustered into several natural communities. The non-intercept regression will then be applied over each natural community to improve the regression estimation by considering primary environmental factors.
- Technique Three (T3) – *Natural Districts-Based Areal Interpolation*: The primary environmental factors (DEM, slope, precipitation, temperature, and distance) and the primary socioeconomic variables will be used to conduct similar multivariate geographic k-medoids clustering analysis and quadratic discriminant analysis to organize MCDs in the study area into several clusters called natural districts. Non-intercept regression models can be applied over each natural district to interpolate POP from MCD to the natural patches.
- Technique Four (T4) – *Road Networks-Based Interpolation*: A set of network characteristics, such as "centrality" (Tocchi et al., 2022), "accessibility" (Zadeh & Rajabi, 2013), "connectivity" (Zhu et al., 2018), or "topology" (Zhang et al., 2015), will be examined to investigate the relationship between transportation network and land-use changes. Road network analysis zones (RNAZs) will be developed to describe the spatial impacts of transportation networks on land-use changes and to divide the study area into several RNAZs (Kasraian et al., 2020; Lao et al., 2016). Non-intercept regression models can be applied over each RNAZ to interpolate POP from MCD to the natural patches.
- Technique Five (T5) – T5 is not the areal interpolation technique, but the validation method. In other words, a validation test will be conducted for T1, T2, T3,

and T4, respectively, by comparing the interpolation results with the observed data in Metropolitan Detroit. Moreover, the multilevel mixed effects regression models included in the design diagram will be illustrated in Chap. 7.

6.1 T1: Natural Patches-Based Areal Interpolation (Non-intercept Regression over Natural Patches)

6.1.1 The Data

The datasets used to conduct the T1 experiment are as follows:

SEM_MCD_Poly – Southeast Michigan MCD (micro civil division) polygon feature layer, downloaded from Michigan's Open Data Portal, https://data.michigan.gov/.

SEMCOG_Landuse2020 – Southeastern Michigan Council of Government (SEMCOG) 2020 land-use polygon feature layer, downloaded from SEMCOG's Open Data Portal, https://semcog.org/data-and-maps/.

semland30 – Southeast Michigan Land Use 2020 Raster Layer at 30 m resolution (1 – agriculture; 2 – commercial; 3 – housing; 4 – industrial; 5 – institutional; 6 – recreational; 7 – vacant; 8 – water), which is converted from SEMCOG_Landuse2020 by using the ArcGIS Pro "Polygon to Raster" tool. A geotiff format of semland30 is also provided.

SEM_MCD_Data – Southeast Michigan MCD (micro civil division) polygon feature layer containing total population, total housing unit, vacant housing unit, manufacturing employment, and the areas of eight land uses. The land-use areas are obtained from semland30 by the ArcGIS Pro "Tabulate Area" tool, which tabulates land-use areas from semland30 to SEM_MCD_Poly. The land-use area is in the square feet unit.

SEM_Ch6_Land_Tract – Southeast Michigan polygon data layer containing both 2019 ACS (American Community Survey) attributes and 2020 land-use area (square miles) data by Census Tract.

SEM_Tract_MCD_T1_Acre – Intersected polygon data layer between SEM_MCD_Data and SEM_Ch6_Land_Tract. The land-use area is the converted Acre unit.

MCD_Land_ACS_Acre – This is the CSV copy of the SEM_MCD_Data, in which the area unit is converted to "acre." This is the input file to the Python file, "Areal_Int_T1_MCD."

Areal_Int_T1_MCD – A python code implements Areal Interpolation Test 1 (T1), which takes MCD_Land_ACS_Acre as an input file and outputs Param_MCD_Land_ACS_Acre as the result.

The above datasets and files are stored in the folder, /The_Book/Chapter_6/T1_Test.

6.1.2 The Interpolation Method and Procedures

T1 includes a set of intercept-suppressed (non-intercept) regression models, taking the form of

$$Y_i = \beta_{ij}\mathrm{NP}_{ij} + \varepsilon, \tag{6.1}$$

where Y_i is a dependent variable; NP_{ij} are the areas of eight natural patches: 1 – agriculture, 2 – commercial, 3 – housing, 4 – industrial, 5 – institutional, 6 – recreational, 7 – vacant, 8 – water; NP_{ij} are the independent variables (i.e., various land-use types) at *i*th MCD and *j*th natural patch; β_{ij} are the coefficients of the independent variables; and ε is the error term.

We have selected four dependent variables to test T1: *TotPop* – total population, *T_Housing* – total housing unit, *Vacant* – total vacant housing unit, and *Manuf* – the number of workers aged 15 and above working in manufacturing industries. *TotPop* is the essential information for planning. *T_Housing* discloses housing market and conveys necessary conditions for economic planning. *Vacant* is a crucial statistic judging the health and direction of the housing market and also reflects the status of urban declining, which is a critical challenge Metropolitan Detroit has been facing for decades. *Manuf* contributes to high-wage jobs, commercial innovation, and trade deficit reduction (Helper et al., 2012). Moreover, Metropolitan Detroit is known for its automobile industry, and manufacturing has long been integral to the Metro Detroit automobile industry (Li & Xie, 2018).

There are 196 valid MCDs in Southeastern Michigan. All records are randomly split into the training and testing datasets at the ratio 1:3. In other words, 66.6% of records fall into the training set, while the remaining 33.3% are in the testing dataset. The fitted regression coefficients derived from the training data are used to predict the dependent variable values by using the testing data. The predicted values are assessed with the observed values in the testing data to obtain a second adjusted R-squared value to judge the performance of the regression model. The results of T1 are reported in Tables 6.1, 6.2, 6.3, and 6.4. The Python code, Areal_Int_T1_MCD, and its input and output files are stored in the folder, /Chapter_6/T1_Test/.

Table 6.1 reveals that adjusted (Adj.) R^2 from the training data is smaller than Adj. R^2 from the testing data. In general, reducing the sample size would increase Adj. R^2. Mathematically, R^2 is calculated by dividing the sum of squares of residuals ($\mathrm{SS}_{\mathrm{res}}$) by the total sum of squared variation ($\mathrm{SS}_{\mathrm{tot}}$) and subtracting it from 1. The sum of squares of explained variation ($\mathrm{SS}_{\mathrm{reg}}$) measures the explained variation, and SSres measures the unexplained variation. $\mathrm{SS}_{\mathrm{reg}} + \mathrm{SS}_{\mathrm{res}} = \mathrm{SS}_{\mathrm{tot}}$.

$$R^2 = 1 - \mathrm{SS}_{\mathrm{res}} / \mathrm{SS}_{\mathrm{tot}} = \mathrm{SS}_{\mathrm{reg}} / \mathrm{SS}_{\mathrm{tot}}.$$

On the other hand, Adj. R^2 measures the proportion of variation explained by only those independent (or predictor) variables that help explain the dependent variable. It penalizes you for adding predictor variables or more samples that do not

Table 6.1 Statistics of TotPop non-intercept regression model

Land use	β Coeff.	*T*-Value	*P*-Value	R^2-Test[a]
Agriculture	−1.053	−4.265	0.000	0.865[b]
Commercial	−2.479	−0.252	0.801	0.909[bc]
Housing	9.239	5.466	0.000	0.000
Industrial	−2.158	−0.876	0.383	0.000
Institution	28.110	2.677	0.008	0.000
Recreation	1.006	0.575	0.566	0.000
Vacant	12.036	5.268	0.000	0.000
Water	−30.960	−6.672	0.000	0.000

[a]R^2 is an abbreviation for the adjusted *R*-squared value; *β Coeff.* indicates β_{ij}, the regression coefficients of the independent variables; *T*-Value is the *t* statistic of the coefficient, reflecting the contribution strength of an independent variable; *P*-Value is the *p* statistic, which reveals if this variable's contribution to the dependent variable is statistically significant. $P < 0.05$ is the common threshold for judging the significance level

[b]The adjusted *R*-squared value from the training data

[c]The adjusted *R*-squared value from the testing data

Table 6.2 Statistics of T_Housing non-intercept regression model

Land use	β Coeff.	*T*-Value	*P*-Value	R^2-Test[a]
Agriculture	−0.611	−4.175	0.000	0.816[b]
Commercial	−5.542	−0.950	0.344	0.892[c]
Housing	4.309	4.299	0.000	0.000
Industrial	−1.071	−0.734	0.465	0.000
Institution	14.893	2.392	0.018	0.000
Recreation	0.415	0.400	0.690	0.000
Vacant	7.699	5.682	0.000	0.000
Water	−17.337	−6.299	0.000	0.000

[a, b, c]These are noted the same as they are in Table 6.1

Table 6.3 Statistics of vacant housing non-intercept regression model

Land use	β Coeff.	*T*-Value	*P*-Value	R^2-Test[a]
Agriculture	−0.196	−3.823	0.000	0.643[b]
Commercial	−4.550	−2.226	0.028	0.783[c]
Housing	0.747	2.128	0.035	0.000
Industrial	−0.321	−0.627	0.532	0.000
Institution	3.399	1.558	0.122	0.000
Recreation	0.136	0.376	0.708	0.000
Vacant	3.072	6.470	0.000	0.000
Water	−5.492	−5.695	0.000	0.000

[a, b, c]These are noted the same as they are in Table 6.1

Table 6.4 Statistics of manufacturing employment non-intercept regression model

Land use	β Coeff.	*T*-Value	*P*-Value	R^2-Test[a]
Agriculture	−0.052	−4.536	0.000	0.934[b]
Commercial	0.589	1.285	0.201	0.892[c]
Housing	0.885	11.235	0.000	0.000
Industrial	0.034	0.293	0.770	0.000
Institution	0.645	1.318	0.190	0.000
Recreation	0.103	1.258	0.211	0.000
Vacant	0.285	2.675	0.008	0.000
Water	−1.705	−7.879	0.000	0.000

[a, b, c]These are noted the same as they are in Table 6.1

help predict the dependent variable.Adj. R^2 can be calculated mathematically in terms of the sum of squares. The only difference between R^2 and Adj. R^2 equation is the degree of freedom.

$$\text{Adj.}R^2 = 1 - \left(\text{SS}_{\text{res}} / df_{\text{e}}\right) / \left(\text{SS}_{\text{tot}} / df_{\text{t}}\right).$$

where df_e is the degrees of freedom $n - p - 1$; n is the total sample size; p is the number of predictors; and df_t is the degrees of freedom $n - 1$.

$$\text{Adj.}R^2 = 1 - \left(\text{SS}_{\text{res}} / n - p - 1\right) / \left(\text{SS}_{\text{tot}} / n - 1\right).$$

When n becomes small, but p remains the same, ($\text{SS}_{\text{res}}/n - p - 1$) can decrease, which increases the Adj. R^2.

Table 6.1 shows that the non-intersect regression model performed well in describing *TotPop*. Both adjusted R-squared values from the training and testing datasets conformed that the land uses could explain 87–90% variations of *TotPop* in Metropolitan Detroit. Moreover, the land uses of housing, vacant, and institution were the primary factors positively affecting population sizes based on *T*-Values. Water and agriculture negatively affected population size. However, commercial, industrial, and recreational land uses were not significantly influencing the total population based on their *P*-Values. These findings are consistent with the common notions in the current literature (Rodrigue, 2020). The regression outcomes for T_Housing were very similar to the results of TotPop (Table 6.2). Logically, the total housing unit is closely related to the total population.

On the other hand, the regression result of vacant housing unit showed several changes (Table 6.3). First, the R-squared values were reduced compared with the regression results of *TotPop* and *T_Housing*. Understandably, the vacancy rate varied among the communities, and the vacancy was a much worse phenomenon in the Detroit downtown area than in remote suburban areas. The relationship between *Vacant* and land uses was less robust over the entire metropolitan region. Furthermore, institutional land use was no longer a significant influencer on *Vacant*, and housing land use reduced its impact on *Vacant*.

The regression of *Manuf* against the land uses displayed several surprising findings (Table 6.4). First, four land uses, industrial, commercial, institutional, and recreation were not significantly impacting manufacturing employment. *T_Housing* and *Vacant* were still the primary factors positively affecting manufacturing employment. Agricultural and water land uses kept their negative influences on employment. This finding confirmed that employment was a statistic closely correlating to population and housing but not to land uses closely related to industries. Second, both Adj. R^2 values were the highest among the four regression models. It explained that manufacturing was a dominant industry across Metropolitan Detroit Region.

The next critical task is how to use the regression results to interpolate the *TotPop*, *T_Housing*, *Vacant*, and *Manuf* from MCD to a finer geographical unit. We used both Census Tract and Block Group as demonstrations.

6.1.3 Accuracy Test at Census Tracts

The accuracy test involves the following steps.

1. Adjusting negative regression coefficients to positive values.
2. Interpolating *TotPop*, *T_Housing*, *Vacant*, and *Manuf* values to Census Tract by using the adjusted regression coefficients.
3. Scaling (adjusting) the interpolated values by their initial values at the MCD scale by intersecting MCDs and Census Tracts.
4. Assessing interpolation accuracies by regressing the interpolated values of *TotPop*, *T_Housing*, *Vacant*, and *Manuf* against the reported values of *TotPop*, *T_Housing*, *Vacant*, and *Manuf* at the Census Tract geography by the Census Bureau.

It should be noted that the use of the reported values of *TotPop*, *T_Housing*, *Vacant*, and *Manuf* by Census Tracts is purposefully designed in this case study to probe into the interpolation accuracy. Usually, such observed data do not exist on the target layer in areal interpolation applications. The validation or accuracy assessment must be done through a survey, field interview, or other available ancillary information.

There are two issues with the current non-intercept regression approach for interpolating variables from one geographic unit to another. The first one is that some of the regression coefficients are negative, while the second is that the land-use areas do not fully explain the variations of these socioeconomic variables. There are missing explanatory variables that we either do not know or cannot access. Scaling techniques are simple but effective approaches to fixing these two concerns. Yuan and her colleagues experimented with the following scaling equations (Yuan et al., 1997):

$$\beta_j^{'} = \beta_j - \beta_l \tag{6.2}$$

Table 6.5 Scaling negative coefficients to positive

β_j	−1.053	−2.479	9.239	−2.158	28.110	1.006	12.036	−30.960
β_j'	29.907	28.481	40.199	28.802	59.070	31.966	42.996	0.000

$$\beta_{ij} = \left(Y_i / Y_i^{'}\right) \times \beta_j^{'} \tag{6.3}$$

where β_j' are adjusted regression coefficients that can be used to estimate (predict) the values of a dependent variable such as *TotPop*. B_l is the lowest negative coefficient. $\beta_j - \beta_l$ will make all coefficients positive. For example, the eight coefficients of TotPop, β_j, are listed as the first row in Table 6.5. B_l is −30.960. The second row are the values of β_j'.

Yuan and her colleagues (Yuan et al., 1997) used the same dataset (the total population over Census Block Groups) to run the non-intercept regression and scale the regression coefficients to get the same values of the initial total population. That scaling consists of a circular calculation, scaling the predicted values to their initial values. In actual applications, this scaling is unrealistic because the initial (observed) values are unavailable in the context of areal interpolation. We suggest a revised version of Eq. (6.3) as follows:

$$\beta_{ijk} = \left(Y_{ik} / Y_i^{'}\right) \times \beta_j^{'} \tag{6.4}$$

where i is the index of target area unit, j is the index of land uses, and k is the index of source area unit. Eq. (6.4) suggests using Y_{ik} rather than Y_i to scale the estimated values. This approach assumes that the initial values of a dependent variable, Y_i, do not exist in the target geography. In contrast, the initial values of this dependent variable, Y_k, exist at the source geography. In our illustration, we assume that TotPop has values at MCD but does not exist at Census Tract. Furthermore, we know that an MCD may include several Census Tracts:

$$\text{Tract}_{km} \in \text{MCD}_k \tag{6.5}$$

where m is the number of Census Tracts within an MCD and m is not a fixed length. It is logical to assume that the sum of the predicted values of *TotPop* over $Tract_{km}$ can be scaled equally to the *TotPop* value at MCD_k:

$$\sum_m \text{TotPop}_{km} = \text{TotPop}_k \tag{6.6}$$

where km are joint indices of m Tracts in MCD k, which can be obtained by intersecting Tracts and MCDs.

Therefore, we name this prediction adjustment as the source–target–zones intersected subtotal adjustment. In the following are the steps for calculating adjusted prediction using this intersected subtotal approach between the source and target layers. All files involved in this process are stored in the folder, /Chapter_6/T1_Test/Accuracy_Assess_from_Tract_by_MCD/:

- The excel file, MCD_Land_ACS_Acre_Non_Neg_Parameters, contains the regression parameters, including the nonnegative regression coefficients.
- The excel file, T1_MCD_Tract_Original_Data, includes the four variables by Census Tract and MCD.
- The excel file, T1_Tract_Obs_Predict_AdjustByTract, contains the original variables and their predicted values by using the non-negative coefficients.
- The excel file, T1_Tract_Obs_Predict_AdjustByTract_with_Formulae, contains formulae to calculate the predicted values of the four variables.
- The excel file, T1_Tract_Obs_Predict_AdjustByMCDandTract, combines the previous two files and is re-sorted by MCD_ID. This file is used to create the following four files to generate the adjusted predicted values based on the subtotals of these four variables on MCD:

  T1_TotPop_Adjust_by_MCD
  T1_THouse_Adjust_by_MCD
  T1_Vacant_Adjust_by_MCD
  T1_Manuf_Adjust_by_MCD

- Next, we are executing the Python code, Adjust_Pred_at_Tract_by_MCD, to obtain the adjusted predictions. The results are saved in the excel file:

  T1_4_Vars_Adjust_by_MCD_Result

- Finally, we can examine the correlations between the adjusted predictions and the original observations to see how well the areal interpolation performs.

We computed the correlations between the adjusted predictions and the original observations for the four variables. The results are reported in Table 6.6. The correlations between the adjusted predictions and the observations are moderate, ranging from 0.342 (T_Housing – total housing), 0.460 (Popu – total population), 0.511 (Vacant – vacant housing) to 0.642 (Manuf – manufacturing employment). Moreover, the correlations between the adjusted predictions of Popu and T_Housing, Popu and Manuf reach 0.978, 0.881. and 0.829, much higher than those between their original observations, 0.881, 0.823, and 0.731. The findings confirm that the areal interpolation results are informative and can be cautiously adopted.

We should point out several warnings here. First, the true sense of accuracy assessment does not exist because no overserved data at the target spatial unit are available for comparison. Second, the areal interpolation is usually conducted between different spatial scales or units. This interpretation assumes consistent land-use densities by various types of land uses. However, the reality is that land uses are often clustered or aggregated in specific locations, and land-use densities dramatically vary over space. Only when the spatial extent differences between the source spatial unit and the target spatial unit are relatively small can we expect good areal interpolation accuracy. We will provide another validation to approve this point. Third, relationships between land use and socioeconomic statistics trained at

Table 6.6 Correlations between adjusted predictions and observations

		Popu	Adjust Popu	T_Housing	Adjust T_Housing	Vacant	Adjust Vacant	Manuf.	Adjust Manuf
Popu	Pearson Cor.	1	0.460**	0.881**	0.393**	−0.127**	−0.137**	0.823**	0.551**
	Sig. (2-tailed)		0.000	0.000	0.000	0.000	0.000	0.000	0.000
	N	1409	1409	1409	1409	1409	1409	1409	1409
Adjust Popu	Pearson Cor.	0.460**	1	0.363**	0.978**	−0.144**	0.295**	0.430**	0.881**
	Sig. (2-tailed)	0.000		0.000	0.000	0.000	0.000	0.000	0.000
	N	1409	1409	1409	1409	1409	1409	1409	1409
T_Housing	Pearson Cor.	0.881**	0.363**	1	0.342**	0.136**	−0.041	0.731**	0.445**
	Sig. (2-tailed)	0.000	0.000		0.000	0.000	0.122	0.000	0.000
	N	1409	1409	1409	1409	1409	1409	1409	1409
Adjust T_Housing	Pearson Cor.	0.393**	0.978**	0.342**	1	−0.062*	0.431**	0.361**	0.829**
	Sig. (2-tailed)	0.000	0.000	0.000		0.020	0.000	0.000	0.000
	N	1409	1409	1409	1409	1409	1409	1409	1409
Vacant	Pearson Cor.	−0.127**	−0.144**	0.136**	−0.062*	1	0.511**	−0.237**	−0.253**
	Sig. (2-tailed)	0.000	0.000	0.000	0.020		0.000	0.000	0.000
	N	1409	1409	1409	1409	1409	1409	1409	1409
Adjust Vacant	Pearson Cor.	−0.137**	0.295**	−0.041	0.431**	0.511**	1	−0.232**	0.041
	Sig. (2-tailed)	0.000	0.000	0.122	0.000	0.000		0.000	0.122
	N	1409	1409	1409	1409	1409	1409	1409	1409
Manuf	Pearson Cor.	0.823**	0.430**	0.731**	0.361**	−0.237**	−0.232**	1	0.642**
	Sig. (2-tailed)	0.000	0.000	0.000	0.000	0.000	0.000		0.000
	N	1409	1409	1409	1409	1409	1409	1409	1409
Adjust Manuf	Pearson Cor.	0.551**	0.881**	0.445**	0.829**	−0.253**	0.041	0.642**	1
	Sig. (2-tailed)	0.000	0.000	0.000	0.000	0.000	0.122	0.000	
	N	1409	1409	1409	1409	1409	1409	1409	1409

** Correlation is significant at the 0.01 level (2-tailed)

* Correlation is significant at the 0.05 level (2-tailed)

a higher (larger) geographic unit hardly remain statistically stable at a lower (smaller) geographic unit. We should be cautious when we adopt the interpolated results.

6.1.4 Accuracy Assessment at Block Groups

We are now conducting a second accuracy assessment of the areal interpolation. We chose the Census Tract as the source spatial unit and the Block Group as the target spatial unit. Since the employment data are unavailable at the Block Group level, we only tested three variables: TotPop, THousing, and Vacant. The accuracy test involved the following steps (the data and Python code files are stored in the folder, /Chapter_6/T1_Test/Accuracy_Assessment_from_BlkGrp_by_Tract/):

- Run the Python code, Areal_Int_T1_Tract, which inputs Ch6_Land_by_Tract_T1_Acre and produces Param_Ch6_Land_by_Tract_T1_Acre. This code executed non-intersect regression models to obtain regression coefficients.
- Reorganize the coefficient excel file, Param_Ch6_Land_by_Tract_T1_Acre, and transform the coefficients as nonnegative.
- Use the nonnegative regression coefficients to predict the values of TotPop, THousing, and Vacant at the Block Group. The formulae and the data were contained in the excel file, sem_blkgrp_data_tractSUM_for_T1_Accuracy_W_Equation.
- Reorganize this file into three files:

  sem_blkgrp_TotPop_tractSUM_for_T1_Accuracy
  sem_blkgrp_Thousing_tractSUM_for_T1_Accuracy
  sem_blkgrp_Vacant_tractSUM_for_T1_Accuracy

- Run the Python code, Aggregate_TotPop_from_BG_to_Tract, Aggregate_THousing_from_BG_to_Tract, and Aggregate_Vacant_from_BG_to_Tract, to get the adjusted prediction values for TotPop, THousing, and Vacant.
- Organize the Python execution results as a single excel file, sem_blkgrp_tract_adjusted 3 vars for_T1_Accuracy, for correlation or regression analysis to assess the model accuracy.
- Run correlation or regression analysis.

The correlation results between the adjusted predictions and the original observations are reported in Table 6.7. TotPop, THousing, and Vacant correlations reached 0.848, 0.811, and 0.770, respectively. The magnitudes of these correlations are very good and significant. This validation confirms that the areal interpolation accuracy could be high when the source spatial unit and the target spatial unit are close in a hierarchy in the spatial structure. The Census Tract is immediately above the Block

Table 6.7 Correlations between the adjusted predictions and the original observations in T1

		BGTotPop	AdjustPop	BGThousing	AdjustHou	BGVacant	AdjustVac
BGTotPop	Pearson Cor.	1	0.848**	0.880**	0.757**	−0.079**	−0.075**
	Sig. (2-tailed)		0.000	0.000	0.000	0.000	0.000
	N	4000	3999	4000	3999	4000	3999
AdjustPop	Pearson Cor.	0.848**	1	0.751**	0.912**	−0.094**	0.004
	Sig. (2-tailed)	0.000		0.000	0.000	0.000	0.812
	N	3999	3999	3999	3999	3999	3999
BGThousing	Pearson Cor.	0.880**	0.751**	1	0.811**	0.194**	0.125**
	Sig. (2-tailed)	0.000	0.000		0.000	0.000	0.000
	N	4000	3999	4000	3999	4000	3999
AdjustHou	Pearson Cor.	0.757**	0.912**	0.811**	1	0.083**	0.232**
	Sig. (2-tailed)	0.000	0.000	0.000		0.000	0.000
	N	3999	3999	3999	3999	3999	3999
BGVacant	Pearson Cor.	−0.079**	−0.094**	0.194**	0.083**	1	0.770**
	Sig. (2-tailed)	0.000	0.000	0.000	0.000		0.000
	N	4000	3999	4000	3999	4000	3999
AdjustVac	Pearson Cor.	−0.075**	0.004	0.125**	0.232**	0.770**	1
	Sig. (2-tailed)	0.000	0.812	0.000	0.000	0.000	
	N	3999	3999	3999	3999	3999	3999

** Correlation is significant at the 0.01 level (2-tailed)

Group in this case study, and a Census Tract usually includes one to a couple of Block Groups. Under such spatial proximity, the areal interpolation result is very significant and dependable.

In summary, we come to three conclusions about areal interpolation. First, the adjusted R-squared value (R^2) of the initial non-intercept regression is a critical statistic to judge the feasibility of an areal interpolation procedure. Only when R^2 is relatively high is it practical to carry out the interpolation. Whether the interpolation outcome is reliable or accurate depends on the spatial disparity between the source and target areas. Greater the disparity, the worse the interpolation, and vice versa. Second, the prediction accuracy is a different concept from traditional regression analysis. The areal interpolation adopts the linear relationship trained at a higher spatial unit to predict unknow values at a lower spatial unit. Too many factors could affect the prediction outcomes because of disparities of spatial patterns. You must be aware of this fact and avoid interpolating an attribute from a high-level areal unit down to a far-away areal unit in the spatial hierarchical structure. Third, it is often unrealistic to validate areal interpolation outcomes because the information on the target spatial unit does not exist. Ancillary data, surveys, interviews, or other sources of data must be used to evaluate the interpolation accuracy. As a result, you must be very cautious to use the interpolation outcomes.

6.1.5 *The Double-Ratio Reallocation*

The purpose of the areal interpolation in this chapter is to create a unified environmental and socioeconomic dataset. As discussed earlier in this chapter, the primary challenge is that environmental data and socioeconomic data are collected over different spatial units. For example, the chapter proposes that the basic spatial unit of a combined environmental and socioeconomic database is the natural patch (NP). There are many types of NPs such as land-use patches, soil patches, plant communities, and forest plots. NPs commonly carry with them specific ecological or environmental characteristics that are collected or tested when they are sampled or surveyed. However, they often do not have socioeconomic statistics. NPs are usually small in size in comparison with statistical enumeration districts or units. That is the foundation of the proposed areal interpolation techniques. The areal interpolation techniques are designed to interpolate the existing socioeconomic statistics from larger enumeration districts to smaller NP-based natural units. An ideal solution is to directly interpolate the socioeconomic statistics from a county, an MCD, or a Census Tract to NPs. However, it is not technically feasible because no ancillary data exist to carry out this interpolation. Through the NP-based areal interpolation, we can disaggregate the available socioeconomic statistics from a large area to a small area that is closer to NPs from the perspective of spatial extent. However, the interpolated socioeconomic statistics are not allocated to NPs at this stage. Therefore, we have designed two techniques to reallocate the interpolated statistics to NPs. The first one is the direct allocation by linking or joining the interpolated small areas with NPs. This technique assumes that the spatial scale of the interpolated spatial unit is closer to the spatial scale of NPs. When two scales are more comparable, we can statistically merge these two datasets together as a unified database. The second method is the double-ratio reallocation, which is a rigorous statistical procedure derived from the non-intercept regression model.

6.1.5.1 Direct Merging

Direct merging assumes that the interpolated spatial unit is finer enough or comparable to the spatial scale of NPs. Under this circumstance, there is no need for further data processing. We can link or join two sets of environmental dataset and socioeconomic dataset (that has been created through the areal interpolation). This chapter's interpolation from the Census Tract to the Block Group is a good example.

The link or join can be performed with a GIS software package. We use ArcGIS Pro as an illustration. This process includes the following steps (all files are in the folder, /Chapter_6/Test_1/Interpolate_to_Natural_Patches/):

1. Join the excel file, sem_blkgrp_tract_adjusted 3 vars for_T1_Accuracy (that includes the adjusted interpolated values of TotPop, THousing, and Vacant at the Block Group level), with SEMBlkGrp shape file (i.e., the GIS data layer for the Block Group). The joint new GIS data layer is called SEM_BG_Interpolated.

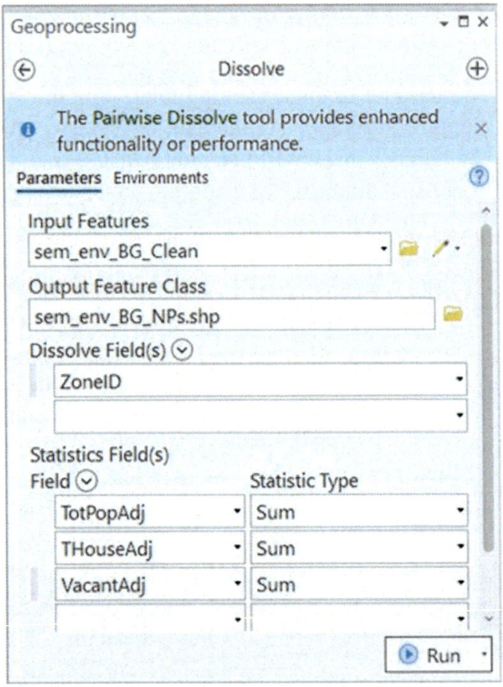

Fig. 6.2 Merging the intersected NPs back to their original patches

2. Union SEMBlkGrp with the shapefile, sem_env (i.e., the shapefile of Southeastern Michigan land-use patches). The purpose of the union is to assign the Block Group index to the land-use patches. The output shape file is called sem_env_BG. After cleaning up, the final shapefile is called sem_env_BG_Clean.
3. Merge the intersected NPs back to their original states by using the GIS function, "Dissolve" (Fig. 6.2). The new shapefile, sem_env_BG_NPs, includes the three newly interpolated variables.

6.1.5.2 Double-Ratio Reallocation

The direct link (join) method is to assign the interpolated values to NPs when the NPs and the interpolated target zones are compatible in scale. However, if the target units are large, direct link or join can lead to rough results. A rigorous statistical approach is preferred for assigning the interpolated values. We illustrate the double-ratio reallocation by using the Census Tracts as the interpolated target units.

There are two sets of ratios involved in the areal interpolation process. The first set is the ratios of the regression coefficient of each NP type to the sum of the coefficients of all eight NPs (named Coef_Ratio). The second set is the ratios of the land area of each NP to the sum of the areas of all NPs in a Census Tract (called Area_Ratio).

Take the Census Tract and the Adjusted TotPop in this Tract as a demonstration. The Coef_Ratio set for Tract 93710100 consists of the following elements for NP_1, NP_2, NP_3, NP_4, NP_5, NP_6, NP_7, and NP_8:

0.052522	0.322968	0.322472	0.041410	0.213737	0.029992	0.000000	0.016899

The sum of the Coef_Ratio set equals to 1.0.

The Area_Ratio set for the same Tract includes the following for the eight types of NPs:

0.419818	0.00237	0.255628	0.017928	0.004676	0.014213	0.259406	0.025961

This set sum is also 1.0. The products of the two sets are the percentages of the population allocated to each NP. However, we must scale the sum of the products to 1.0 as well in order to fully reallocate TotPop in Tract 93710100 to NPs contained in this Tract. Therefore, the sum of the allocated population in this Tract equals to the Adjusted TotPop. The equation is

$$\text{Percent}_{ij} = \text{Coef\_Ratio}_i \times \text{Area\_Ratio}_{ij} \times \left[1 / \sum_{i}^{n} \left(\text{Coef\_Ratio}_i \times \text{Area\_Ratio}_{ij} \right) \right]$$

where Percent_{ij} is the percentage of TotPop allocated to NPi in Tract j, i is the index of NPs, j is the index of the Tracts; and Coef_Ratio_i is the set of the non-intersect all-positive regression coefficients, which are adjusted by elevating negative coefficients to 0.0 or positive values. The Coef_Ratio_i set do not change by the Tract. Area_Ratio_{ij} is the set of the area ratios. The Area_Ratio_{ij} set changes by the Tract.

The final reallocation can be done through the following equation:

$$\text{Population}_{ij} = \text{TotPop}_j \times \text{Percent}_{ij}$$

where Population_{ij} is the population in NP i and Tract j, i is the index of NPs, j is the index of the Tracts; TotPop_j is the total population (the adjusted predicted population) in Tract j; and Percent_{ij} is the percentage of TotPop allocated to NP i in Tract j. The Python code for implementing the double-ratio reallocation, Reallocate_to_NPs, is provided in the folder, /Chapter_6/T1_Test/Interpolate_to_Natural_Patches/. This Python code reads in the excel file, SEM_Tract_LandUse_Percent, and writes out two excel files, SEM_Tract_LandUse_Percent_output and SEM_Tract_LandUse_Percent_result. The former is in the wide format, while the latter is in the narrow format. The narrow format file contains a concatenated field, called "ID'," which is the same as the field, "sem_env_Tract," in the intersected shape file, sem_env_by_Tract (see the next paragraph).

The double-ratio reallocation must be implemented in a GIS software environment like ArcGIS Pro. Let us assume the shapefile, sem_env, contains the NP patches and the shapefile, SEM_Tracts, includes the Tracts bounding polygons. The reallocation involves the following steps:

1. Intersect sem_env and SEM_Tracts to create sem_env_by_Tract.
2. Clean up the sem_env_by_Tract, keeping the needed information.
3. Add a concatenated field, sem_env_Tract, by concatenating sem_env ID with Tract ID (the sem_env_Tract field serves as the link ID for the next step).
4. Join the shape file, sem_env_by_Tract, with the narrow format population table (SEM_Tract_LandUse_Percent_result) by the common fields, "ID" and "sem_env_Tract."
5. It is worth pointing out that the numbers of records in the excel table, SEM_Tract_LandUse_Percent_result, and in the shapefile, sem_env_by_Tract, can be different. The former assumes that each land-use (cover) type only has one patch in each Tract. However, in fact, a Tract could include several land-use patches that are disconnected. When this happens, we can further divide the predicted value to these disconnected patches based on their ratios to the total area of these patches.
6. Export the joined sem_env_by_Tract as the NP_Reallocated_by_Tracts.
7. Merge the intersected NPs back to their original states by using the GIS function, "Dissolve." The new shapefile, NP_Reallocated_Population, is the finally dissolved GIS layer, containing the reallocated population (all the aforementioned GIS files and tables are provided in the folder, /Chapter_6/T1_Test/Interplate_to_Natural_Patches/; the ArcGIS Pro project file, Allocate-to-NPs, implements these processes).

6.2 T2: Natural Communities-Based Areal Interpolation

T1 fitted a non-intercept regression model over the entire source layer (the MCD Census Unit) to predict four socioeconomic variables against eight land-use types. T2 intends to cluster MCDs into several subregions and then fits several non-intercept regression models over these subregions. Because the factors adopted to cluster land-use patches into subregions are all environmental or physical, including digital elevation model (DEM), slope, precipitation, temperature, area, and distance from major transportation lines, we call these subregions "Natura Communities." The rationale for T2 is that dividing the study area into several subregions based on their physical characteristics could improve the prediction accuracy of regression analysis.

Cluster analysis (B. Everitt et al., 2001) aims to discover groups of observations from initially unclassified data. There are several statistical concepts underling the cluster analysis: (1) distance and similarity measures, (2) agglomerative hierarchical techniques, (3) k-means clustering, and (4) k-medoids clustering (Park & Jun, 2009). Cluster analysis requires no prior knowledge or determination of the number of clusters before running a cluster analysis. Cluster analysis has a companion, called Discriminant Analysis (B. S. Everitt & Dunn, 2001). Discriminant analysis works with data that is already classified into groups to derive rules for classifying new (and as yet unclassified) individuals on the basis of their observed variable values. Discriminant analysis can also be used to assess how many clusters will minimize within-group differences and maximize between-group differences (Lan et al., 2022; McLachlan, 2004).

The datasets used in the T2 experiment include:

- sem_env: ArcGIS Shape File contains the environmental attributes for running ArcGIS Pro Spatially Constrained Multivariate Clustering
- semland30: ArcGIS Geotiff File contains the land-use type and area information
- sem_env_data: Excel CSV File contains the environmental attributes for running T2_Discriminant_Analysis Python code to determine how many groups are appropriate
- SEM_MCD_Poly: ArcGIS Shape File contains four variables for areal interpolation

The datasets are stored in the folder, /Chapter_6/T2_Test/. The technical implementation of T2 involves eight steps.

Step 1: Running a discriminant analysis, T2_Discriminant_Analysis, to determine how many clusters should be considered. The result confirms that either six or nine is an acceptable number of clusters. However, the case study only contains 208 MCDs. A minimum of 50 cases is generally regarded as a minimum required sample size to fit a regression analysis (Harris, 1985; Pedhazur & Schmelkin, 1991; Green, 1991; Jenkins & Quintana-Ascencio, 2020). Since the case study only contains 208 MCDs, four have been chosen as the number of clusters in this case study. However, the Python code, T2_Discriminant_Analysis, is provided for future case studies with large-size datasets.

Step 2: Executing a k-medoids clustering using the number of clusters determined in the previous step. The ArcGIS Pro tool is called Spatially Constrained Multivariate Clustering. The input data layer is "sem-env" and the output data layer is "sem-4-clusters" (Fig. 6.3).

Step 3: Sub-setting the four derived clusters into separate sub-data layers.

Step 4: Intersect SEM_MCD_Poly with each sub-data layer, respectively, to join the MCD attribute data with the sub-data layers.

Step 5: Use the ArcGIS Pro tool, Dissolve, to merge the land-use patches within MCD and update the environmental attribute data.

Step 6: Use the tool, Tabulate Area, to tabulate the land-use areas (from semland30m) with the resultant data layers from the previous step.

Step 7: Run non-intercept regression models, respectively.

Step 8: Assess the accuracy of the regression modeling outcomes.

There were four clusters. Large portions of the City of Detroit, Macomb County, and Monroe County along the St Clair Lake and the Detroit River were grouped

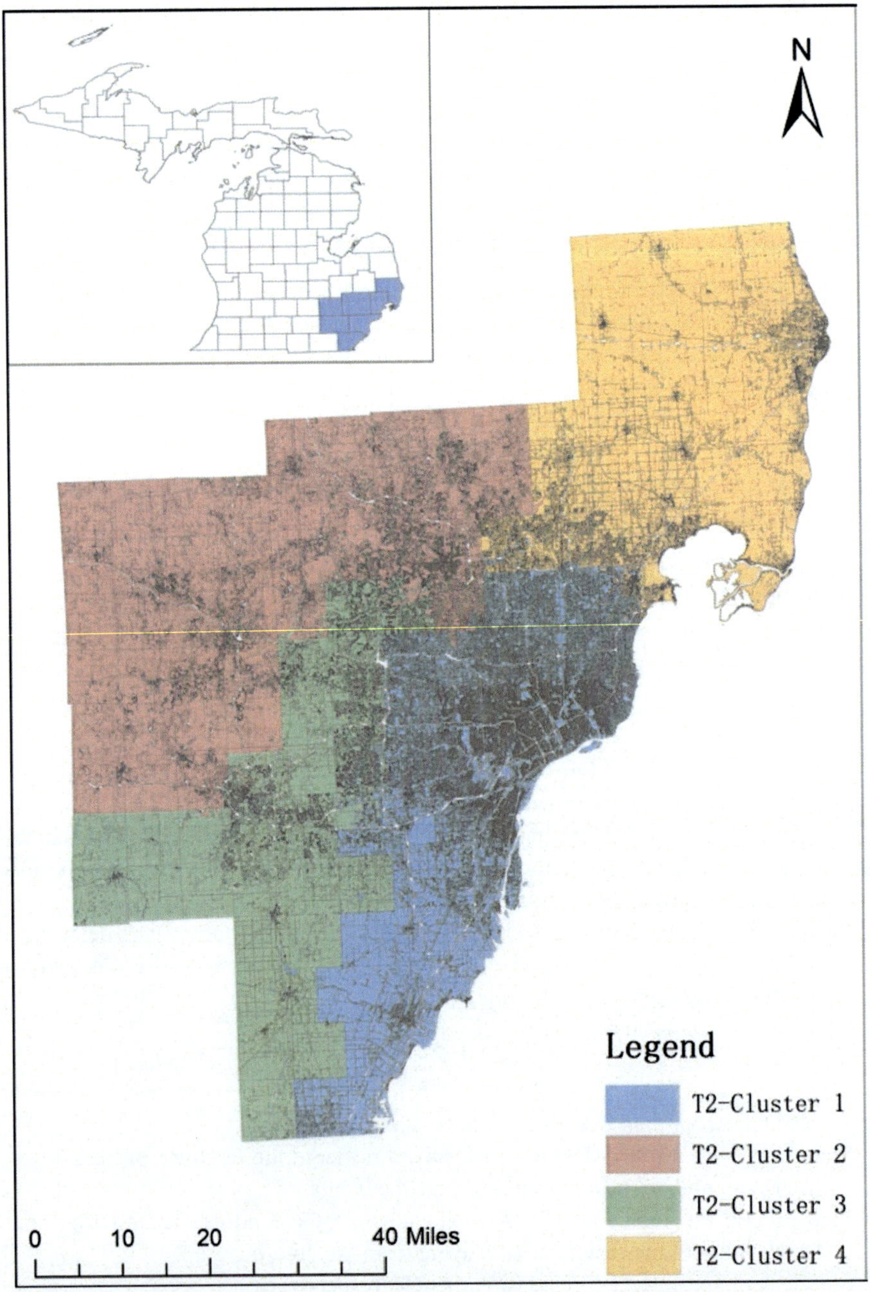

Fig. 6.3 Four nature communities derived from T2 in Southeastern Michigan

Table 6.8 The R^2 value comparison between the T1 and T2 tests

Variable	T1	T2-Cluster 1	T2-Cluster 2	T2-Cluster 3	T2-Cluster 4
TotPop	0.8652	0.8863	0.1679	0.4716	0.2781
THousing	0.8162	0.8952	0.1710	0.4405	0.2560
Vacant	0.6432	0.8807	0.3467	0.4452	0.2405
Manuf	0.9342	0.8268	0.1568	0.3855	0.3106

together as Cluster 1, where the costal lowlands dominate. Large sections of the Livingston County and the western part of Washtenaw County were within Cluster 2, where many hilly areas are present. The western and southern parts of the Washtenaw County and the western section of Monroe County belonged to Cluster 3, which are predominantly flat plains. The hilly areas of St Clair County and Oakland County were grouped as Cluster 4.

The R^2 values of four variables over four clusters were reported in Table 6.8. Cluster 1 derived from Technique 2 (T2) got the highest R^2 values compared to the T1 R^2. The R^2 values of the remaining three clusters were moderate or even lower. This finding was consistent with the research design. When more refined or comparable spatial structures are adopted to organize the data, the corresponding regression analysis outcome should perform better. This finding also indicated that the linear relationship between TotPop, THousing, Vacant and the areas of eight types of land uses was primarily held in the City of Detroit and the river front areas along the River of Detroit.

If these coefficients were used to predict (interpolate) the unknown values, the outcomes should be good as well (Table 6.9). The same procedures as discussed in Sects. 1.3 and 1.4 were followed to conduct the accuracy assessment. All input, output, and Python coding files are provided in the folder, /Chapter_6/T2_Test/.

The correlation results between the adjusted predictions and the original observations for TotPop, THousing, and Vacant reached 0.782, 0.732, and 0.760, respectively. These values are higher than 70.0%, which is pretty good for an areal interpolation application. However, it is interesting to point out that, although the non-intercept regression model in T2-Cluster 1 had higher R^2 values, the prediction correlations were slightly lower than the T1 regression model. The spatial disparity explains this lower interpolation performance. T2-Cluster 1 only occupies one-fourth of the study area (Fig. 6.3), but counts for three-quarters of Block Groups (3252 in Cluster 1 out of the total 4000 Block Groups) (Tables 6.7 and 6.9). The high percentage of Block Groups means Cluster 1 has the most concentrated population, housing, and vacant houses in the study area and contains perceptible internal spatial disparities. As a result, the prediction error is comparably high. This finding iterates what we concluded in the T1 test: the greater the internal inequality, the worse the prediction.

Table 6.9 Correlations between the adjusted predictions and the original observations in T2-Cluster 1

		BGTotPop	AdjustPop	BGThousing	AdjustHou	BGVacant	AdjustVac
BGTotPop	Pearson Cor.	1	0.782**	0.871**	0.676**	−0.091**	−0.101**
	Sig.		0.000	0.000	0.000	0.000	0.000
	N	3252	3252	3252	3252	3252	3252
AdjustPop	Pearson Cor.	0.782**	1	0.690**	0.907**	−0.089**	0.047**
	Sig.	0.000		0.000	0.000	0.000	0.007
	N	3252	3252	3252	3252	3252	3252
BGThousing	Pearson Cor.	0.871**	0.690**	1	0.732**	0.203**	0.117**
	Sig.	0.000	0.000		0.000	0.000	0.000
	N	3252	3252	3252	3252	3252	3252
AdjustHou	Pearson Cor.	0.676**	0.907**	0.732**	1	0.092**	0.289**
	Sig.	0.000	0.000	0.000		0.000	0.000
	N	3252	3252	3252	3252	3252	3252
BGVacant	Pearson Cor.	−0.091**	−0.089**	0.203**	0.092**	1	0.760**
	Sig.	0.000	0.000	0.000	0.000		0.000
	N	3252	3252	3252	3252	3252	3252
AdjustVac	Pearson Cor.	−0.101**	0.047**	0.117**	0.289**	0.760**	1
	Sig.	0.000	0.007	0.000	0.000	0.000	
	N	3252	3252	3252	3252	3252	3252

** Correlation is significant at the 0.01 level (2-tailed)

6.3 T3: Natural Districts-Based Areal Interpolation

T2 used six environmental attributes (DEM, slope, precipitation, temperature, area, and distance) to cluster MCDs into four subregions called Natura Communities because T2 only considered physical ancillary information. T3 is designed to divide the study area into subregions by considering both physical features and socioeconomic characteristics. T3 intends to use comprehensive ancillary data to divide several subregions that could better reveal spatial patterns of regional coupled environmental and socioeconomic disparities. In addition to the six environmental data items used in T2, 17 socioeconomic variables (EDU, Senior, MANUP, UNEMP, BlkPop, NoIncome, GetAid, FHFChild, LOWEDU, BYTR, NoHeat, NoKitch, MULTI, Vacant, POVTY, Rgr25Mi, and YRMV) are added to run the cluster analysis in T3 (Table 6.10). We plan to explore how different spatial structures affect the outcomes of areal interpolation.

Only one dataset is used to run the T3 experiment, sem_mcd_env_land_acs, which is an ArcGIS Shape File containing the complete set of environmental, land

Table 6.10 The variables and explanations

Variable	Explanation
General information	
TotPop	Total population
MHVAL	Median house value
EDU	% Persons with at least a 4-year college degree of the total population
Senior	% Persons aged 65 and above of the total population
Economic	
MANUP	% Manufacturing employees of the workers aged 16 and above
ServiceP	% Service employees (including professional, educational, health care, recreation, and other services) of the workers aged 16 and above
UNEMP	% Unemployed of the total population
Inequality (census tract)	
BlkPop	% Black/African American population of the total population
NoIncome	% Persons aged above 15 not having income of the total population
GetAid	% Persons aged above 15 receiving supplementary aids of the total population
FHFChild	% Female-headed families and subfamilies with own children
LOWEDU	% Persons 25+ years old who did not have a high school diploma or equivalent of the total population
Housing (census tract)	
BYTR	% Structures built more than 30 years ago of the total number of building structures
NoHeat	% Housing units where no heating fuel is used of the total housing units
NoKitch	% Housing units without full kitchen of the total housing units
MULTI	% Multiunit structures of the total housing units
Vacant	% Vacant housing units of the total housing units
Persistent poverty (census tract)	
POVTY	% Below poverty level families of the total families
RGR25	% Renter household heads whose gross rent is 25%+ of last year's income of the total households, which was less than US$ 10,000
YRMV	% Household heads move into a unit less than 10 years ago of the total households
Land-use status	
L1Acre	Agricultural land areas
L2Acre	Commercial land areas
L3Acre	Housing land areas
L4Acre	Industrial land areas
L5Acre	Institutional land areas
L6Acre	Recreational land areas
L7Acre	Vacant land areas
L8Acre	Water surface areas

1 agriculture, 2 commercial, 3 housing, 4 industrial, 5 institutional, 6 recreational, 7 vacant, 8 water

use, and socioeconomic variables. T3 implementation involves three steps: (1) Running the ArcGIS Pro tool, Spatially Constrained Multivariate Clustering, by providing sem_mcd_env_land_acs as the input data layer and generating sem_mcd_T3_4clusters as the output data layer; (2) Sub-setting the four derived clusters

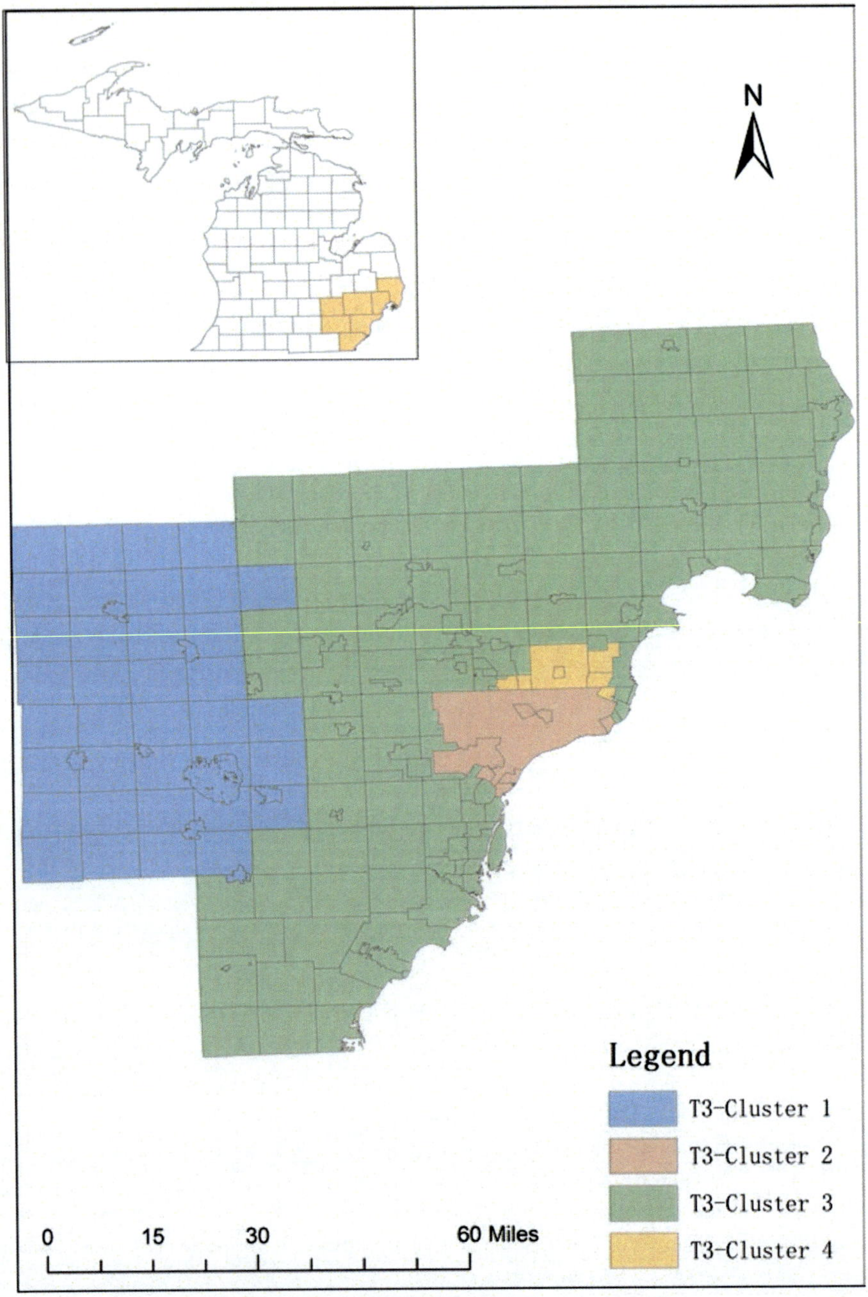

Fig. 6.4 Four nature districts derived from T3 in Southeastern Michigan

Table 6.11 The R^2 value comparison between the T1 and T3 tests

Variable	T1	T3-Cluster 1	T3-Cluster 2	T3-Cluster 3
TotPop	0.8652	0.8700	0.8736	0.9133
THousing	0.8162	0.8676	0.8683	0.9040
Vacant	0.6432	0.8676	0.5954	0.8093
Manuf	0.9342	0.8901	0.8467	0.9003

as four separate sub-data layers; and (3) Joining each cluster data layer with sem_mcd_env_land_acs by matching the Source_ID in the cluster data layer and the Object_ID in the original data layer for clustering. The join function links the attribute data from the original data layer to each clustered data layer for regression analysis.

The T3 classification outcome was shown in Fig. 6.4. The spatial pattern of T3 clustering was dramatically different from the T2 clusters. The socioeconomic variables clearly determined the T3 clustering patterns. Cluster 1 includes MCDs from the Livingston and Washtenaw counties. Cluster 2 coincides with the City of Detroit. Cluster 3 consists of most MCDs in four counties: Macomb, Monroe, St. Clair, and Wayne. Only a couple of MCDs from the Macomb and Oakland counties boarding with the City of Detroit become a separate cluster (Cluster 4). The spatial patterns of T3 clustering reveal three typical urban structural forms: exurban (Cluster 1), suburban (Cluster 3), and urban (Clusters 2 and 4) (Ban & Ahlqvist, 2009; Golding & Winkler, 2020). The City of Detroit and its immediately adjacent MCDs (Clusters 2 and 4) only included a small number of MCDs. They were combined into one cluster (Cluster 2). Since it had few MCDs, the Census Tracts in this combined cluster were used to conduct the noon-intercept regression models.

The Python coding files and their input and output files are included in the folder, /Chapter_6/T3-Test/. The implementation steps are similar to T2. The regression coefficients for the four variables, TotPop, THousing, Vacant, and Manuf, are stored in the output excel files. The R^2 values were reported in Table 6.11. Unlike the T1 model, three T3 models saw high R^2 values except for the variable, Manuf. T3 tests also confirmed that the corresponding regression analysis outcome should have better performance when the data were organized into more comparable spatial structures. We illustrated conducting interpolation accuracy assessments in both T1 and T2 tests. We also discussed how to interpret the interpolation accuracy statistics. These will apply to T3 as well. Therefore, we will omit the interpolation accuracy assessments for T3. Interested readers can use the Python coding files in T2 and the output files in T3 to conduct an accuracy test.

6.4 T4: Road Network Analysis Zone-Based Areal Interpolation

Road network analysis zone (RNAZ) is developed based on the concepts of urban transportation networks and gravity theory. The conceptual and application implications of RNAZ in urban shrinking studies will be discussed in Chap. 7 when this approach is applied to build a hierarchical linear mixed model to examine spatial processes of urban shrinking in Metropolitan Detroit in the United States. It facilitates the understanding when this interpolation approach is introduced in the context of examining spatial processes of urban shrinking.

RNAZ-based areal interpolation has six unique characteristics: (1) consider travel impedance over road networks; (2) treat the source areas as traffic stops; (3) handle ancillary data as attractions to these stops; (4) apply data mining techniques to synthesize attractions into several dimensions or reduce variables into several hypothetic variables; (5) introduce the gravity theory to standardize the attractions; and (6) group the source areas into several spatial clusters by using the standardized gravities. Gravity theory has long been incorporated into urban interaction models and transportation models (Chen, 2009; Guo et al., 2021). The spatial structural organizations derived through this approach are called RNAZ.

On the implementation steps, we will construct a Network Dataset from the Southeastern Michigan Roads (Streets) in ArcGIS Pro. We will obtain the centroids of the source areas (Census Tracts) and assign the attributes of the Census Tracts to these centroids, which we will treat as the attraction nodes. Because we have 25 environmental and socioeconomic variables (Table 6.10), we will reduce these variables into a few hypothetical attraction indicators and interpret their meanings. We will then run a public domain software, Urban Network Analysis (provided in the folder, /Chapter_6/T4_Process/UNA_Toolbox/), to compute the gravity scores of the hypothetical attraction variables through the road networks (Sevtsuk & Mekonnen, 2012; Sevtsuk, 2014). Afterward, we will cluster the source areas into several road network analysis zones (RNAZs). We will run non-intercept regression models over these RNAZs, respectively. Finally, we will use the regression coefficients to interpolate TotPop, THousing, and Vacant over the target areal unit, Block Groups. We are interpolating from Census Tract to Block Group because the published data are available at both units. Therefore, the T4 experiment results can support an accuracy assessment by comparing the interpolated data with the known data.

We have 25 variables (8 environmental and 17 socioeconomic ones). If we just use these variables to cluster the MCDs in the study area, it would be difficult to identify which variable or variables affected the clustering outcome. Therefore, there is a need to reduce 23 variables into fewer dimensions to discern synthetic impacts of these environment-socioeconomic variables on urban clusters. In environmental modeling, many initially observed variables must be grouped into fewer hypothetic (or latent) variables to reveal primary relationships between the dependent and many independent variables (Brown et al., 2013). Principal component analysis (PCA) is a standard statistical method used to classify correlated variables into fewer unobserved variables called factors or latent variables. These latent variables help discern and synthesize relationships and interactions between the initial variables. PCA has been confirmed helpful for identifying water quality indicators and interpreting and analyzing complex data in water quality assessment (Shrestha et al., 2008). PCA can help examine and model complex interrelationships between variables with fewer hypothetical factors (Lewis-Beck, 1994).

Table 6.12 Total variance explained

	Initial eigenvalues			Rotation sums of squared loadings		
Factor	Total	% of variance	Cumulative %	Total	% of variance	Cumulative %
1	9.248	33.027	33.027	6.822	24.363	24.363
2	2.951	10.540	43.567	3.107	11.095	35.458
3	2.340	8.359	51.926	2.763	9.867	45.325
4	1.908	6.814	58.740	2.580	9.213	54.537
5	1.370	4.892	63.632	2.303	8.224	62.761
6	1.071	3.827	67.459	1.315	4.697	67.459
7	0.923	3.296	70.755			
8	Omitted when eigenvalue is less than 1.0					

Extraction method: Principal component analysis

The first critical statistic of PCA is the initial eigenvalue. A factor will be considered significant when its initial eigenvalue is larger than 1.0 (Table 6.12). A rotation method is commonly used to differentiate the associations between the latent variables (factors) and the initially observed variables (Table 6.13). For example, F1 is strongly and positively correlated with the following variables, BLKPOP, GetAid, FHFCLD, NoKitch., VacantH, LOWEDU, POVTY, and UNEMP, but negatively correlated with EDU and MHVAL. All of them reflect sustained poverty (Xie et al., 2018). Therefore, F1 could be named the sustained poverty dimension. Based on its spatial pattern (Fig. 6.5), the centroids (Census Tracts) in the City of Detroit display high values in this dimension, which confirms this finding. F2 is highly correlated with T_Housing, TotPop, Manuf, and housing land use, indicating large volumes of total housing units, population, and manufacturing employment. Thus, this factor describes the economic scale dimension. The economic scale dimension boasts the highest values in the northern suburban areas to Detroit and higher values in almost all suburban areas. The lowest and lower values concentrate in several downtown areas, including Detroit (Fig. 6.5).

F3 highly and positively correlates with MULTI, RGR25, and NoHeat, but negatively with YRMV. The first three variables are the indicators of multiunit housing structures, poor rental households, and no standard heating devices. The last one indicates that these households may have stayed here for a long time (more than 10 years). Therefore, F3 is the low-income rental housing dimension. F4 highly correlates with industrial land use, commercial land use, vacant land use, and agricultural land use, representing the non-housing land-use dimension. F5 highly correlates with water areas and recreational land use, indicating the recreational land-use dimension. F6 highly correlates with the senior population, revealing the senior community dimension.

Afterward, we ran Urban Network Analysis (an ArcGIS toolset; see Sevtsuk & Mekonnen, 2012) to compute the six factors' gravity scores. Two documents in the folder, /Chapter_6/T4_Process/ Areal_Int_T4_Design and /Chapter_6/T4_Process/ UNA_Toolbox/UNA_help, explained how the gravity scores for the six factors were computed. SEM_Gravity_20m (in the folder, /Chapter_6/T4_Process/

Table 6.13 Rotated component matrix[a]

	F1	F2	F3	F4	F5	F6
BLKPOP	0.793	−0.135	0.214	−0.217	−0.060	0.146
GetAid	0.788	−0.073	0.065	−0.123	−0.084	−0.255
FHFCLD	0.774	−0.090	0.281	−0.137	−0.091	−0.082
NoKitch	0.771	−0.182	0.024	−0.169	0.052	0.123
VacantH	0.760	0.146	0.187	−0.172	0.107	0.208
EDU	−0.751	0.168	−0.004	−0.191	0.210	0.254
LOWEDU	0.739	−0.177	0.097	0.020	−0.160	−0.039
POVTY	0.726	−0.174	0.383	−0.101	−0.049	−0.231
UNEMP	0.708	−0.138	0.066	−0.077	−0.039	−0.066
MHVAL	−0.665	0.224	−0.068	−0.057	0.360	0.165
NoIncom	0.571	−0.023	−0.089	−0.136	0.043	−0.377
BTYR	0.359	−0.302	0.067	−0.345	−0.301	0.224
T_Housing	−0.063	0.942	0.103	0.046	0.032	0.137
TotPop	−0.204	0.921	−0.104	0.093	0.056	−0.079
Manuf	−0.314	0.815	−0.203	0.149	0.038	−0.091
L3Acre	−0.290	0.498	−0.311	0.370	0.457	0.139
MULTI	0.006	0.053	0.916	−0.068	−0.106	−0.024
RGR25	0.486	−0.041	0.765	−0.116	−0.136	−0.129
YRMV	−0.414	0.178	−0.666	0.035	0.113	0.066
NoHeat	0.053	−0.063	0.562	0.051	0.023	0.072
L4Acre	−0.008	−0.091	−0.053	0.712	0.168	−0.057
L2Acre	−0.162	0.151	0.146	0.672	−0.095	0.068
L7Acre	−0.038	0.121	−0.162	0.655	0.485	−0.003
L1Acre	0.007	0.003	−0.128	0.633	0.323	0.102
L5Acre	−0.177	0.178	0.014	0.447	0.012	0.000
L8Acre	−0.114	0.030	−0.071	0.143	0.829	0.025
L6Acre	−0.088	0.037	−0.040	0.208	0.807	−0.018
Senior	−0.220	−0.019	−0.054	0.039	0.027	0.850

Extraction method: Principal component analysis
Rotation method: Varimax with kaiser normalization
[a]Rotation converged in seven iterations

UNA_Toolbox/) is the ArcGIS Desktop project file, which includes all data layers created in the process of computing the gravity scores. We then run the ArcGIS Pro tool, "Spatially Constrained Multivariate Clustering," to group the Census Tracts through their centroids into four clusters. The rationale for adopting four clusters was to keep consistent with the previous tests. The clustering result was displayed in Fig. 6.6. Figure 6.6a showed a very compact spatial pattern. The tool, Urban Network Analysis, is using the network travel distance based on the networked route to assess how a place relates to its adjacent places. It is natural to create compact spatial configurations. We generated three additional networked clusters: 8, 12, and 20 to confirm this finding (Fig. 6.6b–d; see the ArcGIS Pro document,

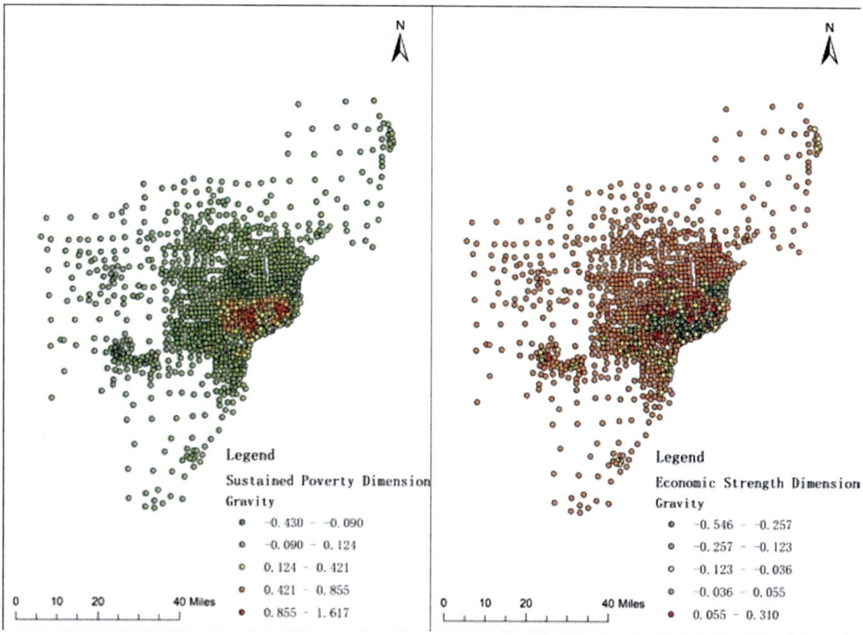

Fig. 6.5 Spatial patterns of Factor 1 – sustained poverty dimension, and Factor 2 – economic scale dimension

T4_Clustering, for the original ArcGIS layers). Only when the clusters reached 20 did the entire study area largely connect except for the northwest quarter.

We used the four clusters to execute the non-intersect regression models. The Python coding files and their input and output files are included in the folder, /Chapter_6/T4_Test. The implementation steps are similar to T2. The regression coefficients for the four variables, TotPop, THousing, Vacant, and Manuf, are stored in the output excel files. The R^2 values were reported in Table 6.14. T4 Cluster 1 was not reported. As we explained before, T4 generated three very compact clusters, 2, 3, and 4. All unconnected (un-networked) areas were included in Cluster 1. In fact, these points or Census Tracts in Cluster 1 were not clustered as can be seen in Fig. 6.5b–d. Compared with the T1 model, the three T4 models saw higher R^2 values except for the variable, Manuf. T4-Cluster 2 was a unique cluster and located in West Wayne County, the fast-growing suburban of Detroit. There are lots of well-maintained residential communities, industries, and businesses. The results of the fine-tuned T4 clusters further confirmed that when the study areas were partitioned into more comparable spatial patterns, the corresponding regression analysis outcome should have better performance. We illustrated how to conduct interpolation accuracy assessments in both T1 and T2 tests. We also discussed how to interpret the interpolation accuracy statistics. These will apply to the T4 outcomes as well. Therefore, we omit the interpolation accuracy assessments for T4. Interested readers can use the Python coding files in T2 and the output files in T4 (in the folder, /Chapter_6/T4_Test/) to repeat this process.

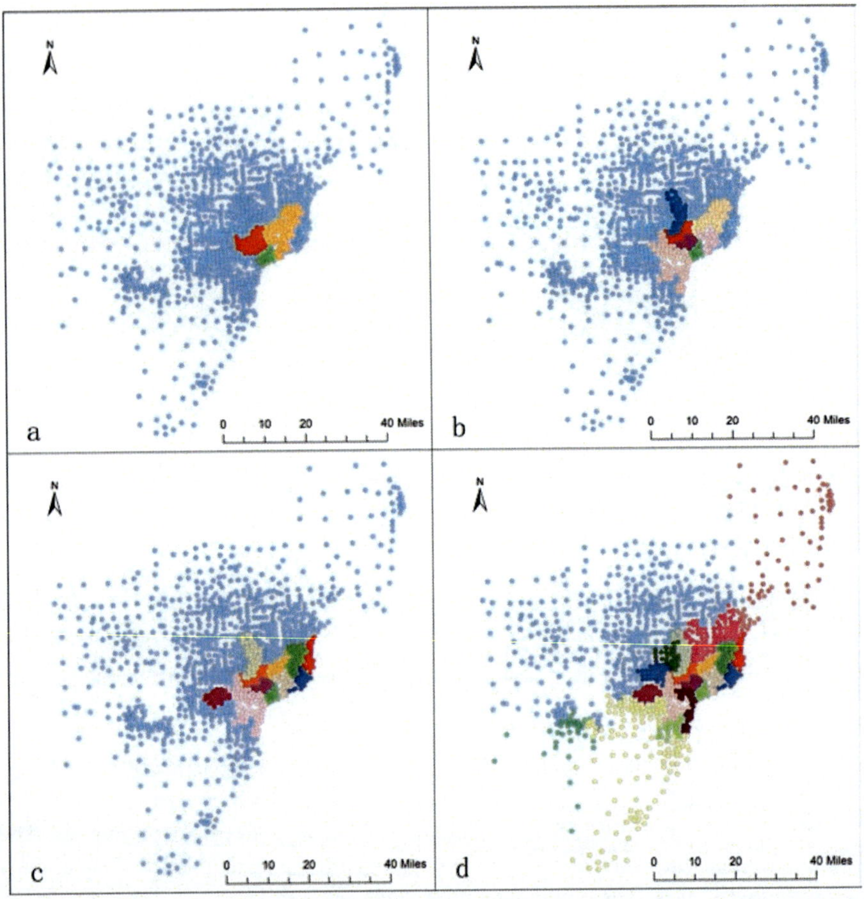

Fig. 6.6 Road network analysis zones: (**a**) 4 zones, (**b**) 8 zones, (**c**) 12 zones, (**d**) 20 zones

Table 6.14 The R^2 value comparison between the T1 and T4 tests

Variable	T1	T4-Cluster 2	T4-Cluster 3	T4-Cluster 4
TotPop	0.8652	0.9669	0.9275	0.9047
THousing	0.8162	0.9591	0.9304	0.9247
Vacant	0.6432	0.8401	0.7228	0.7892
Manuf	0.9342	0.8969	0.7609	0.8555

6.5 Discussion and Conclusion

This chapter explores constructing a unified environmental and socioeconomic database to support a comprehensive analysis of a total socioenvironmental system, which will be examined in the following chapters. The chapter proposes that the natural land-use patches derived from remote sensing (NPs) are the foundation for building such a unified database. The concept of NPs has been commonly used in current literature. However, this chapter presents first efforts to integrate NPs with natural cities, road network analysis zones, and big-data mining analytics for areal interpolation. The chapter designs four types of intermediate zones, NPs, natural communities, natural districts, and road network analysis zones to transform the socioeconomic statistics from a large spatial unit to NPs within a small unit. These experiments confirm that the areal interpolation performances improve when the intermediate zones are constructed with comprehensive and rich information.

The conceptual breakthrough of the NP-based areal interpolation is to use NPs as the basic spatial unit for the unified environmental and socioeconomic database. The socioeconomic data are interpolated (disaggregated) and allocated to NPs, in which no socioeconomic data are initially collected. It is critical to point out that the current land-use-based areal interpolation only regresses total populations in Census Block Groups (CBGs) against various land uses within CBGs (Yuan et al., 1997). This interpolation is applied to validate the reliability of CBGs population counts. It does not intend to disaggregate the population in a CBG to the land-use patches within this CBG. The disaggregation of a socioeconomic variable to natural land-use patches is the core for unifying socioeconomic data with environmental data. The NP-based areal interpolation developed in this chapter is the first effort to interpolate (disaggregate) socioeconomic values from large statistical enumeration areas to small natural land use and land cover patches.

Moreover, the chapter develops two technical innovations to enable the NP-based areal interpolation. The first one is the source–target–zone intersected subtotal adjustment, which adjusts the disaggregation to local constraints and improves the interpolation accuracies. The second one is the double-ratio reallocation, which facilitates to reallocate the adjusted disaggregated information from a socioeconomic enumeration area unit to the NPs within this socioeconomic unit and completes the areal interpolation. It is the first time that these two technical procedures are proposed, named, and implemented with full-documented details in this chapter.

In addition, the natural communities and natural districts-based interpolations are newly developed methods for integrating environmental and socioeconomic ancillary information into areal interpolation to improve interpolation accuracies. Moreover, this chapter innovatively applies the concept of road network analysis zones to empower areal interpolations. In brief, the four areal interpolation methods and their illustrations presented in this chapter are new contributions to the current literature.

Furthermore, the interpolation accuracy depends on spatial variations between the source and target zones. The interpolation accuracy is relatively low when the source and target zones vary significantly in spatial scales. For example, the accuracy test between MCDs and Census Tracts confirms this finding. On the other hand, when the spatial extent variations between the source and the target zones are small, the interpolation outcome accuracy is much improved. The accuracy test between the Census Tracts and the Block Groups validates this conclusion.

In addition, the chapter uses two census geographic units (Census Tracts and Blok Groups) as the target zones so that the interpolation accuracy can be examined. We want to iterate that the socioeconomic statistics must exist in the source zones. However, these statistics do not exist in the target zones. The chapter does not discuss what can be considered as target zones. The general recommendation is that target zone can be a relatively small administrative unit (low on the hierarchy) where no statistics are enumerated. This administrative unit also should be complex enough and usually includes several types of NPs.

It is a limitation at present that it is not always feasible to check the interpolation accuracy. We recommend it is preferable to take a prior accuracy assessment in a pilot area where some validation information exists or can be found. We should be very cautious about the areal interpolation outcomes. We should provide all details about how the interpolation is conducted and point out all limitations you find in the interpolation process.

Finally, we recommend that NPs (the natural patches) will be the ultimate target zones. Direct socioeconomic data disaggregation unto NPs should be the next experiments of spatial (area) interpolation.

References

Anselin, L., & Smirnov, O. (1996). Efficient algorithms for constructing proper higher order spatial lag operators. *Journal of Regional Science, 36*(1), 67–89. https://doi.org/10.1111/j.1467-9787.1996.tb01101.x

Arcaute, E., Hatna, E., Ferguson, P., Youn, H., Johansson, A., & Batty, M. (2015). Constructing cities, deconstructing scaling laws. *Journal of the Royal Society Interface, 12*(102), 20140745. https://doi.org/10.1098/rsif.2014.0745

Ban, H., & Ahlqvist, O. (2009). Representing and negotiating uncertain geospatial concepts – Where are the exurban areas? *Computers, Environment and Urban Systems, 33*(4), 233–246.

Bergs, R. (2018). The detection of natural cities in the Netherlands—Nocturnal satellite imagery and Zipf's law. *Review of Regional Research, 38*(2), 111–140. https://doi.org/10.1007/s10037-018-0122-6

Bronzini, R., & Piselli, P. (2009). Determinants of long-run regional productivity with geographical spillovers: The role of R&D, human capital and public infrastructure. *Regional Science and Urban Economics, 39*(2), 187–199. https://doi.org/10.1016/j.regsciurbeco.2008.07.002

Brown, D. G., Agrawal, A., Sass, D. A., Wang, J., Hua, J., & Xie, Y. (2013). Responses to climate and economic risks and opportunities across national and ecological boundaries: Changing household strategies on the Mongolian plateau. *Environmental Research Letters, 8*(4), 045011. https://doi.org/10.1088/1748-9326/8/4/045011

Burrough, P. A. (1986). *Principles of geographic information systems for land resource assessment* (Monographs on Soil and Resources Survey No. 12). Oxford Science Publications.

Campos, J., Rigotti, J. I., Baptista, E. A., Monteiro, A. M., & Reis, I. A. (2020). Population estimates from orbital data of medium spatial resolution: Applications for a Brazilian municipality. *Sustainability, 12*(9). https://doi.org/10.3390/su12093565

Chen, Y. (2009). Urban gravity model based on cross-correlation function and Fourier analyses of spatio-temporal process. *Chaos, Solitons & Fractals, 41*(2), 603–614. https://doi.org/10.1016/j.chaos.2008.02.030

De Blander, R. (2020). Iterative estimation correcting for error auto-correlation in short panels, applied to lagged dependent variable models. *Econometrics and Statistics, 15*, 3–29. https://doi.org/10.1016/j.ecosta.2020.02.001

Everitt, B. S., & Dunn, G. (2001). *Applied multivariate data analysis* (2nd ed.). Oxford University Press.

Everitt, B., Landau, S., & Leese, M. (2001). *Cluster analysis* (4th ed.). Oxford University Press.

Golding, S. A., & Winkler, R. (2020). Tracking urbanization and exurbs: Migration across the Rural–Urban Continuum, 1990–2016. *Population Research and Policy Review*. https://doi.org/10.1007/s11113-020-09611-w

Green, S. B. (1991). How many subjects does it take to do a regression analysis? *Multivariate Behavioural Research, 26*, 499–510. https://doi.org/10.1207/s15327906mbr2603_7. PMID: 26776715.

Guo, Y., Xi, S., & Zhang, M. (2021). Provincial division of economic zones based on the improved urban gravity model: A case study of Hunan Province, China. *PLoS One, 16*(12), e0261205. https://doi.org/10.1371/journal.pone.0261205. (Accession No. 34936664).

Harvey, J. T. (2002). Estimating census district populations from satellite imagery: Some approaches and limitations. *International Journal of Remote Sensing, 23*(10), 2071–2095. https://doi.org/10.1080/01431160110075901

Harris, R. J. (1985). *A primer of multivariate statistics* (2nd ed.). Academic.

Helper, S., Krueger, T., & Wial, H. (2012). *Why does manufacturing matter? Which manufacturing matters? A policy framework*.

Ippoliti, C., Candeloro, L., Gilbert, M., Goffredo, M., Mancini, G., Curci, G., et al. (2019). Defining ecological regions in Italy based on a multivariate clustering approach: A first step towards a targeted vector borne disease surveillance. *PLoS One, 14*(7), e0219072.

Jenkins, D. G., & Quintana-Ascencio, P. F. (2020). A solution to minimum sample size for regressions. *PLoS One, 15*(2), e0229345. https://doi.org/10.1371/journal.pone.0229345

Jiang, B., & Jia, T. (2011). Zipf's law for all the natural cities in the United States: A geospatial perspective. *International Journal of Geographical Information Science, 25*(8), 1269–1281. https://doi.org/10.1080/13658816.2010.510801

Jiang, B., & Liu, X. (2012). Scaling of geographic space from the perspective of city and field blocks and using volunteered geographic information. *International Journal of Geographical Information Science, 26*(2), 215–229. https://doi.org/10.1080/13658816.2011.575074

Jiang, B., Yin, J., & Liu, Q. (2015). Zipf's law for all the natural cities around the world. *International Journal of Geographical Information Science, 29*(3), 498–522. https://doi.org/10.1080/13658816.2014.988715

Kasraian, D., Raghav, S., & Miller, E. J. (2020). A multi-decade longitudinal analysis of transportation and land use co-evolution in the Greater Toronto-Hamilton Area. *Journal of Transport Geography, 84*, 102696.

Lan, H., Stewart, K., Sha, Z., Xie, Y., & Chang, S. (2022). Data gap filling using cloud-based distributed Markov Chain Cellular Automata framework for land use and land cover change analysis: Inner Mongolia as a case study. *Remote Sensing, 14*(3). https://doi.org/10.3390/rs14030445

Lao, X., Zhang, X., Shen, T., & Skitmore, M. (2016). Comparing China's city transportation and economic networks. *Cities, 53*, 43–50.

Lewis-Beck, M. S. (1994). *Factor analysis and related techniques*. Sage.

Li, Y., & Xie, Y. (2018). A new urban typology model adapting data mining analytics to examine dominant trajectories of neighborhood change: A case of metro Detroit. *Annals of the American Association of Geographers, 108*(5), 1313–1337.

Liu, J., Dou, Y., Batistella, M., Challies, E., Connor, T., Friis, C., et al. (2018). Spillover systems in a telecoupled Anthropocene: Typology, methods, and governance for global sustainability. *Current Opinion in Environmental Sustainability, 33*, 58–69. https://doi.org/10.1016/j.cosust.2018.04.009

Liu, L., Xia, B., Wu, H., Zhao, J., Peng, Z., & Yu, Y. (2019). Delimitating the natural city with points of interests based on service area and maximum entropy method. *Entropy, 21*(5). https://doi.org/10.3390/e21050458

Lo, C. P. (1995). Automated population and dwelling unit estimation from high-resolution satellite images: A GIS approach. *International Journal of Remote Sensing, 16*(1), 17–34. https://doi.org/10.1080/01431169508954369

Martin, D., & Bracken, I. (1993). The integration of socioeconomic and physical resource data for applied land management information systems. *Applied Geography, 13*(1), 45–53. https://doi.org/10.1016/0143-6228(93)90079-G

McLachlan, G. (2004). *Discriminant analysis and statistical pattern recognition*. Wiley.

Montero, G., Tannier, C., & Thomas, I. (2021). Delineation of cities based on scaling properties of urban patterns: A comparison of three methods. *International Journal of Geographical Information Science, 35*(5), 919–947. https://doi.org/10.1080/13658816.2020.1817462

Mur, J., & Angulo, A. (2006). The spatial Durbin model and the common factor tests. *Spatial Economic Analysis, 1*(2), 207–226. https://doi.org/10.1080/17421770601009841

NRC. (2010). *Understanding the changing planet strategic directions for the geographical sciences*. National Academies Press.

Park, H.-S., & Jun, C.-H. (2009). A simple and fast algorithm for K-medoids clustering. *Expert Systems with Applications, 36*(2, Part 2), 3336–3341. https://doi.org/10.1016/j.eswa.2008.01.039

Parrott, L., & Meyer, W. S. (2012). Future landscapes: Managing within complexity. *Frontiers in Ecology and the Environment, 10*(7), 382–389. https://doi.org/10.1890/110082

Pedhazur, E. J., & Schmelkin, L. P. (1991). *Measurement, design, and analysis: An integrated approach*. Erlbaum.

Profillidis, V. A., & Botzoris, G. N. (2019). Chapter 7 – Econometric, gravity, and the 4-step methods. In V. A. Profillidis & G. N. Botzoris (Eds.), *Modeling of transport demand* (pp. 271–351). Elsevier.

Ren, Z., Jiang, B., & Seipel, S. (2019). Capturing and characterizing human activities using building locations in America. *ISPRS International Journal of Geo-Information, 8*(5). https://doi.org/10.3390/ijgi8050200

Rodrigue, J.-P. (2020). *The geography of transport systems*. Routledge.

Rozenfeld Hernán, D., Rybski, D., Andrade José, S., Batty, M., Stanley, H. E., & Makse Hernán, A. (2008). Laws of population growth. *Proceedings of the National Academy of Sciences, 105*(48), 18702–18707. https://doi.org/10.1073/pnas.0807435105

Sarigai, Yang, J., Zhou, A., Han, L., Li, Y., & Xie, Y. (2021). Monitoring urban black-odorous water by using hyperspectral data and machine learning. *Environmental Pollution, 269*, 116166. https://doi.org/10.1016/j.envpol.2020.116166

Seto, K. C., Reenberg, A., Boone, C. G., Fragkias, M., Haase, D., Langanke, T., et al. (2012). Urban land teleconnections and sustainability. *Proceedings of the National Academy of Sciences of the United States of America, 109*(20), 7687–7692. Retrieved from <Go to ISI>://WOS:000304369800033.

Sevtsuk, A., & Mekonnen, M. (2012). Urban network analysis toolbox. *International Journal of Geomatics and Spatial Analysis, 22*(2), 287–305.

Sevtsuk, A. (2014). Networks of the built environment. In D. Ofenhuber & $. C. Ratti (Eds.), *Decoding the city: Urbanism in the big data* (p. 192). Birkhauser.

Sha, Z., Li, R., Li, J., & Xie, Y. (2020). Estimating carbon sequestration potential in vegetation by distance-constrained zonal analysis. *IEEE Geoscience and Remote Sensing Letters, 18*(8), 1352–1356. https://doi.org/10.1109/LGRS.2020.3003448

Shrestha, S., Kazama, F., & Nakamura, T. (2008). Use of principal component analysis, factor analysis and discriminant analysis to evaluate spatial and temporal variations in water quality of the Mekong River. *Journal of Hydroinformatics, 10*(1), 43–56. https://doi.org/10.2166/hydro.2008.008

Shumway, R. H., & Stoffer, D. S. (2017). *Time series analysis and its applications: With R examples Springer texts in statistics* (4th ed., pp. 1 online resource (XIII, 562 pages 148 illustrations, 570 illustrations in color). https://doi.org/10.1007/978-3-319-52452-8

Sun, Z., Wang, X., Wang, Z., Yang, L., Xie, Y., & Huang, Y. (2021). UAVs as remote sensing platforms in plant ecology: Review of applications and challenges. *Journal of Plant Ecology, 14*(6), 1003–1023. https://doi.org/10.1093/jpe/rtab089

Tannier, C., Thomas, I., Vuidel, G., & Frankhauser, P. (2011). A fractal approach to identifying urban boundaries. *Geographical Analysis, 43*(2), 211–227.

Tocchi, D., Sys, C., Papola, A., Tinessa, F., Simonelli, F., & Marzano, V. (2022). Hypergraph-based centrality metrics for maritime container service networks: A worldwide application. *Journal of Transport Geography, 98*, 103225.

Virah-Sawmy, M., Gillson, L., & Willis, K. J. (2009). How does spatial heterogeneity influence resilience to climatic changes? Ecological dynamics in southeast Madagascar. *Ecological Monographs, 79*(4), 557–574. https://doi.org/10.1890/08-1210.1

Webster, C. J. (1996). Population and dwelling unit estimates from space. *Third World Planning Review, 18*(2), 155.

Xie, Y. (1995). The overlaid network algorithms for areal interpolation problem. *Computers, Environment and Urban Systems, 19*(4), 287–306. https://doi.org/10.1016/0198-9715(95)00028-3

Xie, Y., & Ma, T. (2015). A method for delineating a hierarchically networked structure of urban landscape. *Urban Geography, 36*(6), 947–963. https://doi.org/10.1080/02723638.2015.1067410

Xie, Y., Sha, Z., & Yu, M. (2008). Remote sensing imagery in vegetation mapping: A review. *Journal of Plant Ecology, 1*(1), 9–23. https://doi.org/10.1093/jpe/rtm005

Xie, Y., Gong, H., Lan, H., & Zeng, S. (2018). Examining shrinking city of Detroit in the context of socio-spatial inequalities. *Landscape and Urban Planning, 177*, 350–361. https://doi.org/10.1016/j.landurbplan.2018.03.002

Xie, Y., Crary, D., Bai, Y., Cui, X., & Zhang, A. (2019). Modeling grassland ecosystem responses to coupled climate and socioeconomic influences in multi-spatial-and-temporal scales. *Journal of Environmental Informatics, 33*(1). https://doi.org/10.3808/jei.201600337

Yuan, Y., Smith, R. M., & Limp, W. F. (1997). Remodeling census population with spatial information from LandSat TM imagery. *Computers, Environment and Urban Systems, 21*(3), 245–258. https://doi.org/10.1016/S0198-9715(97)01003-X

Zadeh, A. S. M., & Rajabi, M. A. (2013). Analyzing the effect of the street network configuration on the efficiency of an urban transportation system. *Cities, 31*, 285–297.

Zhang, X., Miller-Hooks, E., & Denny, K. (2015). Assessing the role of network topology in transportation network resilience. *Journal of Transport Geography, 46*, 35–45.

Zhu, Z., Zhang, A., & Zhang, Y. (2018). Connectivity of intercity passenger transportation in China: A multi-modal and network approach. *Journal of Transport Geography, 71*, 263–276.

Chapter 7
Empirically Examining Spatial Processes of Urban Declining: Southeast Michigan as the Case Study

7.1 Definitions, Causes, and Dimensions of Urban Decline

Urban decline indicates that an inner city is deteriorating, its housing stock is shrinking, and many social, economic, and environmental problems are accompanying. Many houses become vacant, high levels of unemployment and poverty occur, and public infrastructures deteriorate. People become depressed by the situation and lose aspiration, and some turn to crime to supplement their income. The depressed youth have weaker achievement at schools, and schools started to close. In addition, skilled people migrate out to seek better employment opportunities. Industrial facilities sit empty. Residential, commercial, and industrial properties become abandoned, leading to increased burglary, graffiti, and vandalism (Gamesby, 2022).

For instance, over 60% of the population in Detroit, located in Southeast Michigan, were lost in the past 60 years and 25% during the recent 10 years (Neill, 2015). Detroit has been increasingly abandoned with a vacancy rate 6.5% in 1990, 11.6% in 2000, 18.1% in 2010, and 26.6% in 2020 (ACS, 2022). The city shrinking in Detroit has affected its surrounding regions and towns. Six out of the 15 American ghost towns are located in the Metropolitan Detroit area (Frohlich et al., 2015): Detroit is the No. 4, Highland Park the No. 5, River Rouge the No. 8, Hamtramck the No. 10, Inkster the No. 11, and Ecorse the No. 13. Detroit is the most notorious ghost town in the United States because of the negative impacts from its city shrinking (Frohlich et al., 2015). The neighborhoods in Detroit suffered a lot (McGraw, 2007). Accelerated urban decline in Detroit deteriorated urban structures.

No resources were accessible to curb urban decline and associated challenges. Many buildings were abandoned and became dangerous to the surrounding households and neighboring communities. They had to be demolished. In addition, many

Supplementary Information The online version contains supplementary material available at https://doi.org/10.1007/978-3-031-39594-9_7.

Y. Xie, *Total Socioenvironmental Systems*,
https://doi.org/10.1007/978-3-031-39594-9_7

vacant properties were lost to fire and vandalism. The ruins were man-made and natural causes, illustrated by the following sequences. Political and economic choices caused people to move away from the city. The city's tax revenues were diminished, and houses were vacant. Buildings and infrastructure deteriorated because there was no adequate funding to maintain or repair them. New political and economic decisions focused on demolishing and removing abandoned structures. Bare and open tracts became vacant for dumping wastes or for vegetative growth. For instance, in the Brightmoor neighborhood of Detroit, 31% of the housing units were empty, and there were more than 250 abandoned lots in 2010. Among them, 166 vacant lots had become dumpsites (Ganning & Tighe, 2015). Only 18% of properties were sold at auction for US$ 500 per property between 2002 and 2010 (Ganning & Tighe, 2015).

Urban decline is a global phenomenon (Martinez-Fernandez et al., 2016a, b). More than 25% of large cities worldwide are shrinking (Xie et al., 2018). Urban decline is a hot research subject in advanced capitalist countries such as the United States, the United Kingdom, Germany, France, Australia, and Japan ((Martinez-Fernandez et al., 2016a, b; Danko & Hanink, 2018; Döringer et al., 2020). It is also an emerging research topic in fast-growing countries like China. Currently, China is experiencing rapid urbanization, but hundreds of Chinese cities are suffering a persistent population loss (Hu et al., 2021; Peng et al., 2022).

Why cities shrink and how they can be averted have long been hot topics for a broad array of agencies, disciplines, and organizations. Suburbanization, demographic change, and deindustrialization are commonly regarded as three leading causes of city shrinking (Großmann, et al., 2013). The advance in transportation technology after the Second World War profoundly stimulated suburban development and urban sprawl in the United States. The landmark National Interstate and Defence Highway Act of 1956 signified a long-term commitment by the national government to unite the country through an extensive network of interstates (Gutfreund, 2004). As a result, highways reached almost everywhere.

Meanwhile, cars became affordable for ordinary households, and hence accelerate suburban development. Easily accessible roads and cars sped up migration from city centers to suburbs (Glaeser & Kohlhase, 2003). When people moved out from city centers to suburbs and exurban areas, taxes, financial resources, and political powers came with the flight. Subsequently, the city centers were left with vacant houses, abandoned properties, and unmaintained infrastructures. In addition, housing market was contracting. The shrinking of the housing market led to lower housing prices and higher housing vacancy (Kabisch et al., 2008). These signs of city shrinking in turn became causes of further shrinking, consisting of a vicious circle.

Economic restructuring is a reaction to city shrinking but also became a cause of urban decline. The United States has lost five million manufacturing jobs since 2000 (Long, 2016). When factories and industries were moved to other areas or closed, workers had no choice to move along or faced lay off. The host cities and even the entire regions experienced the outflow of population and capital resources. These outflows sped up the vacancy and abandonment processes of houses, buildings, and facilities, and, eventually led to widespread shrinks (Martinez-Fernandez & Wu, 2007).

Economic globalization has replaced domestic economic restructuring in recent years and become a new driver of urban decline. Global redistribution of production processes and global capital flows have been regarded as an active agent leading to urban shrinking (Martinez-Fernandez & Audirac, 2012). When a city loses its edge in the current competitive international economic environment, it will phase out of the competition, and its economy will shrink and become stagnant. It will witness an outflow of people and capital. It will suffer housing vacancy and infrastructure decaying (Audirac, 2014). Therefore, economic globalization stimulates the mobility of people and capital across countries and regions. When cities are left behind in global competition, their residents lose jobs and must move out to seek new opportunities. This international economic restructuring process leads to the shrinkage of their population, contraction of their housing stocks, and decaying of their infrastructures. In contrast, in places with advantages in the international competitive markets, net gains in people, finance, innovation, and technology are expected (Wiechmann & Pallagst, 2012).

Recently, attention to the economic causes of urban shrinking has shifted to the context and dynamics of shrinking cities. Racial inequality, residential segregation, and gentrification are thought to be primary factors to impact the dynamics of urban shrinking from the perspective of urban political economy (Fol, 2012). Racial segregation has led to fast-growing concentrations of disadvantaged households in poor neighborhoods and aggressive white flight to suburbs (Grossmann et al., 2014). This separation has rapidly decreased property values in city centers and led to unprecedented decaying and shrinking in these areas. Xie and his colleagues (2018) developed a regional logistic categorical regression model to examine the effects of several primary social–spatial inequality factors on urban shrinking in Detroit. The findings confirm that the vacancy odds in the census tracts with the highest concentration of less educated population were 9.01 times higher. The odds were 7.16 times higher in the tracts where there was a good portion of housing structures without a full kitchen. The odds were 7.06 times higher in the tracts with the most concentration of Black population, 5.47 times higher in the tracts with a good portion of multiunit housing structures, and 4.76 times higher in the tracts with the poorest people.

In addition, there is an increasing call for special attention to the spatial dynamics and processes of urban shrinkage rather than being limited to urban centers. For example, Detroit and Cleveland lost half or more of their populations between 1950 and 2010. In 2015, 40.3% and 36.2% of the population were below the poverty line (Pinto & Sablik, 2016). However, in the metropolitan statistical area (MSA) surrounding Detroit, more than one million people were added between 1950 and 2010. Spatial inequalities in poverty, education, and living standards exist in and between cities, which need to be studied by an interdisciplinary team of scientists, planners, and economists (Panori et al., 2019). Spatial inequality trends extend from intra-urban to regional and international scales (Nijman & Wei, 2020). The inner-city inequalities at neighborhood levels and among urban residents have widened and worsened (Li & Xie, 2018). At the inner-city scale, increasing inequality and divergent urban development have caused neighborhood change and spatial polarization (Modai-Snir & van Ham, 2018). This neighborhood change is shaped mainly by poverty/wealth clustering and other spatially explicit processes (Tong & Kim, 2019). Different cities in diverse nations have exhibited different patterns of urban decline attributable to various driving factors (Wang & Hiroatsu, 2019; Batunova & Gunko, 2018; Wolff et al., 2017; Haase et al., 2016). Plenty of evidence suggests that regional inequalities, such as uneven development, urban inequality, spatial disparity, residential segregation, and gentrification, influence the dynamics of shrinking cities. For example, local and regional socioeconomic inequality factors interact across various scales (counties, cities, and census tracts), affecting urban shrinkage in Detroit (Xie et al., 2018).

7.2 Spatial Planning Implication of Urban Decline Studies

Urban planners and policymakers in declining cities understandably want to revitalize their cities. Reversing urban decline trends and restoring declined cities are critical tasks for urban planning agencies in declining cities. First, urban shrinking creates enormous challenges to urban economic growth, and urban and regional planning primarily produces policies and programs that promote economic growth (Batunova & Gunko, 2018). Second, land use and transportation planning are integral elements of urban and regional planning. Urban decline creates many vacant lands and abandoned infrastructures, which planners must deal with to improve city-built landscapes (Deakin, 2020). Urban shrinking is also causing lots of environmental issues. Environmental impact assessment is the primary planning tool planners must use to avoid environmental deterioration and ensure economic growth and quality of life (Daniels, 2022).

From the action point of view, urban renewal is a typical urban planning approach to fight urban decline (Brunner & Kaminski, 2016). Urban revitalization has commonly been regarded as a solution to alleviate urban shrinking. It focuses on mitigating the effects of urban decline by adopting policies for restoring historical landmarks and buildings, demolishing deteriorated and abandoned old buildings, and constructing new infrastructures. The demolition of abandoned buildings in a shrinking city is a priority because this action eradicates many negative socioeconomic and environmental issues (Gao & Ryan, 2020). However, demolition would only eliminate the root of urban decline if this effort is complemented with incentives to redevelop the vacant land left by demolition. In other words, urban revitalization or demolition is a type of reactive planning approach that has received a lot of criticism.

In recent years, the proactive planning approach has gained high popularity. Proactive planning predicts the effects of land use and socioeconomic changes and plans accordingly to achieve sustainable developments. It investigates the causes and consequences of urban decline through a comprehensive analysis of economic indicators. Gu et al. (2019) developed a logistic regression model to examine the housing abandonment issue in shrinking cities. The study used tax delinquency status as a proxy for an urban decline to study relationships between urban shrinking and mixed land uses in Dayton, Ohio. The results confirmed that the tax delinquent residential lots positively affected neighboring single-family residential lots. On the other hand, tax delinquent commercial and industrial lots negatively affected neighboring single-family residential lots (Gu et al. 2019). This finding suggests adopting a mixed land-use design in urban planning can help mitigate the urban decline.

A second proactive urban strategy is to use zoning regulations to impact land uses. Zoning regulations provide guidelines that promote or prevent new developments in a neighborhood. Cho et al. (2012) advocated consideration of housing prices in zoning guidelines. Cho et al. (2012) performed a detailed study on rezoning vacant land into residential, commercial, and industrial uses over 9 years. The findings from the multinomial logit model suggested that a land parcel surrounded by residential zoning is more likely to be rezoned to residential or commercial rather than industrial. The study also identified significant spillover effects of neighboring residential land prices onto neighboring vacant land prices when the vacant land was rezoned to residential. Increased pressure on housing demand is indicated by a rise in the cost of housing in a nearby location. This surge in housing demand is expected to increase demand for residential and commercial construction that complements residential use (Cho et al. 2012).

Another proactive planning practice promotes rightsizing the city's built environment to suit a reduced population size. There are many planning terms analogous to rightsizing, including "creative decline" (Swope 2006), "creative shrinkage" (Oswalt 2006), and "smart decline" (Aalbers & Bernt, 2019). The core of these concepts is "to proactively make plans for a city's possible or likely population shrinkage by adjusting its infrastructure, development patterns, and governmental services to its reduced population" (Alligood, 2008, p. 3). A related practice is allocating empty lands for parks, recreation, or agriculture in Detroit. Although some lovely parks and recreation sites have been established, the primary beneficiaries are suburban residents, outside visitors, or city residents with economic means (Manville & Kuhlmann, 2018). As a result, they fail to contribute much to transforming the urban economy or benefit most of the city's residents.

In recent years, the notion that shrinking has a spatial scale has gotten people's attention. "At what scale do we witness and at what scale do we manage decline?" (Aalbers & Bernt, 2019, p. 4). More critically, there are different scales, such as the description, analysis, and policy scale. In brief, practical solutions to deter urban shrinking or redevelop vacant lands require understanding the spatial effects of urban decline and spatial relationships between declining lots and neighboring communities.

7.3 Methods to Examine Spatial Effects of Urban Decline

Many urban planners and researchers advocate that it is necessary to examine the spatial extent and process of urban shrinking, as well as administrative or organizational structures for implementing urban shrinking prevention policies (Haase et al., 2016). Understanding the spatial scope and dynamics of urban decline requires a comprehensive analysis of spatial pattern, spatial correlation, and spatial interaction of urban decline processes in relationship with their environmental, demographic, and socioeconomic characteristics. On the other hand, responding to urban decline necessitates a deep understanding of complex urban governance processes and hierarchical administrative or policymaking structures (Hospers, 2014). A workable urban decline suppression strategy must be appropriate to each local and regional context (Li et al., 2020). Unfortunately, there is a lack of research on policies preventing or recovering urban decline from spatial perspectives (Gim, 2018). There is an urgent need to analyze the policy effect on urban shrinking in specific localities or regions. We will inevitably develop a hierarchical structure to reflect spatial effects and relationships of urban shrinking and to implement urban decline suppression policies as multilevel (hierarchical) planning strategies (Pallagst et al., 2021). However, few empirical hierarchical studies of urban decline exist in current literature and will be discussed in the following subsections.

7.3.1 *Hierarchical Census Geographic Structure and Logit Regression Model*

Xie and his colleagues (2018) developed a hierarchical logit model to examine the shrinking city of Detroit in the context of socio-spatial inequalities. The study area is the Metropolitan Detroit area, Michigan, including seven counties, Livingston, Macomb, Monroe, Oakland, St. Clair, Washtenaw, and Wayne. Wayne is the largest of these counties, where the City of Detroit is located. In the study area, there are 40 incorporated cities based on the data from Michigan Center for Shared Solutions and Technology Partnerships (MCGI, 2017). The City of Detroit is the largest, with a population of 672,795 in 2016 (US Census Bureau, 2017). The study area also contains around 1400 Census Tracts. The census tracts, incorporated cities, and counties consist of a hierarchical spatial structure with three levels (Fig. 7.1). This design is a hybrid hierarchical structure in which the three models are sequentially linked from the tract-level, city-level, and county-level but not nested in the hierarchy. The three models are individual and each has a specific purpose. Together the three models answer the question how socioeconomic disparity factors affect the urban shrinking in southeastern Michigan, with the City of Detroit as the center. For example, one of the findings is that the odds of becoming vacant in the census tracts with the highest concentration of less educated population were 9.01 times higher than in other census tracts (Xie et al., 2018).

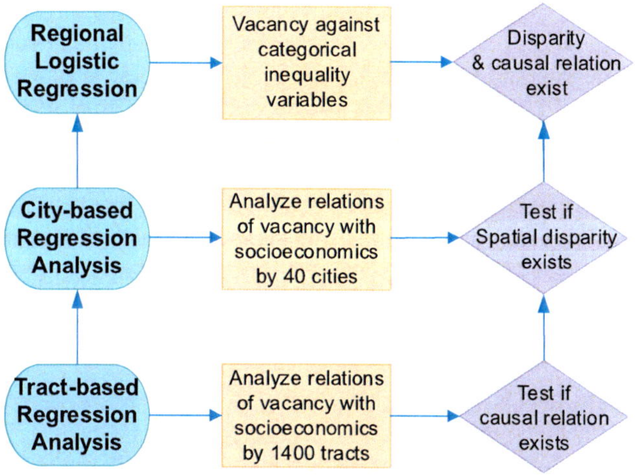

Fig. 7.1 A hierarchical spatial modeling framework: tracts, cities, and region

7.3.2 Spatially Autoregressive Models

The conceptual foundation of the spatially autoregressive models is the first law of geography, which states that everything is related to everything else, but near things are more connected than distant things (Tobler, 1970, 236). The methodological footing of the spatially autoregressive models is the multiple regression analysis, which takes the following form:

$$y_i = \alpha_0 + \beta X + \varepsilon_i \tag{7.1}$$

where y_i is $n \times 1$ matrix of the response variable, n is number of observations for the response variable, X is $n \times p$ matrix of the covariates (the explanatory variables), β is a $p \times 1$ matrix of regression coefficients for X, p is total number of covariates used to fit the model, and ε_i is an error term.

Spatial autoregressive models are also called spatial econometric models (SpEMs) (Anselin, 2003). There are five types of SpEMs: spatial lag model (SLM), spatial error model, spatial lag of X model (SLX), spatial Durbin model (SDM), and spatial Durbin error model (SDEM). They differ on how to handle the spatial lag effects.

The spatial lag model (SLM) takes the following form:

$$y_i = \alpha_0 + \rho W y_i + \beta X + \varepsilon_i \tag{7.2}$$

The term definitions are like Eq. (7.1) except that there is an addition of the spatial lag of the response variable (Wy_i). ρ is the scalar coefficient of Wy_i representing the strength of spatial autocorrelation of y_i. W is a spatial weight matrix that defines the neighboring spatial relationship. The weights of a weight matrix are computed based on several spatial adjacency relationships (Zhou & Lin, 2017): (1) the contiguity of boundaries, for instance, queen contiguity weights and rook contiguity weights are appropriate for polygon features; (2) distances, for instance, distance-band weights and k-nearest neighbor weights are ideal for point features; (3) other user-defined criteria for specific applications (Long, 2021).

The spatial lag model (SLM) examines how the value of the response variable in a spatial unit is affected by the values of the response variable in its neighboring spatial units. This effect is called the spatial autoregressive effect or spatial lag effect.

The spatial error model (SEM) takes the form of

$$y_i = \alpha_0 + \beta X + u_i, u_i = \lambda W_u + \varepsilon_i \tag{7.3}$$

where Wu is the interaction effects among the error terms of the different observations, λ is the scalar coefficient of Wu representing the strength of spatial dependence in error terms, and ε_i is a $n \times 1$ matrix of the error term (LeSage, 2008). SEM examines whether the prediction errors of the response variable are spatially autocorrelated, which is called the spatial error interaction effect (Anselin, 2003).

The spatial lag of X (SLX) model is another type of SpEMs. A linear regression model inspects how the value of the response variable in a spatial unit is affected by the values of the explanatory variables in the same spatial unit, which is called the direct effect of the explanatory variables on the response variable. On the other hand, SLX assumes that the value of the response variable in a spatial unit is affected by the values of the explanatory variables (or called covariates in terms of SpEMs) in its neighboring spatial units (Elhorst & Vega, 2015). In other words, this type of additional effect of the covariates on the response variable is called the spatial spillover effect. SLX can use lagged X to examine spatial interaction effects of covariates on the response variable. SLX can take the following matrix form:

$$y_i = \alpha_0 + \beta X + \Theta WX + \varepsilon_i \tag{7.4}$$

where y_i, α_0, β, X, W, and ε_i are as explained above, Θ is the regression coefficient of WX, representing the effects of neighboring covariates on the response variable of a particular region. Θ is defined as a $p \times 1$ matrix of spatial spillover parameters. The error terms are assumed to be independent and normally distributed with a zero mean and constant variance (Elhorst & Vega, 2015).

The spatial Durbin model (SDM) is an extension of SLX model or SLM model. SDM combines SLX and SLM, thus examining the spatial lags of the covariates and the response variable in a single model (Elhorst & Vega, 2017). SDM can be expressed with a matrix form as

$$y_i = \alpha_0 + \rho Wy + \beta X + \Theta WX + \varepsilon_i \tag{7.5}$$

Another form of SpEMs is the spatial Durbin error model (SDEM). It combines SLX and SEM models and accounts for spatial autocorrelation in covariates and spatial heterogeneity in error term (Elhorst & Vega, 2017). SDEM takes the following matrix form:

$$y_i = \alpha_0 + \beta X + \Theta WX + u_i, u_i = \lambda W_u + \varepsilon_i \tag{7.6}$$

In summary, the five SpEMs can examine four spatial effects: (1) spatial autoregressive effect (SLM), (2) spatial spillover effect (SLX), (3) spatial error interaction effect (SEM), and (4) their mixed effects (SDM and SDEM). However, a nonspatial multiple linear regression model can only examine the direct effect of the covariates (regressors) on the response variable. Therefore, the five SpEMs can statistically reveal spatially explicit processes of the impact of a set of regressors on the response variable.

Moreover, a SpEM model can be structured as hierarchical through its weight matrix. The weight matrix (W) can be constructed as a hierarchical system, for instance, W_1, W_2, W_3, …, W_n. W_1 is the first-order spatial weight matric and W_n is the nth order weight matrix. W_1 can be nested within W_2. The integration of SpEMs and Wnth order weight matrices can reveal a hierarchical order of spatial extents and processes for examining the spatial extent and process of the direct and the spatial spillover effects of the regressors on the response variable. A good application of this combination has been reported for examining the spatial processes of the spillover effects of socioeconomic inequality factors on urban shrinkage in the City of Detroit (Lokhande & Xie, 2023).

7.3.3 Hierarchical Mixed Linear Regression Models

A hierarchical linear mixed model (HLM) has long been developed or adopted to overcome random effects or non-normal distributions common to experimental and observational ecological data (Bolker et al., 2009). Statistically, random effects in ecological and environmental studies are often modeled to examine heterogeneities among individuals, species, communities, regions, and periods (Bolker, 2015). The power of HLM in handling random effects is equivalent to the inclusion of grouping or categorical or indicator variables in the panel data regression model (Xie et al., 2019, 2021; Yang & Xie, 2023). In addition to modeling differences among communities, spatial and temporal dimensions, differentiated effects of organizational, economic, and financial factors can be included in HLM as random effects or covariates of random effects. A generalized hierarchical linear mixed model can take the following form (Hedeker & Gibbons, 2006):

$$y_{ij} = \beta x_{ij} + u_i z_{ij} + e_{ij} \tag{7.7}$$

where y_{ij} is a matrix of continuous responses across the ith subject and the jth repeated measure; β is a vector of regression coefficients for the fixed effect; x_{ij} is the $i \times j$ matrix of covariates for fixed effects; μ_i is a vector of random effects over the j repeated measures; z_{ij} is the $i \times j$ matrix of covariates for random effects; and e_{ij} are the residuals.

In socioeconomic analyses, the concept of the *j*th repeated measure is extended to encompass multiple regions. The repeated measures of a subject over a time series are extended to model subjects of interest over different regions. The temporal hierarchy is introduced as the spatial hierarchy in urban and regional studies. In other words, hierarchical linear modeling (HLM) can examine the impact of multilevel regions. For example, Tong et al. (2021) empirically found that the economic and demographic conditions at the town and county levels affect urban shrinkage differently in Northeast China using a hierarchical linear model. The small population and low-level administrative units at the town level intensified the shrinking phenomena. The Northeast China hierarchical study used two different sets of variables at the town and county levels, respectively. However, the classic HLM models often use the same measurement indicators (Gim, 2018). For example, Greenwald and Boarnet (2001) analyze the same set of variables at different spatial levels (census-block group and zip code levels). They confirm that each variable displays varied influences at a different spatial level. Moreover, Curtis et al. (2012) investigate the poverty-generating process using the same set of variables at different spatial levels.

Nevertheless, at present, no empirical HLM studies exist in the context of urban shrinkage studies. It is confirmed that urban shrinking processes (such as those happening in Metropolitan Detroit) differ on multiple scales (Xie et al., 2018). Regional disparities of urban shrinking should be examined at interregional, intraregional, city, neighborhood, and household scales (Wei, 2015). This chapter develops a comprehensive empirical HLM model to examine spatial processes of urban shrinking to fill the missing literature.

7.3.4 Road Network Analysis Zones (RNAZs): Important Urban Spatial Functions and Interactions

The chapter integrates the concept of networked urban systems to partition the study areas into road network analysis zones (RNAZs) to discern spatial processes of urban shrinking. The concepts of big data and complex networks developed rapidly in the last decade as new approaches in complexity science (Newman et al., 2006; Portugali, 2010; Neal, 2012, 2013; Pijanowşki et al., 2014; Tao et al., 2014). Batty integrates these new concepts into the studies of cities, called *The New Science of Cities* (Batty, 2013). In Batty's vision, cities as complex systems mean that cities are developing like biological organisms evolving in a Darwinian fitness landscape (Szell, 2014). Cities are networked spatial configurations, over which different sets of actions, interactions, and transactions take place. This type of complexity is better interpreted through urban models built on transportation networks and supported with "big data" that are being increasingly collected and available in both spatial and temporal dimensions (Szell, 2014). Batty's new science of cities sets the benchmark that all urban modelers and geo-complexity researchers will have to follow (Thrift, 2013).

Flows and networks are two irreducible and interdependent elements that compose an urban system and define the ways by which the city functions (Batty, 2013, 2018). From the perspective of network analysis used in transportation geography, an urban network consists of nodes (cities, urban centers, or places) and edges (links or transport lines) that enable flows between the nodes. The network can be measured with five groups of interrelated indicators based on the Graph Theory (Diestel, 2005): centrality, clustering, connectivity, accessibility, and hierarchy (Batty, 2013; Neal 2013). Chapter 6 applies these measurements to construct road network analysis zones to represent spatial configurations of urban systems.

This chapter builds a hierarchical linear mixed model (HLM) based on the spatial structures of RNAZs. This new HLM comprises three hierarchical scales – the Census Tracts, the RNAZs, and the Counties – to examine how socioeconomic and inequality factors affect urban shrinkage across different spatial levels. RNAZs are the fundamental spatial configurations to reveal spatial processes of urban growth and decline. This chapter is the first attempt to integrate the concept of networked urban systems to discern spatial patterns and processes of urban shrinking in Metropolitan Detroit, through an HLM model built on RNAZs. Moreover, the chapter investigates how various socioeconomic factors interact across different spatial scales to impact urban shrinkage. These innovative analytical methods will shed new lights on spatial processes and extents of urban shrinking.

7.4 Spatial Processes of Urban Shrinkage in Metropolitan Detroit

7.4.1 The Study Area and Data

The study area is the Metropolitan Detroit area, Michigan, including seven counties: Livingston, Macomb, Monroe, Oakland, St. Clair, Washtenaw, and Wayne. Wayne is the largest of these counties, where the City of Detroit is located. The vacancy is one of the primary characteristics of urban decline and is used as a proxy for urban shrinking in our study. Therefore, the vacancy percentage extracted from the American Community Survey (ACS) in 2019 is the response variable.

Our research goal is to investigate which driving factors directly affect urban decline, which factors exhibit spatial effects on urban shrinking, and how strong the spatial effects extend to different spatial levels. We intend to discover which covariates (regressors) have direct (fixed) effects on urban shrinkage and which have random effects on urban shrinkage. We plan to examine how these regressors interact to affect urban shrinkage and whether regressors' cross-spatial effects on urban shrinkage exist. As discussed in the previous sections, we will also focus on inequality factors in urban shrinkage. Therefore, we follow three principles to select and collect comprehensive datasets to support this exploration.

At first, a shrinking city is witnessing a structural economic transformation and crisis in addition to a long series of losing populations (Wiechmann, 2008). Many shrinking cities are experiencing fiscal problems and deteriorating services (Hummel, 2015). There is an increasing need for welfare and public assistance among the most vulnerable population in shrinking cities (Hoekveld, 2014; Hoekstra et al., 2018). Therefore, we selected a set of persistent poverty variables related to the economic crisis, including the percentages of poverty, no-income, receiving supplementary aids, low-wage renters, and newly moved household heads (Table 7.1).

Moreover, economic crisis often associates with poor quality housing and living conditions. For instance, the variable reflecting housing units without full kitchen out of the total housing units is commonly used to indicate poor housing or living condition (Xie et al., 2018). A full kitchen requires inclusion of three essential items, a sink with a faucet, a stove or range, and a refrigerator for the exclusive use of the occupants of the units (U.S. Department of Housing and Urban Development, 2013). Without a full kitchen is defined as missing one of the three or sharing with others or using portable cooking equipment or icebox instead. 0.17% owner-occupied housing units and 1.0% of renters-occupied housing units are missing at least one of the three required full-kitchen facilities in the nation (Raglin, 2015). No full-kitchen percentage is different from housing vacancy rate. Without full kitchen could very well be a result of overcrowd and limited space (Greer, 1986). However, this variable correlates positively with housing vacancy. Many Federal agencies, such as the Department of Health and Human Services and the Department of Housing and Urban Development, are mandated to collect kitchen data as an essential criterion for many Federal programs, including the Emergency Solutions Grant and the Community Development Block Grants (U.S. Census Bureau, 2014). Without full kitchen is also a crucial socio-spatial inequality index. For example, 40% of the 886 census tracts in Metropolitan Detroit in 2010 had the rate of without full kitchen above 5.0%, 26% of the tracts above 10.0%, and 11% of the tracts above 20.0% (NCDB, 2010). Other variables reflecting poor housing and living conditions include the percentages of no fixed heating equipment housing, multiunit structure housing, and old housing among the total housing units.

Second, one of our focuses is to explore how inequality plays a role in urban shrinkage. We have chosen the following inequality indicators (Xie et al., 2018): the percentage of Black/African American population, the portion of the persons 25+ years old who do not have a high school diploma or equivalent, the percentage of female-headed families and subfamilies with own children out of the total families, and the population density by land area (persons per acre).

Third, we must consider the multiple spatial scales. Our case study constructs one hierarchical spatial system: the Census Tracts, the road network-based analysis zones (RNAZs), and the Counties. This system is based on the outcome of one areal interpolation technique presented in Chap. 6. All the above-selected variables are available at both tracts and RNAZs. In addition, another set of variables is only

Table 7.1 The variables and explanations for hierarchical RNAZ geographics

Variable	Explanation
Responsible variable	
Vacant	% vacant housing units of the total housing units
Level 1: Tract – Inequality, poor quality housing, and poverty	
Inequality (census tract)	
BlkPop	% Black/Afr. Am. Population of the total population, which is recoded as a categorical variable with 1, 2, and 3 based on the dividing values 21.036815 and 61.328790. These values are obtained from the Jenks Natural Break classification method
FHFCLD	% Female-headed families and subfamilies with own children; recoded by Jenks Break values 13.860253–32.394366
Pdens	Population density by land area (persons per acre); recoded by Jenks Break values 7.145280–16.629139
LOWEDU	% Persons 25+ years old who did not have a high school diploma or equivalent of the total population; recoded by Jenks Break values 4.246338–9.697856
Poor quality housing (census tract)	
BYTR	% structures built more than 30 years ago of the total number of building structures
NoHeat	% housing units where no heating fuel is used for the total housing units; recoded as a binary variable: 0 when < mean, 1 when ≥ mean (0.3797)
NoKitch	% housing units without full kitchen of the total housing units; recoded as a binary variable: 0 when < mean, 1 when ≥ mean (6.3089)
MULTI	% multiunit structures of the total housing units; recoded as a binary variable: 0 when < mean, 1 when ≥ mean (19.5344)
Persistent poverty (census tract)	
POVTY	% below poverty level families of the total families; recoded by Jenks Break values 11.257311–24.393807
NoInc	% persons aged above 15 not having income of the total population; recoded by Jenks Break values 12.611301–22.041100
GetAid	% persons aged above 15 receiving supplementary aids of the total population; recoded by Jenks Break values 8.877041–22.950800
RGR25	% renter HHs whose gross rent is 25% + of last year's income of the total households, which was less than US$ 10,000; recoded by Jenks Break values 14.868541–33.451957
YRMV	% household heads move into a unit less than 10 years ago of the total households; recoded by Jenks Break values 16.206898–27.474650
Level 2: Road network analysis zones (RNAZs) – Comprehensive 20 zones	
GRAV1	Gravity score of Factor 1; recoded as 1, 2, 3 by Jenks Break values 5.089560–8.334096.
GRAV2	Gravity score of Factor 2; recoded as 1, 2, 3 by Jenks Break values 1.249804–1.425454
GRAV3	Gravity score of Factor 3; recoded as 1, 2, 3 by Jenks Break values 1.208917–1.652544
GRAV4	Gravity score of Factor 4; recoded as 1, 2, 3 by Jenks Break values 0.725190–0.921762
GRAV5	Gravity score of Factor 5; recoded as 1, 2, 3 by Jenks Break values 0.338070–0.468777

(continued)

Table 7.1 (continued)

Variable	Explanation
GRAV6	Gravity score of Factor 6; recoded as 1, 2, 3 by Jenks Break values 1.481278–1.620512
FAC1	The score of Factor 1; recoded as a binary variable 0: FAC1 < 0, 1: FAC1 > 0
FAC2	The score of Factor 2; recoded as a binary variable 0: FAC2 < 0, 1: FAC2 > 0
FAC3	The score of Factor 3; recoded as a binary variable 0: FAC3 < 0, 1: FAC3 > 0
FAC4	The score of Factor 4; recoded as a binary variable 0: FAC4 < 0, 1: FAC4 > 0
FAC5	The score of Factor 5; recoded as a binary variable 0: FAC5 < 0, 1: FAC5 > 0
FAC6	The score of Factor 6; recoded as a binary variable 0: FAC6 < 0, 1: FAC6 > 0
Level 3: County – Productivity and establishment	
GDP19	Per capita gross domestic product (US$ 1000); recoded as a categorical variable by Jenks Break values 34.088001–50.260000 (1: <34.09; 2: >34.09 and < 52.26; 3: >52.26)
Estab	Number of business and industrial establishments per 1000 persons; recoded as a categorical variable by Jenks Break values 3.167001–25.782000
Employ	% employment over population; recoded as a categorical variable by Jenks Break values 5.372001–42.588000
Wage	Average weekly wage (US$ 100); recoded as a categorical variable by Jenks Break values 9.930001–12.040000

available for the RNAZs (Table 7.1). The RNAZs were constructed through the areal interpolation – Technique 4 based on the gravity theory (see Chap. 6 for details). We grouped 25 variables (8 environmental and 17 socioeconomic) into six dimensions by applying principal component analysis (PCA). We then computed the gravity scores of the six factors based on the road network connectivity and, finally, clustered the census tracts into 20 RNAZs using the gravity scores. As a result, each RNAZ carries six gravity scores and six factor loadings. These scores and loadings comprise a unique set of variables associated with the RNAZs.

The dataset is organized in two formats: (1) the Stata dataset file, Ch7_Trc_RNAZ_Cnty, and (2) the excel file, Ch7_Trc_RNAZ_Cnty_Data. They are saved in the folder/Chapter_7. The HLM models are implemented in Stata. The following analysis sections provide the Stata commands for executing these models.

7.4.2 *The Hierarchical Linear Model Results and Analyses Based on the RNAZs Hierarchical Dataset*

7.4.2.1 A Two-Level Model of Vacancy with RNAZs Effects

We will start with the simplest multilevel model, which allows for RNAZs effects on vacancy without regressors (or covariates or explanatory variables). This two-level null model is written as

Table 7.2 The outputs of the two-level null model at Tract and RNAZ levels. (The Stata command: mixed Vacant RNAZ_ID:, mle variance nostderr)

Mixed-effects ML regression	Number of obs	=	1330
Group variable: RNAZ_ID	Number of groups	=	20
	Obs per group:		
	Min	=	27
	Avg	=	66.5
	Max	=	439
Log-likelihood = −5071.655	Wald chi²(0) = .		
	Prob > chi² = .		

Vacant	Coef.	Std. Err.	z	P > z	[95% Conf. Interval]	
_cons	14.397	1.556	9.26	0.00	11.348	17.446

Random-effects parameters	Estimate	Std. Err.	[95% Conf. Interval]	
RNAZ_ID: Identity				
var(_cons)	45.549	.	.	.
var(residual)	114.883	.	.	.

$$\text{Vacancy}_{ij} = \beta_0 + u_{0j} + \varepsilon_{ij} \tag{7.8}$$

where Vacancy_{ij} is the vacancy percentage of Tract i in RNAZ j, β_o is the overall mean vacancy across RNAZs, μ_{oj} is the effect of RNAZ j on Vacancy_{ij}, and ε_{ij} is the Tract level residual. The RNAZ effects, μ_{oj}, are also called RNAZ (or level 2) residuals. This model outcome is reported in Table 7.2.

LR test vs. linear model: $\text{chibar}^2(01) = 285.26$, $\text{Prob} \geq \text{chibar}^2 = 0.0000$

The overall mean vacancy (across RNAZs) is estimated as 14.397. The vacancy rate for RNAZ j is calculated as 14.397 + μ_{oj}, where μ_{oj} is the RNAZ residual reported in Table 7.3. When an RNAZ has a $\mu_{oj} > 0$, its vacancy rate is higher than the average, while $\mu_{oj} < 0$ for a below-average RNAZ. For example, RNAZ 20 had an estimated residual of −8.697, ranked 1, i.e., the first place from the bottom. For this RNAZ, we estimate a mean score of 14.397–8.697 = 5.7, which is the lowest vacancy rate in Metro Detroit. RNAZ 20 is located in the Northeastern Metro Detroit and primarily within St Clair County (Fig. 7.2). As we explained in Chap. 6, the vacancy rate is the lowest in St Clair County. This two-level null model confirmed this finding. In contrast, the highest residual occurs in RNAZ 5 (ranked 20). The estimated mean score is 14.397 + 13.163 = 27.56, which is the highest vacancy rate. RNAZ 5 locates in the western part of the City of Detroit, where the worst vacancy was observed. The estimated μ_{oj} and ε_{ij} are reported in Table 7.3 and graphed in Fig. 7.3. The vacancy rates in RNAZ 5, 2, 3, 4, 6, and 9 are much higher than the mean vacancy rate. These RNAZs are in or around the City of Detroit. Conversely, the vacancy rates in RNAZ 20, 17, 16, 1, 10, 18, and 14 are much lower than the mean. They are in exurban areas of Metro Detroit (Fig. 7.2).

Table 7.3 Estimates of RNAZ random effects and standard errors

RNAZ_ID	μ_{oj} rank	μ_{oj}[a]	ε_{ij}[b]
20	1	−8.6974	1.2253
17	2	−8.0742	1.7736
16	3	−7.8830	1.0967
1	4	−6.7403	0.5101
10	5	−4.5870	1.9091
18	6	−3.3889	0.8707
14	7	−3.1186	1.6437
13	8	−2.6264	1.8512
12	9	−2.3374	1.9091
8	10	−2.3356	1.4790
11	11	−2.0844	1.4384
15	12	−1.3495	1.8795
7	13	0.3833	1.4516
19	14	0.6156	1.7984
9	15	3.2789	1.9091
6	16	4.2943	1.7984
4	17	8.1653	1.8242
3	18	10.9213	1.9727
2	19	12.4008	1.5714
5	20	13.1631	1.9091

[a]The estimates of the RNAZ random effects using empirical Bayes predictions
[b]The standard errors of the empirical Bayes estimators

It is interesting to examine the spatial pattern of the RNAZ random effects (Fig. 7.4). RNAZs 2–7, 9, and 19 are located in the City of Detroit, and its adjacent areas on the west side, such as Dearborn, Livonia, and Southfield (Fig. 7.2). Urban shrinking is the most serious in the middle and west residential areas of the City of Detroit. The shrinking has spread in three main directions (Fig. 7.4). The primary direction is toward the west and expanded into the city of Livonia; the next direction is the southwest into the city of Dearborn; the third is the northwest into the city of Southfield. We also witness some expansion in the northeast and the south directions.

On the contrary, urban shrinking is not universal in Southeast Michigan. Immediately north of Detroit, two cities of Serling Heights and Clinton consist of a blocking belt deterring urban shrinking expansion from further north. On the southwest corner, the cities of Ann Arbor and Saline do not see much shrinking. In the vast north and west areas, urban shrinking is not a concern. Therefore, we see that urban shrinking in Southeast Michigan displays an apparent spatial process, expanding beyond the City of Detroit and affecting many cities around it. However, the impacts of urban shrinking originated in Detroit show spatial disparities. The shrinking has expanded fast to the west, northwest, and southwest but much slower toward the north. This finding confirms that any action and policy attempting to combat urban shrinking must consider the entire region.

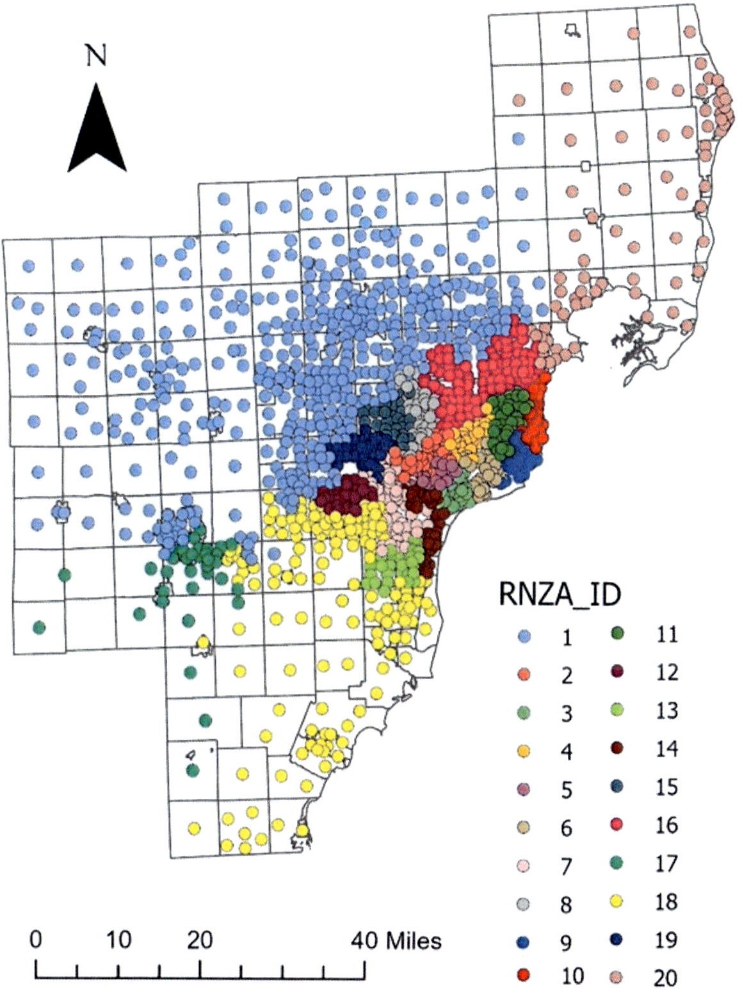

Fig. 7.2 The location map of RNAZs

Now, it is the time to assess how the model is performing. The first statistic we examine is the variance partition coefficient (VPC). The between-RNAZ (level 2) variance in vacancy, *var(_cons)*, is estimated as 45.549 and the within-RNAZ between Tract (level 1) variance, *var(Residual)*, is estimated as 114.883. Thus, the total variance is 45.549 + 114.883 = 160.473. VPC is 0.284 (=45.549/160.473), which indicates that 28.4% of the variance in vacancy can be attributed to the differences between RNAZs.

The second statistic is the likelihood ratio. To test the significance of RNAZ effects, we can conduct a likelihood ratio test comparing the null multilevel model

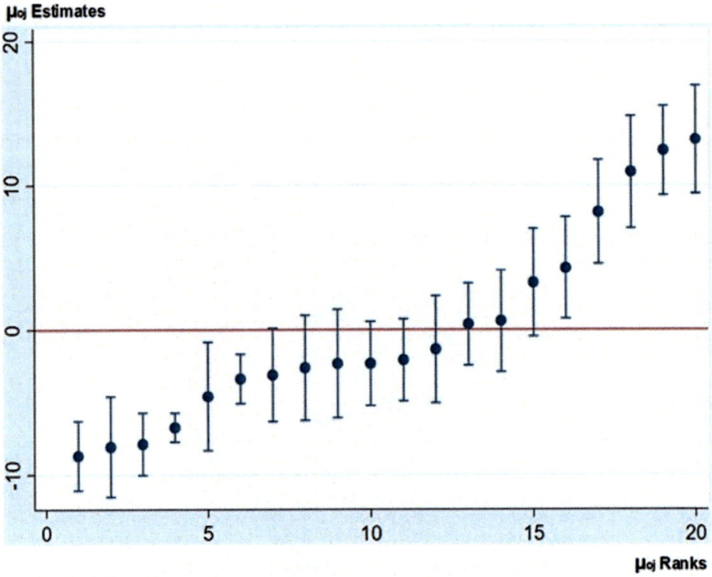

Fig. 7.3 The ranking graph of the RNAZ random effects

with a null single-level model. To fit the null single-level model, we need to remove the random RNAZ effect from Eq. (7.8):

$$\text{Vacancy}_{ij} = \beta_0 + \varepsilon_{ij} \tag{7.9}$$

The log-likelihood of Eq. (7.9) was −5214.285. The likelihood ratio test statistic is calculated as two times the difference in the log-likelihood values for the two models: LR = 2 (−5071.655 to −5214.285) = 142.63 on 1 d.f. (because there is only one parameter difference between the two models). Since the 5% point of a chi-squared distribution on 1 d.f. is 3.84, 142.63 is much higher than this threshold. Therefore, we are confident to conclude that there is overwhelming evidence that RNAZ affects vacancy.

7.4.2.2 A Three-Level Model of Vacancy with RNAZs and Counties Effects

The three-level vacancy model is a direct extension of the two-level vacancy model. Compared with Eq. (7.8), the three-level vacancy variance components model is formulated as follows:

$$\text{Vacancy}_{ijk} = \beta_0 + v_k + u_{jk} + \varepsilon_{ijk} \tag{7.10}$$

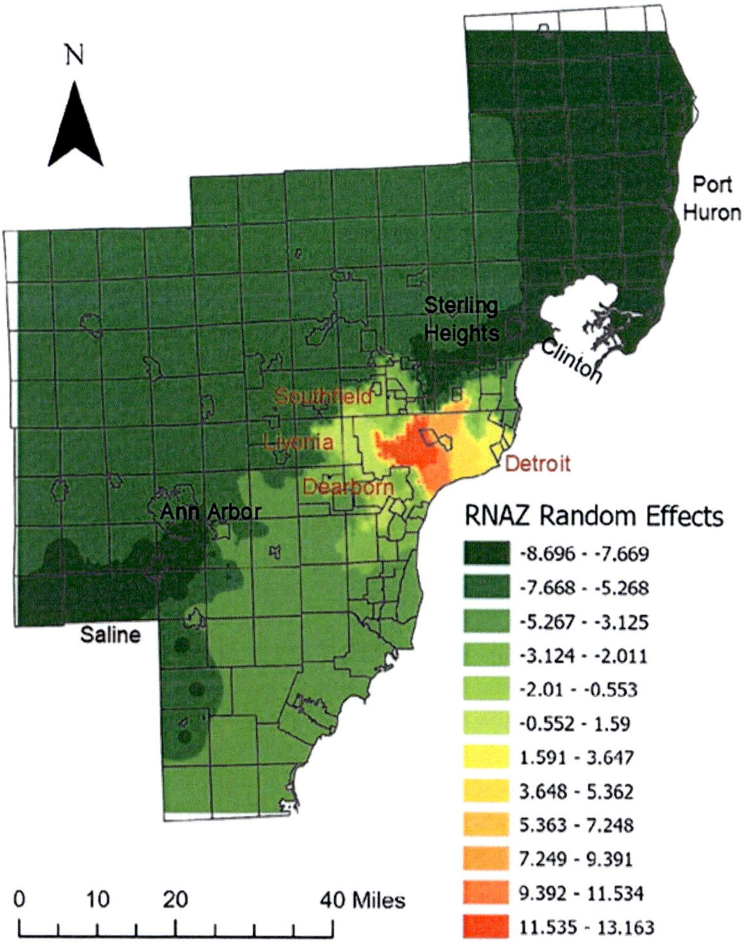

Fig. 7.4 Spatial patterns of RNAZ random effects

where Vacancy_{ijk} is the observed vacancy rate for Tract i, RNAZ j, and County k. β_0 is the mean vacancy across all counties, υk is the effect of county k, μ_{jk} is the effect of RNAZ j, and ε_{ijk} is the tract level residual error term. This model includes only an intercept, two random effects of RNAZ and County, and a Tract level residual error term. The three-level model decomposes the total variance in Tract's vacancy rate into separate variance components over the county, RNAZ, and tract. The county, RNAZ effects, and tract level residual errors are assumed independent and normally distributed with zero means and constant variances. The results are reported in Table 7.4.

The top sub-table of Table 7.4 presents the estimation log-likelihood (in the terminology of the Stata software package) and the general information of the units at level 2 (RNAZ_ID) and level 3 (Cnty_ID). It reports "Log-likelihood = −4975.465."

Table 7.4 The county, RNAZ effects, and the tract level residual errors (the Stata command: mixed Vacant || Cnty_ID: || RNAZ_ID:, mle variance nostderr)

			Observations per group			
Group variable		No. of groups	Minimum		Average	Maximum
Cnty_ID		7	38		190	548
RNAZ_ID		60	1		22.2	240
			Wald $chi^2(0)$	=	.	Log-likelihood
=	−4975.465	Prob > chi^2	=	.		

Vacant	Coef.	Std. Err.	z	$P > z$	[95% Conf. Interval]	
_cons	8.836	1.948	4.54	0	5.0173	12.655

Random-effects parameters	Estimate	Std. Err.	[95% Conf. Interval]	
Cnty_ID: Identity				
var(_cons)	22.130	.	.	.
RNAZ_ID: Identity				
var(_cons)	14.793	.	.	.
var(residual)	97.750	.	.	.

LR test vs. linear model: $chi^2(2) = 477.64$, Prob > $chi^2 = 0.0000$

Note: LR test is conservative and provided only for reference

On the right-hand side, the Wald test "Wald $chi^2(0)$ = .," which compares the current model to a restricted model with only a constant in the fixed part. Currently, this comparison has not been carried out as the current model that includes no fixed effect predictor variables and is equivalent to the restricted model. Stata, therefore, reports a period '.' (Stata's syntax for a missing value) for the test statistic and the associated p-value "Prob > chi^2 = ."

The second sub-table reports the fixed effects. The intercept is 8.836 with a standard error 1.948. Therefore, the mean vacancy rate for counties is estimated to be 8.836. The z-value, p-value, and 95% confidence intervals for this parameter estimate are also given. These statistics tell that the mean vacancy rate of 8.836 is statistically significant.

The third sub-table reports the estimated variance components. The first section of the table is labeled "Cnty_ID: Identity," indicating that this section reports random part parameters at the county level. The between-county variance is estimated as 22.130. The second section of the table labeled "RNAZ_ID: Identity" shows that the estimated within-county-between-RNAZ variance is 14.793. The third section of the table labeled "var(Residual)" exhibits that the estimated within-RNAZ-between-Tract variance is 97.750. Now, we can compute variance partition coefficients (VPCs), which reveal the proportion of the observed response variation at each level of the model hierarchy. These VPCs tell the relative importance of counties, RNAZs, and tracts as sources of variation of urban vacancy rates. The three levels of VPCs are calculated as

$$\text{VPC}v = \sigma_v^2 / \left(\sigma_v^2 + \sigma_u^2 + \sigma_\varepsilon^2\right) = 22.130 / \left(22.130 + 14.793 + 97.750\right) = 0.164 \quad (7.11)$$

$$\text{VPC}u = \sigma_u^2 / \left(\sigma_v^2 + \sigma_u^2 + \sigma_\varepsilon^2\right) = 14.793 / \left(22.130 + 14.793 + 97.750\right) = 0.110 \quad (7.12)$$

$$\text{VPC}\varepsilon = \sigma_\varepsilon^2 / \left(\sigma_v^2 + \sigma_u^2 + \sigma_\varepsilon^2\right) = 97.750 / \left(22.130 + 14.793 + 97.750\right) = 0.726 \quad (7.13)$$

The VPCs confirm that 16.4% of the vacancy variation lies between counties, 11.0% between RNAZs within counties, and 72.6% between tracts with RNAZs. Therefore, the vacancy variation across counties and RNAZs within counties are moderate, and most of the vacancy variations are present between tracts within their RNAZs.

Furthermore, the last two lines report an LR (Log Ratio) test, which compares the current model to a single-level model with no county effects and no RNAZ effects (i.e., linear regression). The LR test statistic is "$\text{chi}^2(2)$ = 477.64" ($\chi^2(2) = 477.64$) with a p-value that is effectively zero, "Prob > chi^2 = 0.0000" ($p < 0.001$, or more precisely $p < 0.0005$). The three-level model, therefore, offers a significantly better fit to the data than the single-level model. The last line of the output states that the "LR test is conservative and provided only for reference." This line means that the reported p-value is an upper bound of the true p-value. Thus, we can reject the null hypothesis even more strongly or confidently.

7.4.2.3 Adding Tract-Level Explanatory Variables and Allowing Intersect to Vary Across RNAZs

This model is also called the random intersect two-level model. Table 7.1 displays three sets of explanatory variables at the tract level: inequality, persistent poverty, and poor-quality housing. We can add them into a random intercept model, respectively. The Stata commands look like the following:

```
mixed Vacant i.BlkPopC i.FHFCLDC i.PdensC i.LOWEDUC || RNAZ_ID:, mle variance nostderr
mixed Vacant i.BlkPopC i.FHFCLDC i.PdensC i.LOWEDUC i.POVTYC i.NoIncC i.GetAidC i.RGR25C i.YRMVC || RNAZ_ID:, mle variance nostderr
mixed Vacant i.BlkPopC i.FHFCLDC i.PdensC i.LOWEDUC i.POVTYC i.NoIncC i.GetAidC i.RGR25C i.YRMVC BTYR NoHeatC NoKitchC MULTIC || RNAZ_ID:, mle variance nostderr
```

Since some of these variables are nonsignificant contributors to the vacancy rate, they are removed from the model. The final Stata command for implementing the random intersect model, which adds the tract-level explanatory variables and allows the model intersect to vary across RNAZs, is the following:

```
mixed Vacant i.BlkPopC i.FHFCLDC i.PdensC i.LOWEDUC i.POVTYC i.NoIncC BTYR NoKitchC || RNAZ_ID:, mle variance nostderr
```

The corresponding equation is written as

$$\begin{aligned} \text{Vacancy}ij = {} & \beta_0 + \beta_1 \text{BlkPopC}_{ij} + \beta_2 \text{FHFCLDC}_{ij} \\ & + \beta_3 \text{PdensC}_{ij} + \beta_4 \text{LOWEDUC}_{ij} + \beta_5 \text{POVTYC}_{ij} \\ & + \beta_6 \text{NoIncC}_{ij} + \beta_7 BTYR_{ij} + \beta_8 \text{NoKitchC}_{ij} + u_{0j} + \varepsilon_{ij} \end{aligned} \tag{7.14}$$

The outcome of Eq. (7.14) is reported in Table 7.5.

First, the random intersect two-level model added some explanatory variables to the two-level null model Eq. (7.8). We can use the LR statistics to judge the performance of this random intersect model Eq. (7.14) in comparison to the null model (Eq. (7.8) and Table 7.2). LR = 2 (−4368.97 to −5071.66) = 1405.37 on 2 d.f. LR is much higher than the chi-square critical value at 95% with 2 degrees of freedom, 5.99. Therefore, there is strong evidence that these covariates strongly and significantly impact the vacancy rate. Moreover, the table's last line reports another LR compared with the standard linear regression model. This second LR test also confirms that the random intersect model is better than its linear regression model.

Second, we examine how the added explanatory variables affected the vacancy rate. All four inequality variables affected vacancy rates. High concentrations of BlkPopC (=3), FHFCLDC (=3), and LOWEDUC (=3) would increase 6.358%, 3.729%, and 4.773% of vacancy, respectively, in comparison to their low concentrations (=1). There are no significant impacts when they are in medium concentrations (=2). On the contrary, both medium and high population densities show preventing effects on vacancy rates (−1.749% and −7.801%, respectively). Two persistent poverty variables, POVTYC and NoIncC, display exacerbating roles to vacancy when they have high values. Two poor-quality housing variables, BTYR and NoKitchC, show accelerating functions to vacancy. Especially, NoKitchC has the most potent exaggerating effect on vacancy (11.475%).

Table 7.5 The outcome of the random intersect two-level model

Mixed-effects ML regression	Number of obs	=	1330
Group variable: RNAZ_ID	Number of groups	=	20
	Obs per group:		
	Min	=	27
	Avg	=	66.5
	Max	=	439
	Wald chi²(14)	=	2557.67
Log-likelihood = −4368.97	Prob > chi²	=	0.0000

Vacant	Coef.	Std. Err.	*z*	*P* > *z*	[95% Conf. Interval]	
BlkPopC						
2	0.775	0.643	1.210	0.228	−0.485	2.035
3	6.358	0.755	8.420	0.000	4.878	7.839
FHFCLDC						
2	0.425	0.555	0.770	0.444	−0.663	1.514
3	3.729	0.872	4.280	0.000	2.020	5.439
PdensC						
2	−1.749	0.437	−4.000	0.000	−2.607	−0.892
3	−7.801	0.805	−9.700	0.000	−9.377	−6.224
LOWEDUC						
2	0.436	0.460	0.950	0.343	−0.466	1.338
3	4.773	0.773	6.170	0.000	3.258	6.289
POVTYC						
2	0.867	0.590	1.470	0.142	−0.289	2.024
3	3.829	0.850	4.510	0.000	2.164	5.495
NoIncC						
2	0.654	0.430	1.520	0.129	−0.189	1.497
3	5.317	0.846	6.290	0.000	3.659	6.974
BTYR	0.033	0.010	3.170	0.002	0.013	0.054
NoKitchC	11.475	0.650	17.670	0.000	10.202	12.748
_cons	3.719	0.939	3.960	0.000	1.878	5.559

Random-effects parameters	Estimate	Std. Err.	[95% Conf. Interval]	
RNAZ_ID: Identity				
var(_cons)	4.175	.	.	.
var(residual)	40.660	.	.	.

LR test vs. linear model: chibar²(01) = 41.70, Prob ≥ chibar² = 0.0000

Third, adding these explanatory variables, the residual, var.(Residual), was dramatically reduced from 114.883 in the null model (Table 7.2) to 40.660 in the random intercept model.

7.4.2.4 Adding the Fixed Effects at Tracts and the Random Effects at RNAZ from the Same Explanatory Variables

This model is also called the random slope two-level model. The Stata command for implementing this model looks like the following:

```
mixed Vacant BlkPopC FHFCLDC PdensC LOWEDUC POVTYC NoIncC NoKitchC
|| RNAZ_ID: BlkPopC FHFCLDC PdensC LOWEDUC POVTYC NoIncC NoKitchC,
covariance(unstructured) mle variance nostderr emonly
```

The corresponding equation is written in the form:

$$\begin{aligned}\text{Vacancy}ij = {} & \beta_0 + \beta_1\text{BlkPopC}_{ij} + \beta_2\text{FHFCLDC}_{ij} + \beta_3\text{PdensC}_{ij} \\ & + \beta_4\text{LOWEDUC}_{ij} + \beta_5\text{POVTYC}_{ij} + \beta_6\text{NoIncC}_{ij} \\ & + \beta_7 BTYR_{ij} + \beta_8\text{NoKitchC}_{ij} + u_{0j} + u_{01}\text{BlkPopC}_{ij} \\ & + u_{02}\text{FHFCLDC}_{ij} + u_{03}\text{PdensC}_{ij} + u_{04}\text{LOWEDUC}_{ij} \\ & + u_{05}\text{POVTYC}_{ij} + u_{06}\text{NoIncC}_{ij} + u_{07} BTYR_{ij} \\ & + u_{08}\text{NoKitchC}_{ij} + \varepsilon_{ij}\end{aligned} \quad (7.15)$$

The outcome of this model is reported in Table 7.6.

At first, we examine the random slopes to test whether the effects of the above set of covariates vary across RNAZs. The null hypothesis for this test is that the intercept variance var.(cons) and the slope variances, var.(BlkPopC), var.(FHFCLDC), var.(PdensC), var.(LOWEDUC), var.(POVTYC), var.(NoIncC), and var.(BTYR), are simultaneously equal to zero. We can use the log-likelihood ratio statistic to test the null hypothesis. The log-likelihood value for the random intersect model is −4368.97 (Table 7.6), while the log-likelihood value for the random slope model is −4327.89 (Table 7.7). Thus, the likelihood ratio test statistic is.

$$\text{LR} = 2(-4327.89 - -4368.97) = 82.16 \,\text{on}\, 2\,\text{d.f.}$$

LR is much higher than the chi-square critical value at 95% with 2 degrees of freedom, 5.99. Therefore, there is solid evidence that the effects of these covariates on vacancy differ across RNAZs.

Moreover, the last two lines of Table 7.7 reveal that the likelihood ratio test, which compares the current model to a single-level model, is statistically significant (Prob > $\text{chi}^2 = 0.0000$). They also state that this LR test is conservative and provided only for reference. This means that the reported p-value is the upper bound for the correct p-value. When the reported p-value is less than 0.05, it is safe to conclude that the two-level model is preferred to the single-level model. This finding provides critical evidence that the random slope model performs well.

Now, let us interpret random slope effects across RNAZs by using BlkPopC as an illustration. The "Coef." and "Random-effects parameter" of BlkPopC are 3.4274

Table 7.6 The outcome of the random slope two-level model

Mixed-effects ML regression	Number of obs	=	1330
Group variable: RNAZ_ID	Number of groups	=	20
	Obs per group:		
	Min	=	27
	Avg	=	66.5
	Max	=	439
	Wald $chi^2(8)$	=	429.730
Log-likelihood = −4327.89	Prob > chi^2	=	0.000

Vacant	Coef.	Std. Err.	z	$P > z$	[95% Conf. Interval]	
BlkPopC	3.4274	0.7273	4.7100	0.0000	2.0019	4.8528
FHFCLDC	0.8696	0.7063	1.2300	0.2180	−0.5147	2.2539
PdensC	−3.4163	0.6730	−5.0800	0.0000	−4.7354	−2.0971
LOWEDUC	1.4673	0.5856	2.5100	0.0120	0.3196	2.6150
POVTYC	1.8471	0.6440	2.8700	0.0040	0.5848	3.1094
NoIncC	1.7381	0.6526	2.6600	0.0080	0.4591	3.0170
BTYR	0.0460	0.0125	3.6600	0.0000	0.0214	0.0706
NoKitchC	11.7247	0.9810	11.9500	0.0000	9.8021	13.6473
_cons	−4.1232	1.2081	−3.4100	0.0010	−6.4910	−1.7554

Random-effects parameters	Estimate	Low	High
RNAZ_ID: Unstructured			
var(BlkPopC)	7.0825	−1.7888	8.6435
var(FHFCLDC)	5.3678	−3.6714	5.4106
var(PdensC)	6.2657	−8.3225	1.4899
var(LOWEDUC)	3.9021	−2.4044	5.3391
var(POVTYC)	4.2537	−2.1953	5.8895
var(NoIncC)	5.3588	−2.7992	6.2753
var(BTYR)	0.0009	−0.0122	0.1042
var(NoKitchC)	9.1796	5.7863	17.6631
var(_cons)	10.7864		

Covariance	Estimate	Covariance	Estimate
cov(BlkPopC,FHFCLDC)	−4.2414	cov(LOWEDUC,POVTYC)	−3.2970
cov(BlkPopC,PdensC)	−2.2077	cov(LOWEDUC,NoIncC)	3.4647
cov(BlkPopC,LOWEDUC)	−1.0798	cov(LOWEDUC,BTYR)	−0.0153
cov(BlkPopC,POVTYC)	1.5105	cov(LOWEDUC,NoKitchC)	−1.4555
cov(BlkPopC,NoIncC)	0.6347	cov(LOWEDUC,_cons)	−3.2294
cov(BlkPopC,BTYR)	0.0483	cov(POVTYC,NoIncC)	−2.7564
cov(BlkPopC,NoKitchC)	4.8064	cov(POVTYC,BTYR)	0.0259
cov(BlkPopC,_cons)	−3.5237	cov(POVTYC,NoKitchC)	2.2800
cov(FHFCLDC,PdensC)	−1.1041	cov(POVTYC,_cons)	1.3065
cov(FHFCLDC,LOWEDUC)	−0.0242	cov(NoIncC,BTYR)	0.0097
cov(FHFCLDC,POVTYC)	−0.8189	cov(NoIncC,NoKitchC)	0.1208
cov(FHFCLDC,NoIncC)	0.7326	cov(NoIncC,_cons)	−6.5096

(continued)

Table 7.6 (continued)

Covariance	Estimate	Covariance	Estimate
cov(FHFCLDC,BTYR)	−0.0138	cov(BTYR,NoKitchC)	0.0780
cov(FHFCLDC,NoKitchC)	−2.2019	cov(BTYR,_cons)	−0.0372
cov(FHFCLDC,_cons)	0.6422	cov(NoKitchC,_cons)	−1.6474
cov(PdensC,LOWEDUC)	−0.2137	var(residual)	35.5796
cov(PdensC,POVTYC)	−0.6799		
cov(PdensC,NoIncC)	−2.6701		
cov(PdensC,BTYR)	−0.0609		
cov(PdensC,NoKitchC)	−5.5663		
cov(PdensC,_cons)	3.9647		

LR test vs. linear model: $chi^2(45) = 243.98$, Prob > $chi^2 = 0.0000$

Note: LR test is conservative and provided only for reference

and 7.0825, respectively. Based on these two critical statistics, we can estimate the BlkPopC random slope effects over $RNAZ_j$ with the following equation:

$$\text{Coef.}(\text{BlkPopC}) \pm 1.96\sqrt{\text{var}(\text{BlkPopC})} \tag{7.16}$$

This equation can be written as.

$$3.4274 \pm 1.96\sqrt{7.0825} \tag{7.17}$$

where 3.4274 is the estimated coefficient of BlkPopC, 1.96 is the 95% coverage interval of a normal curve for $RNAZ_j$, and 7.0825 is the estimated covariance of BlkPopC, var.(BlkPopC). Eq. (7.17) led to two numbers:

$$3.4274 - 1.96\sqrt{7.0825} = -1.7887 \tag{7.18}$$

$$3.4274 + 1.96\sqrt{7.0825} = 8.6435 \tag{7.19}$$

These numbers reveal that the slope ranges of BlkPopC are between −1.789 and 8.644. Thus, assuming a normal distribution, we would expect the middle 95% of RNAZs to have a slope between −1.789 and 8.644 (see Table 7.7 middle section). The other example is NoKitchC. By applying Eq. (7.16), the slope ranges of NoKitchC are between 5.786 and 17.663. Therefore, 95% of RNAZs display the strongest positive associations between vacancy and NoKitchC. BlkPopC is the second strongest positive driver of vacancy. The majority of RNAZs show positive associations between vacancy and BlkPopC.

Furthermore, we will examine what the covariances are telling us. In general, when the covariance is positive, these two variables complement each other to increase the vacancy. For example, cov(BlkPopC,NoKitchC) = 4.806 reveals that BlkPopC highly correlates with NoKitchC, aggravating their effects on vacancy. On the other hand, BlkPopC does not closely correlate with FHFCLDC since cov(BlkPopC,FHFCLDC) is negative −4.241. FHFCLDC is no longer a significant

Table 7.7 Random-effects correlation matrix for level RNAZ_ID

	BlkPopC	FHFCLDC	PdensC	LOWEDUC	POVTYC	NoIncC	BTYR	NoKitchC	_Cons
BlkPopC	1								
FHFCLDC	−0.6879	1							
PdensC	−0.3314	−0.1904	1						
LOWEDUC	−0.2054	−0.0053	−0.0432	1					
POVTYC	0.2752	−0.1714	−0.1317	−0.8093	1				
NoIncC	0.1030	0.1366	−0.4608	0.7577	−0.5773	1			
BTYR	0.6104	−0.2012	−0.8196	−0.2607	0.4227	0.1410	1		
NoKitchC	0.5961	−0.3137	−0.7340	−0.2432	0.3649	0.0172	0.8665	1	
_Cons	−0.4032	0.0844	0.4823	−0.4978	0.1929	−0.8562	−0.3816	−0.1656	1

influencer of vacancy in this random slope model ($P > z - 0.218$). We can also convert the covariance to the correlation coefficient, which can help us interpret their meanings (Table 7.7). When the covariance (or correlation) between an explanatory variable and the intersect (*_Cons*) is negative, it means that RNAZs with a high intercept (above-average vacancy) tend to have a flatter-than-average slope, and vice versa.

7.4.2.5 Examining Interactions across Three Levels

The Detroit case study involves three spatial levels, County, RNAZ, and Tract. They share the common explanatory variables originally processed at the tract level. In addition, the levels of County and RNAZ contain their unique covariates (Table 7.1). Through the above analyses, we have identified that five important variables shared at the three levels – BlkPopC, POVTYC, NoIncC, BTYR, and NoKitchC – significantly impact vacancy in Metropolitan Detroit. Moreover, three gravity scores at RNAZ (G1_AgC, G2_AgC, and G3_AgC) and one covariate at County (GDP19) are statistically critical (Table 7.8). Therefore, we created 17 interaction terms (variables) to represent the cross-interactions between the three spatial scales.

Table 7.8 A random intersect model with the interaction variables

Mixed-effects ML regression	Number of obs	=	1330
Group variable: CLUS_ID	Number of groups	=	20
	Obs per group:		
	Min	=	27
	Avg	=	66.5
	Max	=	439
	Wald $chi^2(8)$	=	2760.24
Log-likelihood = −4345.245	Prob > chi^2	=	0.0000

Vacant	Coef.	Std. Err.	*z*	*P* > *z*	[95% Conf. Interval]	
BlkPxPov	0.9451	0.2037	4.6400	0.0000	0.5458	1.3444
BlkPxNoInc	0.9183	0.1789	5.1300	0.0000	0.5677	1.2689
BlkPxBTYR	0.0127	0.0059	2.1600	0.0310	0.0012	0.0243
BlkPxNoKit	4.5393	0.6029	7.5300	0.0000	3.3576	5.7210
PovtxNoKit	1.9296	0.6971	2.7700	0.0060	0.5633	3.2959
BlkPxGrav2	−0.9111	0.2119	−4.3000	0.0000	−1.3264	−0.4958
NoKitGrav2	−1.6145	0.4644	−3.4800	0.0010	−2.5247	−0.7043
NoKitxGrav1xGDP	0.6808	0.2369	2.8700	0.0040	0.2164	1.1452
_cons	4.6771	0.5836	8.0100	0.0000	3.5333	5.8209

Random-effects parameters	Estimate	Std. Err.	[95% Conf. Interval]	
CLUS_ID: Identity				
var(_cons)	2.530549	.	.	.
var(residual)	39.44935	.	.	.

```
BlkPxPov = BlkPopC*POVTYC
BlkPxNoInc = BlkPopC*NoIncC
BlkPxBTYR = BlkPopC*BTYR
BlkPxNoKit = BlkPopC*NoKitchC
PovtxNoKit = POVTYC*NoKitchC
BlkPxGrav1 = BlkPopC*G1_AgC
BlkPxGrav2 = BlkPopC*G2_AgC
BlkPxGrav3 = BlkPopC*G3_AgC
NoKitGrav1 = NoKitchC*G1_AgC
NoKitGrav2 = NoKitchC*G2_AgC
NoKitGrav3 = NoKitchC*G3_AgC
NoIncGrav1 = NoIncC*G1_AgC
NoIncGrav2 = NoIncC*G2_AgC
NoIncGrav3 = NoIncC*G3_AgC
NoIncxGrav1xGDP = NoIncC*G1_AgC*GDP19C
NoKitxGrav1xGDP = NoKitchC*G1_AgC*GDP19C
BlkPxGrav1xGDP = BlkPopC*G1_AgC*GDP19C
```

For example, BlkPxPov is an interaction variable within the Tract level; BlkPxGrav1 is an interaction variable between two levels, Tract and RNAZ; NoIncxGrav1xGDP is an interaction variable across three levels, Tract, RNAZ, and County.

We used these newly created interaction variables to run a random intersect model between Tract and RNAZ as the start (Table 7.8). Eight out of 17 interaction variables significantly contribute to the urban shrinking in Metropolitan Detroit. Five of them are interaction variables within the Tract level, two are between the Tract and RNAZ levels, and one is across all three levels. For example, two interaction variables, BlkPxGrav2 and NokitGrav2, display strong negative impacts on vacancy. As we know both BlkPopC and NoKitchC are critical driving forces exacerbating vacancy. However, when they interacted with G2_AgC (the gravity score for Factor 2 *Economic Scale Dimension*), they were no longer accelerating vacancy but deterring it. The explanation could be that vacancy is insignificant in the areas with strong economic growth, and more resources are available for community development in these areas.

The interaction across BlkPopC at Tract, Gravity 1 at RNAZ, and GDP at County is also interesting. A high concentration of BlkPopC in a Tract usually associates with high vacancy. G1_AgC, the gravity score of Factor 1 (*Sustained Poverty Dimension*), also relates to the high vacancy. However, when they are in Counties with better GPD, the intensities of their impacts on vacancy are alleviated.

We also used these newly created interaction variables to run a random slope model to compare with the random intersect model. The Stata command running the random slope model is the following:

```
mixed Vacant BlkPxPov BlkPxNoInc BlkPxBTYR BlkPxNoKit PovtxNoKit
BlkPxGrav2 NoKitGrav2 NoKitxGrav1xGDP || RNAZ_ID: BlkPxPov
BlkPxNoInc BlkPxBTYR BlkPxNoKit PovtxNoKit BlkPxGrav2 NoKitGrav2
NoKitxGrav1xGDP, covariance(unstructured) mle variance nost-
derr emonly
```

Table 7.9 A random slope model with the interaction variables

Mixed-effects ML regression	Number of obs	=	1330
Group variable: CLUS_ID	Number of groups	=	20
	Obs per group:		
	Min	=	27
	Avg	=	66.5
	Max	=	439
	Wald chi²(8)	=	1159.05
Log-likelihood = −4301.12	Prob > chi²	=	0.0000

Vacant	Coef.	Std. Err.	*Z*	*P* > *z*	[95% Conf. Interval]	
BlkPxPov	0.8916	0.2829	3.1500	0.0020	0.3372	1.4461
BlkPxNoInc	0.7440	0.3294	2.2600	0.0240	0.0985	1.3896
BlkPxBTYR	0.0119	0.0063	1.8900	0.0590	−0.0004	0.0243
BlkPxNoKit	6.2621	1.1371	5.5100	0.0000	4.0334	8.4908
PovtxNoKit	0.8215	1.3235	0.6200	0.5350	−1.7725	3.4154
BlkPxGrav2	−0.7713	0.2338	−3.3000	0.0010	−1.2295	−0.3131
NoKitGrav2	−2.5154	0.6664	−3.7700	0.0000	−3.8216	−1.2092
NoKitxGrav1xGDP	0.6897	0.3867	1.7800	0.0740	−0.0682	1.4476
_cons	4.8562	0.5637	8.6200	0.0000	3.7515	5.9609

Random-effects parameters	Estimate	Low	High
CLUS_ID: Unstructured			
var(BlkPxPov)	0.6896	−0.7360	2.5193
var(BlkPxNoInc)	1.2996	−1.4904	2.9784
var(BlkPxBTYR)	0.0001	−0.0089	0.0327
var(BlkPxNoKit)	13.0995	−0.8318	13.3560
var(PovtxNoKit)	19.1210	−7.7492	9.3921
var(BlkPxGrav2)	0.3295	−1.8964	0.3538
var(NoKitGrav2)	1.8868	−5.2076	0.1769
var(NoKitxGrav1xGDP)	0.5971	−0.8248	2.2043

LR test vs. linear model: chi²(45) = 104.66, Prob > chi² = 0.0000

Note: LR test is conservative and provided only for reference

The outcome of the random slope model is reported in Table 7.9. We first assess the performance of the random slope model by comparing it with the random intersect model. We can use the log-likelihood ratio statistics LR = 2 (−4301.12 to −4345.25) = 88.26 on 2 d.f., which is much higher than the chi-square critical value at 95% with 2 degrees of freedom, 5.99. Therefore, we can confidently conclude that these interactions' effects on vacancy differ across RNAZs.

How much the slopes of these variables depart are reported in the last panel of Table 7.9. The interaction slope between BlkPopC and NoKitchC (BlkPxNoKit) varies by almost 14 units. The cross-two-level interaction slopes, NoKitGrav2

(NokitchC at Tract level and G2_AGC at RNAZ level), are nearly 5.0 units apart. The cross-three-level interaction slopes, NoKitxGrav1xGDP (NoKitchC at Tract, G1_AGC at RNAZ, and GDP19 at County), are 3.0 units apart.

Finally, the LR test vs. linear model is $chi^2(45) = 104.66$ with $Prob > chi^2 = 0.0000$ with a note: LR test is conservative and provided only for reference. This LR test further confirms that the two-level random slope model performs much better than the linear model.

The covariances between these interactions are converted to correlations (Table 7.10). Generally, their random slopes synchronize when the correlations between two covariances are positive. If the correlations between a pair of covariances are negative, their random slopes often negate each other. However, greater attention should be given to the covariance between these interactions with the intercepts (_Cons). When the covariance or correlation is negative, it indicates that the intercept is high, but the slope is relatively less steep. This phenomenon often means spatial clustering. If the covariance or correlation is positive, it reveals that the intercept is low, but the slope is steep, indicating a spatial dispersion.

Table 7.10 Random-effects correlation matrix for level RNAZ_ID

	BlkPxPov	BlkPxNoInc	BlkPxBTYR	BlkPxNoKit	PovtxNoKit
BlkPxPov	1				
BlkPxNoInc	−0.6584	1			
BlkPxBTYR	−0.5665	0.6214	1		
BlkPxNoKit	−0.3719	−0.3408	−0.1933	1	
PovtxNoKit	0.7085	−0.1094	−0.2209	−0.8138	1
BlkPxGrav2	0.0152	−0.6874	−0.5489	0.7940	−0.4493
NoKitGrav2	−0.5477	0.5221	0.2690	−0.0921	−0.1975
NoKitxGrav1xGDP	−0.4543	0.3355	0.5119	0.0219	−0.4750
_cons	0.3654	−0.4067	−0.6285	−0.0237	0.3702
	BlkPxGrav2	NoKitGrav2	NoKitxGrav1xGDP	_cons	
BlkPxPov					
BlkPxNoInc					
BlkPxBTYR					
BlkPxNoKit					
PovtxNoKit					
BlkPxGrav2	1				
NoKitGrav2	−0.1834	1			
NoKitxGrav1xGDP	−0.1840	0.0286	1		
_cons	0.2228	0.1741	−0.5777	1	

7.5 Discussion and Conclusion

The case study of Metropolitan Detroit urban shrinking through the hierarchical linear mixed model has revealed several significant findings. First, it is the first empirical study of spatial processes and characteristics of urban shrinking by applying a hierarchical linear mixed model. It has two distinct innovations compared to the current approaches. The logistic linear models of examining spatial inequality of urban shrinking treat hierarchical urban spatial structures as a set of individual layers and apply models at each layer to analyze the similarities and differences between them (Xie et al., 2018). These models indirectly identify spatial connections between different spatial structures by analyzing the model outcomes to deduce how these models at different spatial units affect each other. The spatial structures are often composed of existing census enumeration units, such as block groups, tracts, towns, cities, and counties.

The spatial econometric models (SpEM) use spatial weight matrices to analyze spatial relationships. SpEM can use different orders of spatial adjacency matrices or spatial distance belts to model spatial structures. For example, SpEM spatial structure grows outwards from a basic spatial unit like a block group to the first group of block groups that touch the original block group, then the second group of block groups that touch the first group, and so on so forth (Lokhande & Xie, 2023). One challenge that SpEM face is determining what this spatial structure tells about physical, socioeconomic, or policy implementation.

The distinction of the hierarchical linear mixed model (HLM) is that its spatial structure is a nested structure of spatial units at different administrative levels, statistical enumeration levels, or researcher-defined hierarchical levels. Therefore, the HLM spatial structure has explicit meanings in planning, policy, administration, or purposeful research. For example, the spatial structure of this case study is RNAZs, which are grouped from census tracts based on six dimensions of socioeconomic factors that affect urban shrinking. The term nested spatial structure is the unique characteristic of HLM deviating from other spatial structures. The spatial unit, RNAZs, is constructed from the networked urban system theory with abundant (big) data and directly supports urban and regional planning practices.

The second feature of HLM is the term mixed. In our case study, the first spatial level of the Detroit Vacancy HLM models is the census tract, in which the dependent variable, Vacancy, responds to tens of explanatory variables. We know causal relationships exist between them based on the literature and hence model their relationships as fixed effects. In the HLM terminology, this is a random intercept model. On the other hand, we plan to explore how their relationships vary across hierarchical spatial structures. Using the random slope model, we can add the level 2 and level 3 spatial units and model their relationships as random effects. In other words, we examine how their relationships vary across spaces as random effects. Therefore, we can investigate the causal relationships between urban shrinking and its driving forces through the fixed model (the random intersect model) and the spatial processes and interactions of urban shrinking by using the random model (the random slope model). Thus, we call this research design and approach a hierarchical linear mixed model.

The Detroit Vacancy HLM model got compelling findings. First, the model confirms the vacancy is driven by three groups of driving factors reflecting inequality, persistent poverty, and poor housing. The fixed effect counts 72.6% vacancy variation within RNAZs between tracts. The random effects explain 16.4% of the vacancy variation between counties and 11.0% within counties between RNAZs. The LR tests confirm that the random effects at both County and RNAZ levels are statistically significant. Since the case study only contains seven counties, we analyzed the spatial pattern of the random effects by RNAZs. Urban shrinking was the most severe in Detroit's middle and western residential areas. Urban shrinking displayed three expansion axes: the primary axis toward the west, the city of Livonia; the secondary axis toward the northwest, the city of Southfield; and the third toward the southwest, the city of Dearborn. In addition, urban shrinking extended toward the northeast but soon stopped in the cities of Clinton and Sterling Heights. The spatial process and dynamics of urban shrinking in Metropolitan Detroit were clearly revealed.

NoKitchC is the strong driving force for aggravating urban shrinking based on the fixed effects. BlkPopC, LOWEDUC, POVTYC, and FHFCLDC are driving factors exacerbating urban shrinking in decreasing order. However, there is a surprising finding that population density is often regarded as a deteriorating factor intensifying vacancy (Hummel, 2020). Our case study suggests population density (PdensC) becomes relatively low in a sustained shrinking city like Detroit. Moreover, high population densities are found in university towns and senior residential areas. As a result, high population density becomes a deterring factor in urban shrinkage.

NoKitchC and BlkPopC are the two most important explanatory variables among the random effects. Both have high intercepts but less steep slopes, indicating spatial clustering effects on the vacancy rate. In other words, both variables vary significantly across RNAZs and also display extensive spatial clusters. The effects of the remaining variables on vacancy also diverge across RNAZs, but are less clustered compared to the spatial patterns of NoKitchC and BlkPopC. Moreover, the covariance analysis confirms that cov(NoIncC,_cons), cov(BlkPopC,_cons), cov(LOWEDUC,_cons), and cov(NoKitchC,_cons) show negative values, which indicate that these variables have high intercepts but gentle slopes, displaying extensive spatial clusters. On the other hand, cov(PdensC,_cons), cov(POVTYC,_cons), and cov(FHFCLDC,_cons) have positive covariances. Their intercepts are lower, but their slopes are uprising quickly, showing more spatial disparities.

Furthermore, the Detroit Vacancy HLM examined interactions between the vacancy driving factors and the cross-level (cross-space) interactions. There exist both within-tract interactions and cross-tract-RNAZ and cross-tract-RNAZ-county interactions. The within-tract interactions significantly impact urban shrinking (the interaction random intercept model). The cross interactions confirm that the effects on urban shrinking display distinct spatial variations (the interaction random slope model). The interactions, BlkPxBTYR, NoKitxGrav1xGDP, BlkPxNoInc, and BlkPxNoKit, show high intercepts with gentle slopes. In other words, these interactions show extensive spatial patterns, reflecting spatial clustering effects. On the contrary, PovtxNoKit, BlkPxPov, BlkPxGrav2, and NoKitGrav2 display low intercepts with steep slopes. The spatial patterns of these interactions are more dispersed.

In summary, the Detroit Vacancy HLM effectively reveals the causal relationships between urban shrinking and inequality, persistent poverty, and poor housing factors. It has depicted spatial processes of urban shrinking in the context of the City of Detroit with its adjacent towns. It has differentiated that the driving factors, NoIncC, BlkPopC, LOWEDUC, and NoKitchC, are active in forming spatial clusters and aggravating urban shrinking. Moreover, the combinations between BlkPopC, BTYR, NoIncC, and NoKitchC, and the cross-level interactions of BlkPopC at tract, Sustained Poverty Dimension at RNAZ, and GPD at county level cause spatial clusters.

These findings have significant regional and urban planning implications for averting urban shrinking. As revealed in this study, the spatial process and expansion dynamics of urban shrinking recommend that all affected cities (Detroit, Livonia, Southfield, and Dearborn) collaborate to make policies to avert urban shrinking as intercity cooperation. The case study has identified that the vast areas in St Clair County, Oakland County, Livingston County, and Washtenaw County are less affected by urban shrinking, contrasting with the above cities. A regional approach must be considered to stop urban shrinking expansion and help avert urban shrinking in the four cities. Moreover, the case study has found that the towns of Clinton and Sterling Heights are adjacent to the City of Detroit, but they have effectively fenced off urban shrinking intrusion. We suggest conducting more in-depth studies and interviews in these cities to learn and summarize how they have tried to control urban shrinking. Furthermore, the case study has found several driving factors, and their interactions are the primary forces driving the spatial expansion of urban shrinking. More studies will be conducted to identify how these factors spread urban shrinking and to find effective measures to stop their spatial processes.

However, there are a few issues warranting further explorations. First, HLM supports many types of examinations, as introduced in this chapter. Although this chapter intends to provide comprehensive coverage of HLM and conduct a meaningful application of HLM in studying Detroit urban shrinking, there are still some missing pieces on the HLM methodology or about Detroit urban shrinking analysis. For instance, the random slope model can be conducted at three levels, tract, RNAZ, and county. However, in the Detroit case study, the three levels of the random slope model did not give significant outcomes based on the LR test. Therefore, this analysis was excluded from this chapter.

Second, the three levels have their unique variables. A random slope intersect model could be run across three levels with different sets of explanatory variables. However, the three levels of random intercept model did not generate meaningful outcomes based on the LR test. Moreover, this chapter purposefully designed 20 RNAZs as the second spatial level to get interpretable findings about the spatial process and dynamics of urban shrinking. However, 20 RNAZs and five counties are too few cases for running random intercept models. As a result, most of the explanatory variables at these two levels are statistically nonsignificant. Therefore, the random intercept models were excluded as well from this chapter.

Third, a different spatial structure could be examined in this chapter, such as tracts, towns, and counties. However, in the study area of Southeast Michigan, there are 200 towns. The intuitive judgment is that too many units at a high spatial level may make the analysis too cloudy and hard to interpret, and that is a future experiment that should be tested. Moreover, we will conduct more empirical HLM urban vacancy models in other metropolitan cities in the United States or other countries to further refine the models. We can also adopt the Detroit Vacancy HLM model in different research fields.

References

Aalbers, M. B., & Bernt, M. (2019). The political economy of managing decline and rightsizing. *Urban Geography, 40*(2), 165–173. https://doi.org/10.1080/02723638.2018.1524654

ACS. (2022). *American Community Survey – Detroit City of Michigan*. https://www.census.gov/programs-surveys/acs. Accessed 18 July 2022.

Alligood, L. S. (2008). *Creative shrinkage: In search of a strategy to manage decline* (Master of Community Planning thesis). Portland, OR: Portland State University. http://rave.ohiolink.edu/etdc/view?acc_num=ucin1212081537.

Anselin, Luc. 2003. Chapter 14: Spatial econometrics. In Badi H. Baltagi(Ed.), *A companion to theoretical econometrics* (pp. 310–330). Blackwell Publishing Ltd. http://web.pdx.edu/~crkl/WISE/SEAUG/papers/anselin01_CTE14.pdf.

Audirac, I. (2014). Shrinking cities in Latin America – An oxymoron? In H. Richardson & C. W. Nam (Eds.), *Shrinking cities: A global perspective*. Routledge.

Batty, M. (2013). *The new science of cities*. MIT Press.

Batty, M. (2018). *Inventing future cities*. MIT Press.

Batunova, E., & Gunko, M. (2018). Urban shrinkage: An unspoken challenge of spatial planning in Russian small and medium-sized cities. *European Planning Studies, 26*(8), 1580–1597. https://doi.org/10.1080/09654313.2018.1484891

Bolker, B. M. (2015). Linear and generalized linear mixed models. In G. A. Fox, S. Negrete-Yankelevich, & V. J. Sosa (Eds.), *Ecological statistics* (pp. 378–379). Oxford University Press.

Bolker, B. M., Brooks, M. E., Clark, C. J., Geange, S. W., Poulsen, J. R., Stevens, M. H. H., & White, J. S. S. (2009). Generalized linear mixed models: A practical guide for ecology and evolution. *Trends in Ecology and Evolution, 24*, 127–135.

Brunner, D., & Kaminski, V. (2016). *Urban studies and sprawl (concepts, elements & issues)*. Academic Studio.

Cho, S., Kim, J., Roberts, R. K., & Kim, S. G. (2012). Neighborhood spillover effects between rezoning and housing prices. *The Annals of Regional Science, 48*, 301–319. https://doi.org/10.1007/s00168-010-0401-9

Curtis, K. J., Voss, P. R., & Long, D. D. (2012). Spatial variation in poverty-generating processes: Child poverty in the United States. *Social Science Research, 41*(1), 146–159.

Daniels, T. (2022). *The environmental planning handbook: For sustainable communities and regions* (2nd ed.). Routledge Taylor & Francis Group.

Danko, I. J. J., & Hanink, D. M. (2018). Beyond the obvious: A comparison of some demographic changes across selected shrinking and growing cities in the United States from 1990 to 2010. *Population, Space and Place, 24*, e2136.

Deakin, E. (2020). *Transportation, land use, and environmental planning*. Elsevier.

Diestel, R. (2005). *Graph theory*. Springer-Verlag.

Döringer, S., Uchiyama, Y., Penker, M., & Kohsaka, R. (2020). A meta-analysis of shrinking cities in Europe and Japan-towards an integrative research agenda. *European Planning Studies, 28*, 1693–1712.

Elhorst, J. P., & Vega, S. H. (2015). The SLX model. *Journal of Regional Science, 55*(3), 339–363. https://doi.org/10.1111/jors.12188

Elhorst, J. P., & Vega, S. H. (2017). The SLX model: Extensions and the sensitivity of spatial spillovers to W. *Papeles de Economía Española, 152*, 34–50.

Fol, S. (2012). Urban shrinkage and socio-spatial disparities: Are the remedies worse than the disease? *Built Environment, 38*(2), 259–275.

Frohlich,T.C.,Stebbins,S.,Sauter,M.B.,&Comen,E.(2015).15American.GhostTowns.November18, 2015 7:04 pm EST. http://247wallst.com/special-report/2015/11/18/15-american-ghost-towns.

Gamesby, R. (2022). World cities- urban decline, Coolgepgraphy.co.uk. https://www.coolgeography.co.uk/A-level/AQA/Year%2013/World%20Cities/Decline/Urban_Decline.htm. Accessed 18 July 2022.

Ganning, J. P., & Tighe, J. R. (2015). Assessing the feasibility of side yard programs as a solution to land vacancy in U.S. shrinking cities. *Urban Affairs Review, 51*(5), 708–725. https://doi.org/10.1177/1078087414555435

Gao, S., & Ryan, B. D. (2020). Can neighborhood planning in shrinking cities achieve demolition goals? A conformance and performance evaluation of neighborhood action plans in Youngstown, Ohio. *Journal of Planning Education and Research*, 1–22. https://doi.org/10.1177/0739456X20940782

Gim, T. H. T. (2018). SEM application to the household travel survey on weekends versus weekdays: The case of Seoul, South Korea. *European Transport Research Review, 10*(1), 1–16.

Glaeser, E., & Kohlhase, J. (2003). *Cities, regions, and the decline of transport costs*. NBER Working Paper Series, National Bureau of Economic Research.

Greenwald, M. J., & Boarnet, M. G. (2001). Built environment as determinant of walking behavior: Analyzing nonwork pedestrian travel in Portland, Oregon. *Transportation Research Record, 1780*(1), 33–41.

Greer, W. (1986, January 9). Living without a kitchen (or almost). *New York Times*. http://www.nytimes.com/1986/01/09/garden/living-without-a-kitchen-or-almost.html?mcubz=1.

Großmann, K., Bontje, M., Haase, A., & Vlad Mykhnenko, V. (2013). Shrinking cities: Notes for the further research agenda. *Cities, 35*(2013), 221–225.

Grossmann, K., Haase, A., Arndt, T., Cortese, C., Rumpel, P., Rink, D., et al. (2014). How urban shrinkage impacts on patterns of socio-spatial segregation: The cases of Leipzig, Ostrava, and Genoa. In C. C. Yeakey, V. S. Thompson, & A. Wells (Eds.), *Urban ills: Post recession complexities to urban living in global contexts* (Vol. 1). Lexington Books.

Gu, D., Newman, G., Kim, J. H., Park, Y., & Lee, J. (2019). Neighborhood decline and mixed land uses: mitigating housing abandonment in shrinking cities. *Land Use Policy, 83*, 505–511. https://doi.org/10.1016/j.landusepol.2019.02.033

Gutfreund, O. D. (2004). *Twentieth century sprawl: Highways and the reshaping of the American landscape* (p. 2004). Oxford University Press.

Haase, A., Bernt, M., Großmann, K., Mykhnenko, V., & Rink, D. (2016). Varieties of shrinkage in European cities. *European Urban and Regional Studies, 23*(1), 86–102. https://doi.org/10.1177/0969776413481985

Hedeker, D., & Gibbons, R. D. (2006). *Longitudinal data analysis*. Wiley.

Hoekstra, M. S., Hochstenbach, C., Bontje, M. A., & Musterd, S. (2018). Shrinkage and housing inequality: Policy responses to population decline and class change. *Journal of Urban Affairs, 42*(3), 333–350.

Hospers, G. (2014). Policy Responses to Urban Shrinkage: From Growth Thinking to Civic Engagement. *European Planning Studies 22*(7). https://doi.org/10.1080/09654313.2013.793655

Hoekveld, J. J. (2014). Understanding spatial differentiation in urban decline levels. *European Planning Studies, 22*(2), 362–382.
Hu, Y., Wang, Z., & Deng, T. (2021). Expansion in the shrinking cities: Does place-based policy help to curb urban shrinkage in China? *Cities, 113*. https://doi.org/10.1016/j.cities.2021.103188
Hummel, D. (2015). Right-sizing cities in the United States: Defining its strategies. *Journal of Urban Affairs, 37*(4), 397–409.
Hummel, D. (2020). The effects of population and housing density in urban areas on income in the United States. *Local Economy*. https://doi.org/10.1177/0269094220903265
Kabisch, S., Steinführer, A., Haase, A., Großmann, K., Peter, A., & Maas, A. (2008). *Demographic change and its impact on housing*. Final report for EUROCITIES. European Commission, Leipzig and Brussels.
LeSage, J. P. (2008). An introduction to spatial econometrics. *Translation-Industrial Economics Journal, 123*, 19–44. https://doi.org/10.4000/rei.3887
Li, Y., & Xie, Y. (2018). A new urban typology model adapting big-data analytics to examine dominant trajectories of neighborhood change: A case of metro detroit. *Annals of the American Association of Geographers*. https://doi.org/10.1080/24694452.2018.1433016
Li, Y., Kong, X., & Zhu, Z. (2020). Multiscale analysis of the correlation patterns between the urban population and construction land in China. *Sustainable Cities and Society, 61*.
Lokhande, T., & Xie, Y. (2023). *Spatial Spillover Effects of Urban Decline in Southeast Michigan*. Submitted to Journal of Applied Geography for consideration of publication.
Long, H. (2016). U.S. has lost 5 million manufacturing jobs since 2000. *CNN Money News*. http://money.cnn.com/2016/03/29/news/economy/us-manufacturing-jobs/
Long, J. (2021). *GeoDa Documentation*. https://geodacenter.github.io/documentation.html.
Manville, M., & Kuhlmann, D. (2018). The social and fiscal consequences of urban decline: Evidence from large American cities, 1980–2010. *Urban Affairs Review, 54*(3). https://doi.org/10.1177/1078087416675741
Martinez-Fernandez, C., & Audirac, I. (2012). Shrinking cities: Urban challenges of globalization. *International Journal of Urban and Regional Research*. https://doi.org/10.1111/j.1468-2427.2011.01092.x
Martinez-Fernandez, C., & Wu, C. T. (2007). *Shrinking cities: Urban management and environmental legacy*. Refereed Conference Proceedings of the II International Congress on Environmental Management and Planning.
Martinez-Fernandez, C., Weyman, T., Fol, S., Audirac, I., Cunningham-Sabot, E., Wiechmann, T., et al. (2016a). Shrinking cities in Australia, Japan, Europe and the USA: From a global process to local policy responses. *Progress in Planning, 108*, 1–38.
Martinez-Fernandez, C., Weyman, T., Fol, S., Audirac, I., Cunningham-Sabot, E., Wiechmann, T., & Yahagi, H. (2016b). Shrinking cities in Australia, Japan, Europe and the USA: From a global process to local policy responses. *Progress in Planning, 108*, 1–38.
MCGI (Michigan Center of Geographical Information) (2017). MI Geographic Framework Cities (v14a) Geographic. *Theme*. https://www.mcgi.state.mi.us/mgdl/?rel=thext&action=thmname&cid=9&cat=MI+Geographic+Framework+Cities+%28v14a%29. Accessed 28 Jan 2017.
McGraw, B. (2007). Life in the ruins of Detroit. *History Workshop Journal, 63*(1), 288–302.
Modai-Snir, T., & van Ham, M. (2018). *Inequality, reordering and divergent growth: Processes of neighbourhood change in Dutch Cities*, IZA Discussion Papers, No. 11883, Institute of Labor Economics (IZA), Bonn.
NCDB. (2010). Neighborhood change databases – Tract data for 1970-80-90-00-10, Version 2.1, Geolytics, Inc.
Neal, Z. P. (2012). Structural determinism in the interlocking world city net-work. *Geographical Analysis, 44*, 162–170.
Neal, Z. P. (2013). *The connected city – How networks are shaping the modern metropolis and modern life*. Routledge.
Neill, W. J. V. (2015). Carry on shrinking?: The bankruptcy of urban policy in Detroit. *Planning Practice and Research, 30*(1), 1–14. https://doi.org/10.1080/02697459.2014.997462

Nijman, J. & Wei, Y. D. (2020). Urban inequalities in the 21st century economy. *Applied Geography, 117*, Article 102188. https://doi.org/10.1016/j.apgeog.2020.102188

Newman, M., Barabási, A. L., & Watts, D. J. (2006). *The structure and dynamics of networks*. Princeton University Press.

Oswalt, Philipp (Ed.). (2006). *Shrinking cities*, Vol. 2: *Interventions. Ostfildern*: Hatje Cantz Publishers.

Panori, A., Mora, L., & Reid, A. (2019). Five decades of research on urban poverty: Main research communities, core knowledge producers, and emerging thematic areas. *Journal of Cleaner Production, 237*. https://doi.org/10.1016/j.jclepro.2019.117850, 117850.

Pallagst, K., Fleschurz, R., Nothof, S., & Uemura, T. (2021). Shrinking cities: Implications for planning cultures? *Urban Studies, 58*(1), 164–181.

Peng, X., Zhou, Y., Fu, X., & Xu, J. (2022). Study on the spatial-temporal pattern and evolution of surface urban heat Island in 180 shrinking cities in China. *Sustainable Cities and Society, 84*, 104018. https://doi.org/10.1016/j.scs.2022.104018

Pijanowski, B. C., Tayyebi, A., Doucette, J., Pekin, B. K., Braun, D., & Plourde, J. (2014). A big data urban growth simulation at a national scale: Configuring the GIS and neural network based land transformation model to run in a high performance computing (HPC) environment. *Environmental Modelling and Software, 51*, 250–268.

Pinto, S., & Sablik, T. (2016). Understanding Urban Decline – Federal Reserve Bank of Richmond 2016 ANNUAL REPORT. https://www.richmondfed.org/-/media/richmondfedorg/publications/research/annual_report/2016/pdf/article.pdf. Accessed 18 July 2022.

Porutgali, J. (2010). *Complexity, cognition and the city*. Springer.

Raglin, D. A. (2015). *Plumbing and kitchen facilities in housing units*. U.S. Census Bureau, Washington DC 20233-0001.

Swope, C. (2006). *Smart decline*. Governing. Retrieved from http://www.governing.com/mag/November-2006.html.

Szell, M. (2014). Connecting paradigms. *Science, 343*, 970.

Tao, S., Corcoran, J., Mateo-Babiano, I., & Rohde, D. (2014). Exploring bus rapid transit passenger travel behaviour using big data. *Applied Geography, 53*, 90–104.

Thrift, N., 2013. The cover commentary on the book. In Batty (Ed.), *The new science of cities*. The MIT Press.

Tobler, W. R. (1970). A computer movie simulating urban growth in the Detroit region. *Economic Geography, 46*, 234–240. https://doi.org/10.2307/143141

Tong, X., & Kim, J. H. (2019). Concentration or diffusion? Exploring the emerging spatial dynamics of poverty distribution in Southern California. *Cities, 85*, 15–24. https://doi.org/10.1016/j.cities.2018.11.022

Tong, Y., Liu, W., Li, C., Zhang, J., & Ma, Z. (2021). Understanding patterns and multilevel influencing factors of small town shrinkage in Northeast China. *Sustainable Cities and Society, 68*.

U.S. Census Bureau. (2014). *American community survey: Handbook of questions and current Federal uses*. http://www.census.gov/acs/www/Downloads/congress/ACS_Federal_Uses.pdf

U.S. Census Bureau. (2017). *QuickFacts*. https://www.census.gov/quickfacts/fact/table/detroitcitymichigan/PST045216.

U.S. Department of Housing and Urban Development. (2013). *American Housing Survey: Housing adequacy and quality as measured by the AHS*. https://www.census.gov/content/dam/Census/programs-surveys/ahs/publications/HousingAdequacy.pdf

Wang, Y., & Hiroatsu, F. (2019). Sustainable urban regeneration for shrinking cities: A case from Japan. *Sustainability, 11*(5), 1–14. https://doi.org/10.3390/su11051505

Wei, Y. D. (2015). Spatiality of regional inequality. *Applied Geography, 61*, 1–10.

Wiechmann, T. (2008). Errors expected – Aligning urban strategy with demographic uncertainty in shrinking cities. *International Planning Studies, 13*, 431–446.

Wiechmann, T., & Pallagst, K. (2012). Urban shrinkage in Germany and the USA: A comparison of transformation patterns and local strategies. *International Journal of Urban and Regional Research, 36*(2), 261–280.

Wolff, M., Fol, S., Roth, H., & Cunningham-Sabot, E. (2017). Is planning needed? Shrinking cities in the French urban system. *Town Planning Review, 88*(1), 131–145. https://doi.org/10.3828/tpr.2017.10

Xie, Y., Gong, H., Lan, H., & Zeng, S. (2018). Examining shrinking city of Detroit in the context of socio-spatial inequalities. *Landscape and Urban Planning, 177*, 350–361. https://doi.org/10.1016/j.landurbplan.2018.03.002

Xie, Y., Crary, D., Bai, Y., Cui, X., & Zhang, A. (2019). Modeling grassland ecosystem responses to coupled climate and socioeconomic influences from multi-spatial-and-temporal scales. *Journal of Environmental Informatics, 33*(1), 37–46.

Yang, X., & Xie, Y. (2023). Exploring socio-ecological interactions meddling ecological recovery policies: A case study in Inner Mongolia, China. In R. R. Thakur, S. D. Brunn, B. Thakur, & S. K. Thakur (Eds.), *Environment, development and culture in south and East Asia – Local, regional and international perspectives (Vol-1: Environment)*. Springer Nature.

Zhou, X., & Lin, H. (2017). Spatial weights matrix. In S. Shekhar, H. Xiong, & X. Zhou (Eds.), *Encyclopedia of GIS* (p. 79). Springer. https://doi.org/10.1007/978-3-319-17885-1_1307

Chapter 8
Modeling Grassland Deterioration Considering Spatial Variations and Time-Lag Effects at a Finer Geographical Unit

8.1 Introduction

Chapter 3 illustrates how to build temporally compatible total socioenvironmental systems (TSES) datasets. Chapter 4 uses the time-series datasets created through various types of temporal data preprocessing to explore time-series causal relationships between grassland deterioration and a set of socioenvironmental driving factors. Chapter 4 confirms that different combinations of time-series data preprocessing and panel regression analyses disclose varied capabilities, revealing causal relationships hidden in complex socioenvironmental datasets. In other words, the outcomes from different panel models with distinctive time-series filtering methods are not consistent and sometimes contradict one another. For example, we found out that per capita GDP (*gdppc*), grain production density (*gpsk*), highway length per square kilometers (*lohsk*), grassland ratio (*Grass*), and the urban land ratio (*Urban*) showed inconsistent impacts on grassland productivity (enhance vegetation index – EVI – as a proxy) among different models. They disclosed varied impacts on EVI in strength or direction in three different panel models. We pointed out in Chap. 4 that the data at a coarse geographical scale, county, may have caused these variations. This finding indicates a need to obtain finer-scale data to explain why these models generate inconsistent results.

This scaling issue reveals a significant challenge that the investigations face concerning the integrated, total, or coupled human and natural systems (Zhou et al., 2022a). Some environmental or ecological datasets, such as plant and soil samples, come from field observations. Some come from remotely sensed data, including precipitation, temperature, and land use and land cover (LULC) data. The former is usually point-based data, while the latter is a raster or grid. However, socioeconomic statistics are enumerated in administrative areas. The point samples or raster

---

Supplementary Information The online version contains supplementary material available at https://doi.org/10.1007/978-3-031-39594-9_8.

Y. Xie, *Total Socioenvironmental Systems*,
https://doi.org/10.1007/978-3-031-39594-9_8

images must be interpolated or resampled over the boundaries of the administrative area. The most common administrative unit globally is the county (Xie et al., 2019). A county usually covers a large piece of stretched land (huge spatial extent), but its spatial scale is small. Equivalently, its spatial resolution for image-source data is coarse. Therefore, it loses resolution to resample high-resolution grid-based or image-based environmental data to (large area) administrative boundaries. In other words, the current practice of matching environmental data with socioeconomic statistics loses spatial details, which becomes a severe data limitation in integrated environmental and socioeconomic studies.

One solution to this challenge is to collect or survey socioeconomic data during the sampling process of environmental field data. However, surveying or interviewing socioeconomic data or information has a high price tag. Furthermore, the frequencies and contents of the survey and interview data vary widely across time and space, constituting severe limitations to integrated spatial and temporal environmental and socioeconomic studies (Bretschger & Pittel, 2020). There is an urgent need to develop new methods and techniques for integrating various sources and types of data into a unified dataset (Bai et al., 2016). Only integrated data assimilation can fill critical gaps between environmental and socioeconomic data to advance our understanding of how coupled environmental and socioeconomic systems work. Therefore, this chapter introduces a new solution to this challenge, spatially interpolating the county-based data to the township-based data and constructing a newly synthesized fine-resolution socioenvironmental data. Afterward, we will fit a panel regression model to analyze the model outcomes compared with the results of the panel models in Chap. 4. We will discuss new insights obtained through the finer resolution data fitted panel model.

8.2 Spatially Interpolate Key Variables from Counties to Towns

Chapter 4 recommends that the empirical mode decomposition (EMD) and the hybrid EMD and support vector regression (SVR-A)-filtered data are good choices for experimenting with panel regression models. Since the SVR-A filtering algorithm is an improved variation of the EMD decomposition, this chapter chose the EMD and SVR-A-filtered datasets for further exploration with a focus on EMD. The IMAR (Inner Mongolia Autonomous Region of China) dataset includes the response variable EVI (enhanced vegetation index) and 30 explanatory variables, including 2 climate factors, 20 socioeconomic indicators, and 8 land use and land cover (LULC) variables. These variables were filtered with the EMD technique from 2001 to 2017. To approve the concept and reduce the preprocessing data time, we took a subset of the EMD-filtered dataset, including all 31 variables in 4 years: 2001, 2006, 2011, and 2017.

The research question is whether the panel regression model fitted with fine-resolution data will help reveal the causal relationship between ecological deterioration, climate change, and socioeconomic transformation in IMAR. We know we can obtain high-resolution ecological, climate, and LULC data. The challenge is to get socioeconomic statistics at finer (or small area, or low level) administrative units. Moreover, some socioeconomic statistics are only meaningful at relatively high-level administrative units. For instance, our study area, IMAR, is a primary animal husbandry region with a relatively sparse population and less commercial intensity. Several data items, such as professional and technical employment, health and medical professionals, and tertiary industry output value, are not very meaningful at the township level. Based on the literature review (Li and Xie, 2013; Li et al., 2013, Chen et al., 2018, Xie et al., 2018; Lan et al., 2022; Zhou et al., 2022a, b), six variables are identified as the key local variables, including population density (*DPop*), percent of rural population to total population (*prpop*), per capita gross domestic product (*gdppc*), net income of farmers and pastoralists (*iofp*), the number of the livestock per square kilometers (*lssk*), and grain production per square kilometers (*gpsk*). In other words, these six variables show noticeable spatial variations and change dramatically between different towns. We include two sets of these six variables in the new dataset. The first set is at the town level, while the second set is at the county level. The remaining 14 socioeconomic variables are kept at the county level. We also include three primary land uses, percentage of grassland area to total land area (*Grass*), percentage of cropland to total grassland land (*CropPG*), and percentage of urban land area to total land area (*Urban*), at both town and county levels. Finally, EVI, the response (dependent) variable, is calculated at the town level. The variables at the town level have the prefix "*T_*." The variables at the county level keep their original names. The rationale for including these nine variables at both town and county levels is that these variables displayed inconsistent impacts on grassland productivity from the county-based panel models (see Chap. 4). Including them at both town and county levels aims to see how these variables behave differently at town and county levels, influencing grassland conditions.

However, the variables at the town level do not exist, and we must generate them with the spatial (or areal) interpolation method illustrated in Chap. 6. We apply the natural patches-based areal interpolation (also called the non-intercept regression over natural patches). In other words, we treat the EMD-fitted values of *EVI*, *DPop*, *prpop*, *gdppc*, *iofp*, *lssk*, and *gpsk* over counties in 2001, 2006, 2011, and 2017 as the dependent variables. We treat the areas of various LULC types within counties in 2001, 2006, 2011, and 2017 as the independent variables, respectively. We then apply the regression coefficients obtained at the county unit to the LULC data at the town level to predict the values of *EVI*, *DPop*, *prpop*, *gdppc*, *iofp*, *lssk*, and *gpsk* by towns. This process is also called the areal interpolation of county data to town data.

The land use and land cover (LULC) data were derived from the NASA MCD12Q1 LULC Data (see Table 3.2 in Chap. 3). The urban land areas in the NASA MCD12Q1 Data product were the values in 2000 and no further updates were provided. Therefore, the urban land data was replaced with the impervious land surface dataset of China (Gong et al., 2019). When extracting the LULC data by the town boundaries, we separated the croplands with the planted grassland for

Table 8.1 The independent land use variables

Land code	Land class
1.	Forests
2.	Shrubs
3.	Grasslands
4.	Wetlands
5.	Croplands
6.	Urban
7.[a]	*Farmed Grass*
8.[a]	*Ice*
9.	Barren
10.	Water

[a]Excluded from the areal interpolation because of lots of missing records

harvest (*Farmed Grass*), because *Farmed Grass* was missing in most of towns. In addition, the land cover, *Ice*, was missing in most of towns. We excluded them in the areal interpolation processing (Table 8.1).

The data preprocessing steps and accuracy assessments are described in the final section of the chapter. The input and output datasets are stored in the folder, /Chapter8/DataProcess/2001_2017. The final combined town–county panel datasets are stored in the folder, /Chapter8/DataProcess/TownBasedData.

8.3 Town–County Combined Panel Regression Model Analysis

We developed a town–county combined panel regression model. T_EVI (the EMD-filtered enhanced vegetation index by town) is the response variable. The model included nine variables extracted by town boundaries through the areal interpolation or direct resampling by using the town boundaries. The same set of nine variables were also processed over the county boundaries (Table 8.2). We collected this set of variables at both town and county boundaries to analyze how these variables affected grassland deterioration differently at two scales. The model also contained 16 variables collected at the county boundaries only. The time series included 2001, 2006, 2010, and 2017. The time series was very short; thus, we did not involve the lagged effects in the model.

The town–county mixed model can be formed as follows:

$$Y_{it} = \alpha + \sum_{m}^{m=1} X_{itm}\beta_{itm} + \sum_{n}^{n=1} X_{jtn}\beta_{jtn} + \delta_i + \gamma_t + \varepsilon_{it} \tag{8.1}$$

Table 8.2 List of variables in the town–county combined panel data analysis[a]

Short name	Long name	Unit
T_EVI	EMD-filtered enhanced vegetation index by town in 2001, 2006, 2011, 2017	EMD-filtered values
T_DPop	EMD-filtered population density by town in 2001, 2006, 2011, 2017	EMD-filtered values
T_gdppc	EMD-filtered per capita gross domestic product (GDP) by town in 2001, 2006, 2011, 2017	EMD-filtered values
T_prpop	EMD-filtered percent of rural population to total population by town in 2001, 2006, 2011, 2017	EMD-filtered values
T_iofp	EMD-filtered net income of farmers and pastoralists by town in 2001, 2006, 2011, 2017	EMD-filtered values
T_lssk	EMD-filtered the number of livestock per square kilometer by town in 2001, 2006, 2011, 2017	EMD-filtered values
T_gpsk	EMD-filtered grain production per square kilometers by town in 2001, 2006, 2011, 2017	EMD-filtered values
T_Grass	Grassland area (percentage to total land area)	% in town
T_CropPG	Crop area + planted grassland for harvest (percentage to grassland)	% in town
T_Urban	Urban area (percentage to total land area)	% in town
PRE	Precipitation (summer precipitation of each period in the growing season)	Mm
TEM	Temperature (averaged temperature of each period in the growing season)	°C
DPop	The population density in a county	100/sq.km
gdppc	Per capita gross domestic product (GDP) in a county	1000 yuan pc
prpop	Percent of rural population to total population in a county	%
iofp	Net income of farmers and pastoralists in a county	%
lssk	The number of livestock per square kilometers in a county	100 head/sq.km
gpsk	Grain production per square kilometers in a county	Ton/sq.km
Grass	Grassland area (percentage to total land area)	% in county
CropPG	Crop area + planted grassland for harvest (percentage to grassland)	% in county
Urban	Urban area (percentage to total land area)	% in county
gova	The gross output value of farming, forestry, animal husbandry, and fishery – Share of GDP in a county	%
Paa	Percent of the arable area to total land area in a county	%
fai	Fixed assets investment – Share of GDP in a county	%
lgr	Local government revenue – Share of GDP in a county	%
lohsk	The total length of highways per square kilometers in a county	Km/sq.km)
HMP	Health and medical professionals per 10,000 people in a county	Person/10,000 people
HPB	Hospital beds per ten thousand people in a county	Bed/10,000 people

(continued)

Table 8.2 (continued)

Short name	Long name	Unit
MHT	Middle and high school teachers per 10,000 people in a county	Person/10,000 people
PTE	Professional and technical employment per 10,000 people in a county	Person/10,000 people
PcUDI	Per capita disposable income of permanent urban residents in a county	1000 yuan pc
PIO	Primary industry output value – Share of GDP in a county	%
SIO	Secondary industry output value – Share of GDP in a county	%
TIO	Tertiary industry output value – Share of GDP in a county	%
TCRV	Total consumer retail value – Share of GDP in a county	%

[a] "T_" indicates town-based

Table 8.3 The results of the Hausman test

Hausman test: Ho: Difference in coefficients not systematic	
chi^2 (29) =	220.88
Prob > chi^2	0.000

Extracted from the Fixed_or_Random_Effect_Model_Test result file (stored in the Chap. 8 folder)

where Y_{it} is the dependent variable (EMD-fitted EVI index values), i is the index of towns, and t is the index of Years. X_{itm} is the m-th independent variable over town i at Year t; β_{itm} is the coefficient of the m-th independent variable; X_{jtn} is the n-th independent variable over county j at Year t; β_{jtn} is the coefficient of the n-th independent variable; α is the intercept and the overall constant of the model; δ_i indicates the cross-sectional effect; γ_t is the temporal-specific effect; and ε_{it} represents the errors that vary across sections throughout the studied period.

We first ran a Hausman test (Baltagi, 2005; Li et al., 2013) to assess which type of estimation our panel regression model should apply. The null hypothesis of the Hausman test is that the differences of estimated coefficients between the fixed and the random models are not systematic, which means the random effect. The alternative is that the differences of estimated coefficients between the two models are systematic, which indicates the fixed effect. In the Hausman test, the chi^2 was relatively small for the panel data analysis (PDA) model, with more than 29 independent variables and 1400 sections (Table 8.3). The chi^2 statistic was also highly significant (Prob > chi^2 = 0.000). Therefore, we accepted the Hausman test null hypothesis that the model estimation coefficients were not systematic and hence exhibited the random effect. In other words, the estimation method for the combined town–county panel model ought to be the random-effect approach.

Next, we applied three random-effect panel regression models: the panel regression model with the covariates (the covariates model), the panel regression model with both covariates and *year* (the time-series model), and the panel regression model with both covariates and *towns* (the cross-sections model). We can see that the model statistics of the covariates model and the time-series model were very

Table 8.4 The statistics of three random-effects panel models

Random-effects GLS regression		Number of obs. = 5608
Group variable: Towns		Number of towns = 1402
Observations (obs.): 1402 × 4 = 5608		Number of years = 4
The covariates model	*The time-series model*	*The cross-section model*
R-sq:	R-sq:	R-sq:
Within = 0.5245	Within = 0.5301	Within = 0.5336
Between = 0.6889	Between = 0.6874	Between = 1.0000
Overall = 0.6134	Overall = 0.6154	Overall = 0.7952
Wald $chi^2(34) = 7638$	Wald $chi^2(37) = 7731$	Wald $chi^2(1435) = 16{,}196$
Prob > $chi^2 = 0.0000$	Prob > $chi^2 = 0.0000$	Prob > $chi^2 = 0.0000$

similar. The within and overall R-sq (R-squared values) were improved slightly with the time-series model (Table 8.4). However, the between R-sq was not improved. The time-series model included four time series, 2001, 2006, 2011, and 2017. Since *year* is a categorical variable, the model only generated three coefficients for *year* because 2001 was the default base category (Table 8.5). Based on the $P > z$ values, the difference between 2006 and 2001 was not statistically significant, the time-series effects between 2011 and 2001 were significant at 10% confidence level, and the temporal results between 2017 and 2001 were highly significant. Moreover, the time-series effects were adverse because the coefficients were negative. For example, the EMD-filtered EVI would decrease 5.81 units from 2001 to 2017 when other covariates remained unchanged (Table 8.5). This finding confirms that grassland deterioration in IMAR got worsened over time.

The panel regression model with covariates and *towns* (the cross-section model) disclosed the best performance (Table 8.4). The between R-sq. rose to 1.0, which confirmed that the selection of a random-effects model was correct. Adding *town* into the model explained all cross-section (i.e., the spatial) variations. The overall R-sq. (0.795) was dramatically improved compared to the other two models. Therefore, we conducted an in-depth analysis of the coefficients of this cross-section model to reveal the causal relationships, strengths, directions, and underlying environmental and socioeconomic implications compared with the current literature.

The coefficients of the three-panel models are reported in Tables 8.5 and 8.6, respectively. The coefficients of the covariates were analogous to some degrees among the three models. However, since adding towns to the model dramatically improved the model performance, we will provide a detailed analysis of the town-included panel model in the following sections.

The coefficients of the cross-section panel model disclosed several interesting findings (Table 8.6). First, five socioeconomic variables at the town level, *T_prpop*, *T_iofp*, *T_DPop*, *T_gpsk*, and *T_gdppc*, showed very strong and positive causal relationships with T_EVI based on their coefficients and $P > z$ values. On the contrary, their counterparts at the county level, *iofp*, *prpop*, *gpsk*, *gdppc*, and *DPop*, displayed substantial negative impacts on T_EVI. However, their impact intensities changed

Table 8.5 The coefficients of the covariates and the time-series panel regression models

The covariates model					The time-series model			
T_EVI[a]	Coef.	$P > z$	Coef.	$P > z$	Coef.	$P > z$	Coef.	$P > z$
County level			Town level[b]		County level		Town level[b]	
DPop[c]	−0.0916	0.000	1.0917	0.000	−0.1150	0.000	1.1214	0.000
gdppc	−0.0023	0.000	0.0247	0.000	−0.0017	0.001	0.0246	0.000
prpop	−0.0112	0.000	0.1876	0.000	−0.0073	0.000	0.1879	0.000
iofp	−0.1842	0.000	2.2328	0.000	−0.1818	0.000	2.2102	0.000
lssk	0.0234	0.009	−0.2346	0.001	0.0246	0.006	−0.2357	0.001
gpsk	−0.0018	0.002	0.0538	0.000	−0.0009	0.130	0.0537	0.000
Grass	0.0032	0.013	−0.2388	0.054	0.0030	0.018	−0.2239	0.071
CropPG	−0.0002	0.010	0.4580	0.000	−0.0002	0.023	0.4476	0.000
Urban	−0.0023	0.691	−0.4449	0.002	0.0049	0.402	−0.4471	0.002
PRE	0.0448	0.111			0.1109	0.000		
TEM	−0.9566	0.000			−1.0851	0.000		
gova	0.0033	0.138			0.0033	0.137		
fai	0.0006	0.343			0.0007	0.288		
HMP	−0.0034	0.009			−0.0008	0.583		
HPB	0.0057	0.000			0.0071	0.000		
lgr	−0.0096	0.365			−0.0071	0.515		
lohsk	−0.0443	0.718			−0.0037	0.977		
MHT	0.0075	0.001			0.0034	0.154		
Paa	0.0180	0.010			0.0073	0.311		
PcUDI	0.0000	0.303			0.0000	0.247		
PIO	0.0027	0.437			−0.0015	0.666		
PTE	−0.0001	0.626			−0.0002	0.198		
SIO	0.0016	0.000			0.0022	0.000		
TCRV	−0.0049	0.014			−0.0020	0.322		
TIO	0.0075	0.000			0.0095	0.000		
				Year[d]				
				2006	−0.0257	0.573		
				2011	−0.1198	0.069		
				2017	−0.4524	0.000		
_cons	2.2564	0.000		_cons	2.1022	0.000		
sigma_u	0.6007			sigma_u	0.6031			
sigma_e	1.0052			sigma_e	0.9994			
rho	0.2631			rho	0.2670			

[a]T_EVI is the areal interpolated EMD-filtered EVI values from counties to their corresponding towns within the counties. T_EVI is the dependent variable

[b]The independent variables available at the town level also exist at the county level for two-level panel regression models

[c]The nine variables available at the town and county levels are named differently. Suppose that there is a prefix T in front of the town variables but the variable names for the town level were omitted due to the table width limitation

[d]"Year" is a categorical variable and its first value 1 (indicating 2001) was treated as the base category for comparison

Table 8.6 The coefficients of the cross-section panel regression model

Variables	Coef.	$P > z$	T_EVI[a]	Coef.	$P > z$
Independent variables common to county and town[b]					
DPop	−0.1382	0.044	T_DPop	1.1791	0.000
gdppc	−0.0018	0.003	T_gdppc	0.0213	0.000
prpop	−0.0119	0.000	T_prpop	0.1837	0.000
iofp	−0.1950	0.000	T_iofp	2.0140	0.000
lssk	0.0171	0.100	T_lssk	−0.1546	0.061
gpsk	−0.0026	0.002	T_gpsk	0.0511	0.000
Grass	0.0099	0.055	T_Grass	0.3892	0.136
CropPG	0.0001	0.580	T_CropPG	0.8234	0.005
Urban	0.0330	0.000	T_Urban	−1.5155	0.003
Independent variables by county only[c]					
PRE	−0.1215	0.003	*Township variations*		
TEM	−1.6464	0.018	*Towns*[d]		
gova	0.0004	0.908		–	–
fai	−0.0002	0.754	The model constant		
HMP	−0.0002	0.881	_cons	1.5149	0.270
HPB	0.0075	0.000	Other model parameters		
lgr	−0.0229	0.091	sigma_u	0.0000	
lohsk	−0.1512	0.335	sigma_e	1.0052	
MHT	−0.0036	0.316	rho	0.0000	
Paa	0.0407	0.004			
PcUDI	0.0000	0.966			
PIO	0.0081	0.060			
PTE	−0.0001	0.816			
SIO	0.0016	0.000			
TCRV	−0.0087	0.002			
TIO	0.0090	0.002			

[a]T_EVI are the areal interpolated EMD-filtered EVI values from counties to their corresponding towns within the counties. T indicates town. T_EVI is the dependent variable

[b]The nine independent variables available at the town level also exist at the county level for two-level panel regression models

[c]These variables are only available at the county levels

[d]There are more than 1400 towns in IMAR. Therefore, we did not list the coefficients of towns in the panel model. We included the categorical variable, towns, in the model to confirm that the suitable panel model for this dataset should be the random-effects panel model

from town to county levels. *T_prpop* and *T_iofp* switched their positions to *iofp* and *prpop* at the county level. *T_DPop* dropped its influence from the third strongest position at the town level to the fifth negative at the county level. Second, *T_lssk* displayed a marginally negative impact on *T_EVI* (at the 10% significance level) at the town level. Still, *lssk* showed a slightly positive effect (at the 10% significance level) on *T_EVI* at the county level. Third, *T_Urban* exhibited a negative causal relationship with *T_EVI* at the town level, but *Urban* showed a positive impact on *T_EVI* at the county level. Fourth, grassland (*T_Grass*) disclosed a nonsignificant impact on *T_EVI* at the town level but revealed a weak positive impact on *T_EVI* at the county level (*Grass*). Fifth, cropland (*T_CropPG*) displayed a significant impact on *T_EVI* at the town level but a nonsignificant influence on *T_EVI* at the county level (*CropPG*). Sixth, both climate factors (PRE and TEM) showed moderate negative impacts on T_EVI.

We also examined the causal relationships between *T_EVI* and the socioeconomic variables only available at the county level. First, the production-related indicators, such as the primary industrial output (*PIO*), the secondary industrial output (*SIO*), and the tertiary industrial output (*TIO*), showed positive impacts on *T_EVI*. Second, the variables concerning income and revenue disclosed nonsignificant impacts on *T_EVI*. These variables included *gova* (gross output value of farming, forestry, animal husbandry, and fishery), *fai* (fixed asset investment), *lgr* (local government revenue), *PcUDI* (urban resident dispensable income), and *TCRV* (total consumer retail value). However, *TCRV* revealed a negative impact on *T_EVI*. Third, the model contained four social development indicators, hospital beds (*HPB*), health and medical professionals (*HMP*), middle and high school teachers (*MHT*), and professional and technical employment (*PTE*). They displayed nonsignificant impacts on *T_EVI*, except that *HPB* revealed a significantly positive impact on *T_EVI*. In addition, highway development (*lohsk*) showed a nonsignificant relationship with *T_EVI*. Finally, arable land (*Paa*) displayed a moderate positive impact on *T_EVI*.

Furthermore, we compared the relationships between the grassland productivity (EVI) and the county-level socioeconomic variables among three models: the town–county two-level model developed in this chapter and the two recommended models from Chap. 4, the EMD-fitted dependent-lagged model and the EMD-SVR-A-fitted dependent-lagged model (Table 8.7). It is critical to point out that the data filtering methods differ among the three models, and there is no foundation for a direct comparison. However, the two-level (town and county) panel model is built with the data interpolated to the finest spatial unit of town. The outcomes of this model reveal significant variations between the town-based causal relationships and the county-derived causal relationships. Therefore, it is worth comparing its county-level causal relationships with the findings of the other county-based panel models.

This three-model comparison exhibited three apparent differences. First, the three models confirmed that the nine independent variables interpolated to the town level were the most important variables affecting grassland productivity. Second, the town–county two-level model displayed a significant deviation from two county-based models in terms of the strengths and directions of these variables on grassland productivity. Significantly, the findings from the two-level model were dramatically different from the EMD-SVR-A hybrid model. Third, the remaining socioeconomic and LULC variables only available at the county level behaved differently among the three models. Many were not significant factors affecting grassland productivity in the two-level model, but almost all were substantial influencers in the EMD model.

Table 8.7 The coefficients of the covariates by county: Town–County Model vs County-Alone Models

	Town–County Model		(County-based) EMD Model		(County-based) SVR Hybrid Model	
T_EVI	Coef.	$P > z$	Coef.	$P > z$	Coef.	$P > z$
Independent variables that were included in the Town–County levels						
DPop	−0.1382	0.044	0.0101	0.000	0.0014	0.004
gdppc	−0.0018	0.003	0.0001	0.000	0.0000	0.000
prpop	−0.0119	0.000	0.0001	0.020	0.0001	0.000
iofp	−0.1950	0.000	0.0019	0.000	−0.0028	0.000
lssk	0.0171	0.100	0.0002	0.004	−0.0002	0.019
gpsk	−0.0026	0.002	0.0003	0.000	−0.0001	0.000
Grass	0.0099	0.055	0.0004	0.000	−0.0001	0.003
CropPG	0.0001	0.580	0.0000	0.000	0.0000	0.453
Urban	0.0330	0.000	0.0013	0.000	0.0006	0.000
Independent variables only at the county level						
PRE	−0.1215	0.003	0.0042	0.000	−0.0071	0.000
TEM	−1.6464	0.018	0.0274	0.000	0.0221	0.000
gova	0.0004	0.908	0.0006	0.000	0.0000	0.375
fai	−0.0002	0.754	0.0000	0.247	0.0000	0.508
HMP	−0.0002	0.881	0.0002	0.000	0.0000	0.129
HPB	0.0075	0.000	0.0001	0.070	0.0000	0.440
lgr	−0.0229	0.091	0.0005	0.007	−0.0006	0.004
lohsk	−0.1512	0.335	0.0333	0.000	0.0090	0.002
MHT	−0.0036	0.316	0.0001	0.014	0.0000	0.975
Paa	0.0407	0.004	0.0017	0.000	−0.0039	0.000
PcUDI	0.0000	0.966	0.0000	0.000	0.0000	0.000
PIO	0.0081	0.060	0.0005	0.000	0.0000	0.637
PTE	−0.0001	0.816	0.0000	0.000	0.0000	0.340
SIO	0.0016	0.000	0.0000	0.376	0.0000	0.428
TCRV	−0.0087	0.002	0.0003	0.000	−0.0003	0.000
TIO	0.0090	0.002	0.0001	0.178	−0.0003	0.000
			L1.EVIemd		**L1.EVIsvr_a**	
			1.0023	0.000	1.0017	0.000
_cons	1.5149	0.270	0.0910	0.000	0.0093	0.461

8.4 Discussion

The two-level random-effects panel model makes the first attempt to examine causal relationships between grassland deterioration and climate factors, LULC changes, and socioeconomic driving forces from a fine spatial unit, town. Socioeconomic statistics are not available at such a fine administrative unit. Therefore, the current literature lacks empirical or quantitative analyses of integrated (coupled, synthetic, or total) socioenvironmental systems at preferred satisfactory resolutions. Existing empirical studies of the total socioenvironmental systems at coarse resolutions often lead to contradictory findings (Zhou et al., 2022b).

Let us take IMAR as an example. Two schools are giving different interpretations of the driving effects of environmental and socioeconomic factors on grassland deterioration. One school argues that fast socioeconomic development in the IMAR of China negatively affects grassland productivity. Rapid economic growth in IMAR in the past several decades has been achieved at the expense of its natural environments, resources, and grasslands (Robinson et al., 2017). Unprecedented GDP increase has converted large pieces of grasslands into urban construction, grain plantation, and mining industries (Hu & Nacun, 2018).

Furthermore, large-scale urban expansion and intensified coal mining have accelerated large-scale road construction. Therefore, the increased impervious surfaces, concrete roads, and high road densities directly damaged grasslands and worsened grassland deterioration (Sha & Li, 2020). In addition, rapid economic growth accelerated rural population growth. Increased population exacerbated the burden on natural resources and stimulated the desire for expanding socioeconomic growth, leading to more intensified industrial development, grazing, and agricultural production. The consequences consist of a vicious cycle. All these economic activities accelerated the conversion of grasslands into urban, residential, and agrarian usages (Hu & Nacun, 2018). The increased industrial output and agricultural output values intensified the demand for more grassland conversion, denser grazing density, and higher agricultural production. Uncontrolled conversion of grasslands into urban and arable lands would eventually break fragile ecosystem resilience and cause irreversible ecological damages (Robinson et al., 2017; Zhou et al., 2022a).

The other school suggests that GDP can exhibit a reversed causality to grassland productivity. The GDP's positive impact on EVI indicates that healthy economic development means more financial resources available to support sophisticated grassland management and utilization (Xie et al., 2019). Especially, in the degraded grassland areas, easy access to financial resources, markets, and transportation can create more economic opportunities, diversify income sources, and help grasslands recover in Inner Mongolia of China (Deng et al., 2011). High GDP and industrial output values indicate more available financial resources to fund large-scale construction projects, such as urban infrastructures, railroads, highways and sand-dune prevention greenbelts (Gao et al., 2016). In addition, more resources reduce dependences on grassland and positively impact grassland recovery through direct protection, providing diversified economic opportunities or improving market access (Robinson et al., 2017).

However, empirical analyses at coarse resolutions failed to dispute these contradicting arguments. The two-level panel regression model developed in this chapter provided an evident clarification for this controversy. The model pointed out that *T_prpop*, *T_iofp*, *T_DPop*, *T_gpsk*, and *T_gdppc* disclosed strong positive causal relationships with *T_EVI* at the local (town) scale. On the other hand, *iofp*, *prpop*, *gpsk*, *gdppc*, and *DPop* negatively affected T_EVI at the regional (county) scale. This finding means that high GPD, a high income of farmers and pastoralists, high grain production, and a high population at the local (town) scale can positively impact grassland productivity. This finding indicates that more available resources help diversify revenue sources and reduce dependence on grasslands at the local (town) scale.

On the contrary, these driving factors have negative causal relationships with grassland productivity at the regional (county) scale. In other words, when economic activities are intensified across the entire county or region, they become damaging agents to ruin grassland resources. This finding supports the popular view of the impacts of rapid economic development on natural resources. The different results from the town and the county variables add new insights concerning the environmental Kuznets curves (Chang et al., 2021; Chen et al., 2020). In the early stage, rapid economic development and population growth, in general, bring with them environmental deterioration. However, sound and accessible economic and financial resources could alleviate environmental stresses. Even in the early stage of economic development, such environmental benefits of economic maturity could be observed in endowed local communities. Thus, the two-level panel model clarifies where these controversies come from. The town–county model also confirms that geographical scale is important for examining total socioenvironmental systems.

Next, there is controversy about the impact of the grazing density on grassland productivity. It is generally agreed that intensive grazing speeds up grassland deterioration. However, several studies reveal that grazing positively affects grassland productivity. The geographical explanation indicates that higher livestock densities occur in fertile grasslands with higher EVI values (Xie et al., 2021). Our two-level panel model found that grazing density negatively impacted grasslands locally but had a positive effect regionally. This finding is more logical. Overgrazing deteriorates grasslands, which is reflected by the town-level panel coefficient. However, grazing mainly concentrates on highly productive or fertile grasslands. Consequently, a positive association between livestock and grassland productivity is found at the regional (county) scale.

In addition, the impact of urban development on grasslands discloses another dichotomy. The two-level panel model pointed out that extensive urban development significantly negatively impacted grassland at a local scale. However, from the regional perspective, urban development would improve the regional economy, bring in more resources, diversify economic opportunities and reduce its negative impact on grasslands.

Moreover, current literature holds a dialectic view on the impacts of crop plantation and arable land on grasslands. Crop plantation increases landscape greenness based on remote sensing images (Zhou et al., 2022b). Still, a high percentage of arable land means an extensive conversion of grasslands to agricultural lands, accelerating grassland deterioration (Tang et al. 2016). Our two-level panel model confirmed that crop plantation increased landscape greenness at the local scale but showed a nonsignificant impact at the regional scale. Moreover, a large percentage of arable land at the regional scale was positively correlated with grasslands. This finding has two indications. First, easy access to arable land supports diversified economic activities and hence reduces pressure on grasslands. Second, from the remote sensing point of view, crop plants correlate with higher EVI values.

It is also a nuance that our two-level panel model identified moderate and negative impacts of two climate factors, PRE and TEM, on grassland productivity. Most literature recommends that precipitation is a critical climate factor that positively determines grassland productivity (Li et al., 2013). In our model, T_ EVI values were linearly transformed through the EMD residual transformation. The EMD-transformed EVI data diminished the direct impacts of the climate factors on EVI but significantly augmented the influences of the socioeconomic variables (Zhou et al., 2022b). In addition, our model is the covariates plus *towns* random-effects model. When the random regional effects were counted for, the fixed effects of PRE and TEM on T_EVI were adverse. This finding was consistent with the fixed effects of the *year* on T_EVI, which illustrated negative relationships with T_EVI. This finding indicated that PRE and TEM had declined over the years. Therefore, our model confirms that grassland productivity deteriorated during the study period from 2001 to 2017.

Furthermore, three additional findings need to be discussed. First, primary, secondary, and tertiary industrial output values positively improved grassland productivity. This finding confirms that IMAR and China have experienced rapid industrial development since the economic reforms. The trickle-down economic effects became noticeable (Niu, 2012; Tan et al., 2018; Lo-Iacono-Ferreira et al., 2022). Second, three of the four social development indicators did not significantly impact T_EVI except for the hospital beds (HPB). This finding indicated that the trickle-down effect was primarily confined to the production sectors but was slow in social development, including education and health. Third, *fai* indicated the state's fixed asset investment share in GDP. *fai* came from the Central Government and continued to increase in IMAR, especially after 2010, when the economic compensation policy was launched. However, the compensation policy did not achieve its initial goal of restoring a large chunk of deteriorated grasslands.

On the contrary, the conversion of natural grasslands to agricultural, industrial, and urban land use continued or, at best, slowed down (Shang et al., 2019). The increased *fai* failed to reverse the environmental deterioration trends in IMAR, and our model found that *fai* was a nonsignificant influencer of T_EVI. In other words, the increased state investment in IMAR did not achieve its intended goal.

8.5 Conclusion

The town–county random-effects panel model is the first empirical econometric model to examine total socioenvironmental systems at the town level in IMAR. It is also the first fine-scale panel regression model to study coupled human–nature systems. The areal interpolating socioeconomic statistics from the county to the town level makes this two-level modeling possible. The inclusion of town-based socioenvironmental data enables the model to delve into fine details of the causal relationships between grassland deterioration, climate and LULC changes, and socioeconomic transformations. As a result, the model can disclose subtle interactions between grassland productivity, climate, LULC, and socioeconomic factors. This two-level model can reveal local dynamics and regional responses. In other words, the model can differentiate local and regional impacts of climate, LULC, and socioeconomic factors on grasslands deterioration. As a result, this model clarifies several confusions about the effects of socioeconomic development, climate factors, and LULC changes on grasslands. The findings confirm that the interactions between ecosystem functions (grassland productivity) and climate, LULC, and socioeconomic factors vary at different geographic scales. We must examine the dynamics of total (or coupled, or integrated) socioenvironmental systems with fine-resolution data and at multiple geographical scales. The areal data interpolation and the two-level panel data modeling approaches can be applied to conduct empirical studies of integrated human–natural systems in other regions, countries, or coupled systems.

However, there are several limitations that we need to address and further investigate as soon as possible. First, there is no adequate methods to validate the interpolation outcomes from counties to towns. Second, it will be interesting to directly interpolate the county-based socioeconomic statistics from counties to land use (or cover) patches. This type of interpolation will enable the building of land parcels or natural patches based on empirical econometric models to study total socioenvironmental systems. Third, the data preprocessing (areal interpolation) takes time since no existing tools exist. Although this chapter has developed prototype tools to implement the illustrated data interpolation, there is a need to make these tools robust. In this chapter, we only interpolated seven variables over 4 years. Ideally, we would like to have all variables interpolated yearly for a complete dataset for analysis.

8.6 The Steps of Preprocessing the Dataset Needed to Run the Town-Based Panel Regression Model

8.6.1 The Input Data Files

1. The four EMD-filtered county data files (2001, 2006, 2011, and 2017) are stored in the folder, Chapter_8\DataProcess\Datasets:

    IMAR_County_Data_2001
    IMAR_County_Data_2006
    IMAR_County_Data_2011
    IMAR_County_Data_2017

2. The four county-based land use files are stored in the folder, \Chapter_8\DataProcess\Datasets:

    IM_County_Lands2001
    IM_County_Lands2006

IM_County_Lands2011
IM_County_Lands2017

3. The four datasets of county data and land-use file:
   Join IMAR_County_Data_20xx and IM_County_Lands20xx to create IM_County_Data_Land_20xx

   IM_County_Data_Land_2001
   IM_County_Data_Land_2006
   IM_County_Data_Land_2011
   IM_County_Data_Land_2017

   - These four files are the input files for the areal interpolation are stored in the folder, \Chapter_8\DataProcess\2001_2017, under each year subfolder.

4. The four town-based land use files are stored in the folder, Chapter_8\DataProcess\Datasets:

   IM_Town_Land_2001
   IM_Town_Land_2006
   IM_Town_Land_2011
   IM_Town_Land_2017

5. The four town-based land-use ratio files:

   IM_Town_Land_Ratio_2001
   IM_Town_Land_Ratio_2006
   IM_Town_Land_Ratio_2011
   IM_Town_Land_Ratio_2017

   - These files are the input files for predicting the values at the town level and are stored in the folder, \Chapter_8\DataProcess\2001_2017, under each year subfolder.

8.6.2 Run the Areal Interpolation (Non-intercept Regression)

1. Run the non-intersect regression Python code, **IMAR_Areal_Interpolation** (in the folder, \Chapter_8\DataProcess\2001_2017) to interpolate EVIemd, DPop, gdppc, iofp, lssk, prpop, and gpsk, respectively, by each year, creating:

   Param_IM_County_Data_Land_2001
   Param_IM_County_Data_Land_2006
   Param_IM_County_Data_Land_2011
   Param_IM_County_Data_Land_2017

   - The Python codes are provided in the folder, \Chapter_8\DataProcess\2001_2017.

- The input and output files are stored in the folder, \Chapter_8\DataProcess\2001_2017, under each year subfolder.
- The file locations are applied to all the following processes.

2. Adjust the coefficients as nonnegative adjusted coefficients (see Chap. 6 for details); Use the adjusted nonnegative coefficients and four town-based land-use files (Chap. 6: Step 1–4) to predict the seven variables, EVIemd, DPop, gdppc, iofp, lssk, prpop, and gpsk and save the results in the following files:

   IM_Towns_Data_Prediction_2001 (under each year subfolder, \Chapter_8\DataProcess\2001_2017\)
   IM_Towns_Data_Prediction_2006
   IM_Towns_Data_Prediction_2011
   IM_Towns_Data_Prediction_2017

- Bring this IM_Towns_Data_Prediction_20xx into ArcGIS (Pro or Desktop); Join it with IMAR_Towns_in_County with TID; Afterwards, join IM_County_Data_Land_20?? with IMAR_Towns_in_County with CountyID (Now IMAR_Towns_in_County should include the predicted Town-based EVIemd and DPop and County-based original seven variable values.)
- Export the joint IMAR_Towns_in_County as a new shape file and as a new dbase file TownPred_CountyTotal_20xx, which should include the seven predicted values at Towns and the seven original values at Counties. We will use the county values to adjust the predicted town values to improve the prediction accuracy. In fact, we have done the above ArcGIS or ArcGIS Pro functions and exported the files with the same names:

   IM_Towns_Data_Prediction_2001 (under each year subfolder, \Chapter_8\DataProcess\2001_2017\)
   IM_Towns_Data_Prediction_2006
   IM_Towns_Data_Prediction_2011
   IM_Towns_Data_Prediction_2017

8.6.3 Adjust the Towns Predictions Controlled by the County Totals

1. Reorganize the above four files by years into seven files based on each of the seven variables (under each year subfolder, \Chapter_8\DataProcess\2001_2017\):

- PREPRPOP.XLSX
- PRELSSK.XLSX
- PREIOFP.XLSX
- PREGPSK.XLSX
- PREGPDPC.XLSX

- PREEVIEMD.XLSX
- PREDPOP.XLSX

2. Run Python code, Aggregate_EVIemd_Towns_to_Counties, to adjusted **Towns** predictions controlled by County totals, creating:

   - PredDPop_CountyTotal_2001_Outcome.XLSX
   - PredEVIemd_CountyTotal_2001_Outcome.XLSX
   - Predgdppc_CountyTotal_2001_Outcome.XLSX
   - Predgpsk_CountyTotal_2001_Outcome.XLSX
   - Prediofp_CountyTotal_2001_Outcome.XLSX
   - Predprpop_CountyTotal_2001_Outcome.XLSX
   - Predlssk_CountyTotal_2001_Outcome.XLSX

3. After you get the adjusted seven variable values, you shall combine (join) these adjusted values with IM_Towns_Land_Ratio_2001 as.

   - IM_Towns_Land_Ratio_Reallocate_2001

8.6.4 Reallocate the Adjusted Town Predictions to Each Land-Use Parcel Within a Town

1. Run the Python code, Intersect_Based_Reallocate_Sort (in \Chapter_8\DataProcess\2001_2017\), to match Town-IDs.
2. Run the Python code, Intersect_Based_Reallocate_Allocate (in \Chapter_8\DataProcess\2001_2017\), to disaggregate the predictions to each land-use patch.

8.6.5 Combine the Seven Predicted Town-Based Variables with the Remaining Variables at the County Level

The aforementioned GIS processes and results are stored in the folder, \Chapter8\DataProcess\TownBasedData\

1. Township-based dataset

   - Four years interpolated data: IM_Town_Land_Ratio_Reallocate_2001 (TownID – TID)
   - The case number of towns in 2017 are fewer than the case numbers of towns in 2001, 2006, and 2011, which needs to be joined to 2017 to keep the same

2. County-based dataset

   - Four years aggregated county data: IMAR_Aggregated_Data_2001. ... _2017

3. Join IMAR_Aggregated_Data_200# with IMAR_Towns_in_County by CID
   Export the joined results to IM_CntyTownData2001, ..., IM_CntyTownData2017, respectively
4. Make sure the town lists in Step 1 and Step 3 are consistent
5. Merge Steps 1 and 3 results into IM_CntyTownData_4Y:
   - Make sure to have TID and Time (Year) fields to define the panel, also a variable like TSlice (Year) to model "Repeated Measurements"
6. Reorganize IM_CntyTownData_4Y into the panel dataset:
   - Town_Econ_EVI_4YPanel_Sorted.xlsx (This file is the input file for the panel regression models)
   - The panel regression model, test, and result are stored in the folder, /Chapter_8 (under this root directory).
   - IM_Town_Econ_EVI_4YPanel.dat – the Stata Dataset for panel regression analysis
   - EVI_stationarity_test.do – the Stata Do-file for testing variable's stationarity
   - EVI_fixed_or_random_test.do – Stata Do-file for testing either fixed-effect or random-effect model
   - EVI_random_effect_model.do – the Stata Do-file for running the selected random-effect two-level panel regression model
   - "T_" is the prefix for the variables interpolated by towns in the analysis sections. However, in the datafiles (both excel and Stata), these variables are prefixed as "Ip" (i.e., "interpolated").

8.6.6 Accuracy Assessments of the Areal Interpolation (Non-intercept Regression)

We did areal interpolation of seven variables over 4 years. The R-squared values of the areal interpolation results were reported in Table 8.8. We made a through discussion on the areal interpolation accuracies in Chap. 6. However, there were some distinctive findings from the areal interpolation results in this IMAR grassland deterioration case study. First, the interpolation (estimation) accuracies of *EVI* were stable. The *EVI* R-squared values were within 0.6–0.7 for both training and testing datasets over the 4 years. Second, the R-squared values of rural population, *prpop*, were largely in the range of 0.45–0.66. Both findings confirmed that IMAR is mainly a grassland region, and most people live in the countryside, engaging in animal husbandry or farming. Third, the two socioeconomic variables, *gdppc* and *iofp*, disclosed relatively stable R-squared values, mostly ranging from 0.45 to 0.70, with both training and testing datasets over 4 years. This finding indicated that the socioeconomic indicators showed less spatial disparities and hence relatively constant estimation (interpolation) accuracies.

Table 8.8 R-squared values of areal interpolations (non-intercept regression)

Year	EVI	DPop	prpop	gdppc	iofp	lssk	gpsk
2017[a]	0.6671	−0.0152	0.5026	0.5324	0.6680	0.2867	0.2750
2017[b]	*0.7238*	*0.6062*	*0.4992*	*0.6037*	*0.4549*	*0.6737*	*0.3170*
2011	0.6202	0.0018	0.4756	0.4756	0.4927	0.2068	0.3055
2011	*0.7028*	*0.5027*	*0.6617*	*0.6617*	*0.4554*	*0.0346*	*−0.0694*
2006	0.6354	−0.0047	0.4910	0.6075	0.5588	0.1917	0.2797
2006	*0.7061*	*0.5815*	*0.4517*	*0.7708*	*0.0166*	*−0.2666*	*−0.0451*
2001	0.5954	0.0038	0.4949	0.5711	0.6867	0.1192	0.1076
2001	*0.6064*	*0.3897*	*0.3600*	*0.1286*	*0.2317*	*−0.3370*	*−0.2429*

[a]Non-italic and right-adjusted entries are R-squared values from the training data
[b]Italic and left-adjusted entries are R-squared values from the testing data. The regression was done over 89 counties in Inner Mongolia Autonomous Regions of China. The split of the training and testing is two-thirds for training and one-third for testing

Fourth, the remaining three variables, *DPop*, *lssk*, and *gpsk*, disclosed some abnormal R-squared values. Several R-squared values were negative, which is rare and leads to question how a linear regression generates negative R-squared values. Statistically, linear regression models can produce negative R-squared values when either the intercept or the slope are constrained so that the "best-fit" line (given the constraint) fits worse than a horizontal line (Motulsky, 2011). The regression model we adopted was the non-intercept regression. In other words, the intercept was forced to be 0.0. A negative R-squared value usually means that the chosen model poorly fits the data. Why did this happen in our case study? IMAR has 89 counties (also called banners in the Mongolian language). Among them, 36 are designated as agricultural counties, 33 as herding counties, and 20 as mixed agriculture-herding counties. In other words, 36 counties heavily engage *gpsk*, 33 counties focus on *lssk*, and the remaining 20 involve both *gpsk* and *lssk*. Furthermore, we randomly split the dataset into the training and testing sub-datasets. As a result, an ill-structured sub-dataset could happen. When *lssk* was regressed primarily against agricultural land use, or *gpsk* against grassland land use, the model fit would be inferior, leading to negative R-squared values. In addition, IMAR has extensive natural lands, and relatively sparse urban land uses, although it witnessed fast urban development (Xie et al., 2022). An odd combination of population density and varied land uses could happen when the training and testing sub-datasets were randomly picked. A final note was that statistical relationships between these variables (*DPop*, *lssk*, and *gpsk*) and land uses were not linear. As a result, the linear regression models performed very poorly.

As discussed in Chap. 6, the areal interpolation accuracy assessment was only available at the source data layer (the county in this application), but not feasible at the target data layer (the town). Because we apply the county observations or statistics as the total that adjust and control the interpolated values to the town level, we can deduce that the interpolations (predictions) from the county to town shall not be far off from the accuracy assessment made at the county level.

References

Bai, X., Surveyer, A., Elmqvist, T., Gatzweiler, F. W., Güneralp, B., Parnell, S., Prieur-Richard, A.-H., Shrivastava, P., Siri, J. G., Stafford-Smith, M., Toussaint, J.-P., & Webb, R. (2016). Defining and advancing a systems approach for sustainable cities. *Current Opinion in Environmental Sustainability, 23*, 69–78. https://doi.org/10.1016/j.cosust.2016.11.010

Baltagi, B. H. (2005). *Econometric analysis of panel data*. Wiley.

Bretschger, L., & Pittel, K. (2020). Twenty key challenges in environmental and resource economics. *Environmental and Resource Economics*, 1–26. https://doi.org/10.1007/s10640-020-00516-y

Chang, H. Y., Wang, W., & Yu, J. (2021). Revisiting the environmental Kuznets curve in China: A spatial dynamic panel data approach. *Energy Economics, 104*, 105600.

Chen, T., Xie, Y., Liu, C., Bai, Y., Zhang, A., Mao, L., & Fan, S. (2018). Trend analysis of relationship between primary productivity, precipitation and temperature. *ISPRS International Journal of Geo-Informatics, 7*, 214.

Chen, H., Zhang, X., Wu, R., et al. (2020). Revisiting the environmental Kuznets curve for city-level CO_2 emissions: Based on corrected NPP-VIIRS nighttime light data in China. *Journal of Cleaner Production, 2020*(268), 121575.

Deng, X., Huang, J., Huang, Q., Rozelle, S., & Gibson, J. (2011). Do roads lead to grassland degradation or restoration? A case study in Inner Mongolia, China. *Environment and Development Economics, 16*(6), 751–773. https://doi.org/10.1017/S1355770X11000180

Gao, L., Kinnucan, H. W., Zhang, Y., & Qiao, G. (2016). The effects of a subsidy for grassland protection on livestock numbers, grazing intensity, and herders' income in inner Mongolia. *Land Use Policy, 54*(2016), 302–312.

Gong, P., Li, X. C., & Zhang, W. (2019). 40-year (1978–2017) human settlement changes in China reflected by impervious surfaces from satellite remote sensing. *Science Bulletin, 64*(2019), 756–763.

Hu, Y., & Nacun, B. (2018). An analysis of land-use change and grassland degradation from a policy perspective in Inner Mongolia, China, 1990–2015. *Sustainability, 2018*(10), 4048.

Lan, H., Stewart, K., Sha, Z., Xie, Y., & Chang, S. (2022). Data gap filling using cloud-based distributed Markov chain cellular automata framework for land use and land cover change analysis: Inner Mongolia as a case study. *Remote Sensing, 2022*(14), 445. https://doi.org/10.3390/rs14030445

Li, S., & Xie, Y. (2013). Investigating coupled impacts of climate change and socioeconomic transformation on desertification by using multi-temporal Landsat images: A case study in central Xilingol, China. *IEEE Geoscience and Remote Sensing Letters, 10*(5), 1244–1248. https://doi.org/10.1109/LGRS.2013.2257158

Li, S., Xie, Y., Brown, D. G., Bai, B., Jinhua, J., & Judd, K. (2013). Spatial variability of the adaptation of grassland vegetation to climatic change in Inner Mongolia of China. *Applied Geography, 43*, 1–12. https://doi.org/10.1016/j.apgeog.2013.05.008

Lo-Iacono-Ferreira, V. G., Garcia-Bernabeu, A., Hilario-Caballero, A., & Torregrosa-López, J. (2022). Measuring urban sustainability performance through composite indicators for Spanish cities. *Journal of Cleaner Production, 359*, 131982.

Motulsky, H. 2011. *When is R squared negative?* https://stats.stackexchange.com/questions/12900/when-is-r-squared-negative. Accessed 25 Aug 2022.

Niu, W. (2012). The connotation cognition of sustainable development theory: Commemorating the 20th anniversary of the United Nations Rio conference on environmental development. *Chinese Resources and Environment, 22*(05), 9–14.

Robinson, B. E., Li, P., & Hou, X. (2017). Institutional change in social-ecological systems: The evolution of grassland management in Inner Mongolia. *Global Environmental Change, 47*(2017), 64–75.

Sha, Z., & Li, R. (2020). Assessment of human-related driving forces for reduced carbon uptake using neighborhood analysis and geographically weighted regression: A case study in the grassland of Inner Mongolia, China. *Applied Science, 2020*(10), 7787. https://doi.org/10.3390/app10217787

Shang, C., Wu, T., Huang, G., & Wu, J. (2019). Weak sustainability is not sustainable: Socioeconomic and environmental assessment of Inner Mongolia for the past three decades. *Resources, Conservation and Recycling, 141*, 243–252.

Tan, Y., Jiao, L., Shuai, C., et al. (2018). A system dynamics model for simulating urban sustainability performance: A China case study. *Journal of Cleaner Production, 199*, 1107–1115.

Tang, J., Davy, A. J., Jiang, D., Musa, A., Wu, D., Wang, Y., & Miao, C. (2016). Effects of excluding grazing on the vegetation and soils of degraded sparse-elm grassland in the Horqin Sandy Land, China. *Agriculture Ecosystem & Environment, 235*, 340–348.

Xie, Y., Zhang, Y., Lan, H., Mao, L., Jin, H., & Zeng, S. (2018). Investigating long-term trends of climate change and their spatial variations caused by regional and local environments through data mining. *Journal of Geographical Sciences, 28*(6), 802–818.

Xie, Y., Crary, D., Bai, Y., Cui, X., & Zhang, A. (2019). Modeling grassland ecosystem responses to coupled climate and socioeconomic influences from multi-spatial-and-temporal scales. *Journal of Environmental Informatics, 33*(1), 37–46. https://doi.org/10.3808/jei.201600337

Xie, Y., Fan, S., & Zhou, C. (2021). Examining ecosystem deterioration using a total socioenvironmental system approach. *Science of the Total Environment, 784*, 147171.

Xie, Y., Liu, C., Chang, S., & Jiang, B. (2022). Urban Sustainability: Integrating Socioeconomic and Environmental Data for Multi-Objective Assessment. *Sustainability 14*, 9142. https://doi.org/10.3390/su14159142

Zhou, C., Xie, Y., Zhang, A., Liu, C., & Yang, J. (2022a). Spatiotemporal analysis of interactions between seasonal water, climate, land use, policy, and socioeconomic changes: Hulun-Buir steppe as a case study. *Water Research, 117937*. https://doi.org/10.1016/j.watres.2021.117937

Zhou, C., Liang, J., & Xie, Y. (2022b). Investigating causal relationships between grassland deterioration and climate and socioeconomic changes through time-series computational learning. *Journal of Cleaner Production, 132963*. https://doi.org/10.1016/j.jclepro.2022.132963

Chapter 9
Innovations, Limitations, and Future Directions of TSES

It is time to summarize what total socioenvironmental system (TSES) intends to accomplish from the perspectives of the grand challenges that socioenvironmental system (SES) modeling faces. TSES develops a new set of SES modeling approaches and techniques based on the socioenvironmental coevolution concept. TSES creates several formal representations of complex socioenvironmental systems by integrating data assimilation techniques, quantitative models, and qualitative considerations.

TSES mainly addresses five of the eight grand challenges identified in socioenvironmental systems modeling (Elsawah et al., 2020):

1. Breaking silos between disciplines
2. Integrating the human dimension
3. Dealing with scales
4. Combing quantitative, qualitative, and data
5. Elevating the adoption of TSES models and impacts on policy

Elsawah et al. (2020) have done a seminal work identifying the eight grand challenges in socioenvironmental modeling based on a thorough literature review. They point out the nature of these challenges and recommend possible ways to overcome them. In the following sections, we will briefly summarize what TSES has done in the five challenges, what is currently lacking, and what forward thoughts are (Fig. 9.1).

9.1 Breaking Silos Between Disciplines

The primary theme of TSES is the "totality" of socioenvironmental systems, which aims to break the silos between disciplines. The book makes the first effort to establish that ecological–socioeconomic coevolution is the theoretical foundation for studying total socioeconomic and environmental systems. The book particularly

Y. Xie, *Total Socioenvironmental Systems*,
https://doi.org/10.1007/978-3-031-39594-9_9

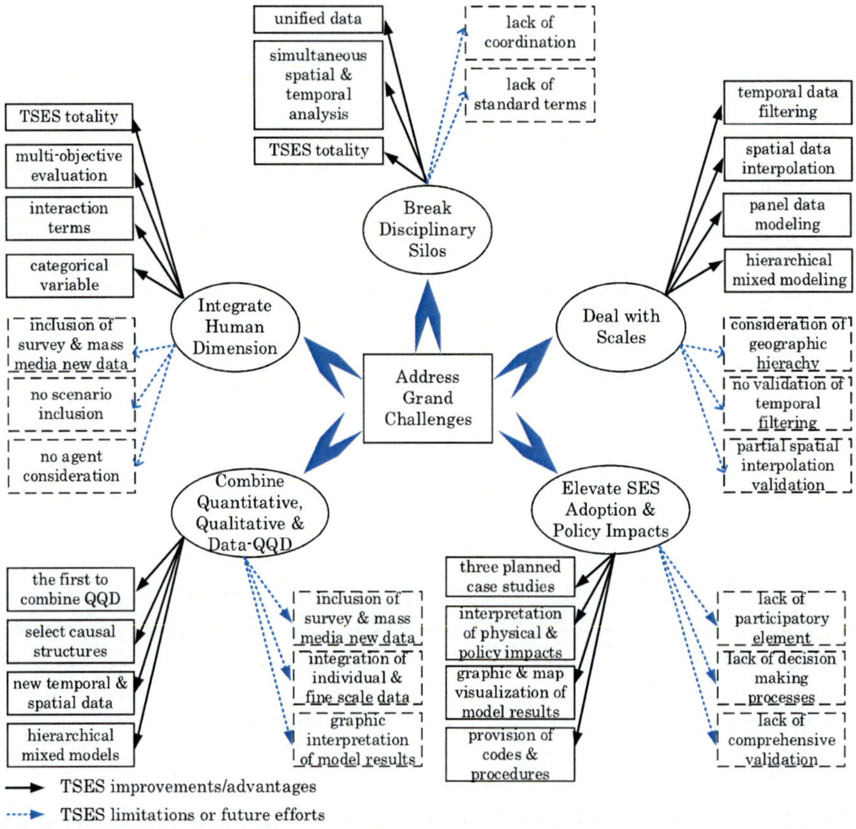

Fig. 9.1 The main innovations and limitations of the TSES modeling framework

discusses the coevolution concepts and applications from the angles of geography, ecological modeling, environmental analysis, sustainable development, and urban planning because the critical TSES applications of this book focus on these fields. The book makes several contributions: (1) affirming that the coevolution theory breaks silos constructed by individual disciplines and can provide a common ground for these disciplines to communicate and collaborate; (2) offering a comprehensive literature review, developing a set of analytical methods, and conducting several case studies to establish and validate TSES as a systematic and viable solution to examine socioeconomic and environmental interactions in the context of sustainability science; (3) conducting the first extensive experiments for constructing a unified socioeconomic and environmental database by spatially and temporally matching and merging two divergent socioeconomic and environmental datasets in space and time; and (4) making the first effort to design and develop a suite of tools to conduct natural patch-based areal interpolation to transform socioeconomic and environmental datasets across nonconformable geographic boundaries.

However, TSES is developed by the author through collaboration with his coauthors and the literature review. As a result, there is a lack of broad coordination with scholars, researchers, and practitioners in socioenvironmental system modeling. TSES may only cover limited elements of SES and report a partial view of SES modeling. Furthermore, TSES has made no efforts to standardize the terminologies in SES modeling, which is strongly recommended as a critical forward way to bridge epistemologies across disciplines (Elsawah et al., 2020).

9.2 Integrating the Human Dimension

TSES emphasizes the consideration of the totality of socioenvironmental systems, which includes the critical human dimension. TSES has made two noticeable efforts to integrate the human dimension. The first is to consider the human dimension when forming or constructing TSES empirical models. For example, Chap. 4 develops a causal framework to examine grassland deterioration from four major components of a grassland ecosystem, global climate change, grassland deterioration, ecological/environmental consequences, and socioeconomic adaptations. Second, Chap. 5 presents a cross-lagged structural equation model with panel data (also named cross-lagged panel model, CLPM). The formulation of the CLPM for urban sustainability (CLPM-US) builds on the literature review and qualitative analysis of urban sustainability and its underlying causal structure in China. Primarily, three constructs – STA (the state investment), NAT (natural resource consumption), and CP (clean production) – are included in the CLPM-US model to investigate how the centralized government in China has employed policy incentives and investment preferences to promote sustainable urban growth. Although exploring CLPM-US is far from exhaustive, it is an excellent effort to use qualitative analysis to determine the urban sustainability causal structures for the 278 prefecture cities in China.

The other effort is to use categorical variables to represent qualitative considerations. Using categorical variables to express qualitative information in statistical analysis is common (Xie et al., 2005). For instance, incorporating the Grazing Policy in the grassland panel models in Chap. 4 helps us understand how different production structures affect grassland productivity. Furthermore, integrating interaction terms between categorical and numerical variables extends the qualitative examination capability in statistical analysis. For example, the hierarchical linear mixed model in Chap. 7 has designed several interaction terms to examine the impacts of socioeconomic factors on urban vacancy rates across three geographical scales (Census Tracts, Road Network Analysis Zones – RNAZs, and counties). Moreover, TSES integrates multi-objective optimization (MOO) to assess and validate the urban sustainability rankings in Chap. 5. MOO allows the evaluation of trade-offs between different choice sets, which often falls into the human decision-making domain.

However, TSES is an empirical modeling framework, as opposed to scenario-based simulation models, agent-based models, cellular-automata models, artificial neural network models, genetic algorithms, and evolutionary computational models (de Smith et al., 2021). These modeling paradigms adequately integrate human dimensions or incorporate different forecasting scenarios. One improvement from the TSES perspective is integrating survey data, individualized data, and new data sources from mass media into the TSES modeling framework.

9.3 Dealing with Scales

TSES has designed four procedures to deal with scales: (1) time-series filtering to deal with temporal scale discrepancies between socioeconomic and environmental data; (2) spatial interpolation to match up spatial units between socioeconomic and environmental data; (3) panel data analysis enables cross-scale modeling; and (4) hierarchical linear mixed modeling supports cross-scale data analysis (Fig. 9.1).

TSES filters trends of change from the environmental data containing both seasonal and annual changes and the socioeconomic and LULC data that only display yearly changes. Current literature about synchronizing socioeconomic and environmental datasets into a temporally compatible one is sporadic. Chap. 3 illustrates how to generate temporally consistent socioeconomic and environmental datasets. This chapter has made several improvements. First, the chapter synthesizes a comprehensive review of time-series trend extraction methods. It provides detailed algorithmic foundations and illustrations of these methods, filling a gap in the current literature. Second, the chapter analyzes the advancement of the support vector regression (SVR) model for filtering trends of change compared with the traditional ordinary least squares (OLS) regression method. Third, the chapter integrates a thorough analysis of stationarity tests, including both uni-section and panel time-series data, bringing fragmented literature together. Fourth, the chapter provides a complete set of well-organized datasets, including the original, intermediate, and final processed data and the corresponding Python codes.

Moreover, Chap. 3 argues that the trend extraction for filtering temporal noise is a kind of variable standardization (VS). VS is an accepted practice in statistical analysis, which is the rationale for implementing time-series data mining to match temporal rhythms between socioeconomic and environmental data. The common notion is that the variable standardization does not change the initial relationship between the original (unstandardized) data but only removes outliers and noises in the data, thus augmenting their linear relationship.

However, the rigorous statistical verification of this statement is missing in the current literature. As we can see from the illustrations in Chap. 4, different combinations of time-series data mining techniques and panel regression models do generate significantly different regression coefficients and signs for the same variables in various combinations. As a result, the physical and socioeconomic interpretations of the impacts of these variables on grassland deterioration vary dramatically. Although we can explain these differences from the current literature, we cannot validate these differences and explanations from the field data or local scholars and practitioners. This shortcoming reflects the lack of cross-disciplinary cooperation. There is a need for seeking such coordination with experts who can verify or assess how the model outcomes match with location conditions.

Next, TSES develops four areal interpolation methods to match geographic units between socioeconomic and environmental data. Environmental and LULC data are collected through field sampling or remote sensing techniques, and they are in the formats of points or pixels and thus have high spatial resolutions. On the contrary, socioeconomic data are enumerated over an administrative area. Most countries publish socioeconomic statistics at the county level, and the county is usually a large area. When matching environmental data with socioeconomic data, the common practice is to resample the fine-scale data over the coarse scale. As a result, we lose spatial resolutions in the resampling process. The fundamental task of areal interpolation is to disaggregate socioeconomic data available in a large area unit to a small area, where socioeconomic statistics usually do not exist.

This book develops a comprehensive and systematic areal interpolation schema to support TSES-based analyses in Chap. 6. This chapter proposes that socioeconomic data are commonly enumerated over administrative areas but are seldom collected simultaneously with environmental samples. In other words, no comprehensive or complete socioeconomic statistic data accompany environmental data. A critical task of constructing a unified socioenvironmental database is to disaggregate the socioeconomic statistical data from large administrative areas to small environmental area units. The chapter further recommends that small but congruent natural land use and land cover patches are the basic environmental area unit. These smallest natural land use and land cover patches are basic natural patches (NPs) for spatial interpolation. Therefore, the areal interpolations are based on natural patch, including four algorithms: (1) natural patches-based areal interpolation; (2) natural communities-based areal interpolation; (3) natural districts-based areal interpolation; (4) road network analysis zones (RNAZs)-based areal interpolation.

Chapter 6 has experimented with these four interpolation methods using a case study examining spatial patterns of urban decline in Metropolitan Detroit, USA. The case study confirms all the four interpolations perform well, and the interpolation accuracies improve when more ancillary information is incorporated into the areal interpolation algorithms. The chapter also discusses, with illustrations, which factors impact the interpolation accuracy. The NP-based areal interpolations demonstrate two technical breakthroughs: (1) the source–target–zone intersected subtotal adjustment, which adjusts the disaggregation of socioeconomic statistical data by applying local constraints and improves the interpolation accuracies, and (2) the double-ratio reallocation, which facilitates to reallocate the adjusted disaggregation from a socioeconomic area unit to the NPs within this socioeconomic unit and completes the areal interpolation.

In addition, the natural communities and natural districts-based interpolations are newly developed methods for integrating environmental and socioeconomic ancillary information into areal interpolation to improve interpolation accuracies. Moreover, this chapter innovatively applies the concept of road network analysis zones to empower areal interpolations. In brief, the four areal interpolation methods and their illustrations presented in this chapter are new contributions to the current literature.

The challenge is how to validate the areal interpolation results when the data in small areas do not exist. The book (Chap. 6) purposefully uses two pairs of spatial units (Micro Civil Divisions (M.C.D.s) vs. Census Tracts (C.T.s) and C.T.s vs. Block Groups (B.G.s)) to validate the areal interpolation outcomes. Both validations use the available data in C.T.s and B.G.s to prove the concept of areal interpolation. The interpolation results from the two pairs of spatial units are acceptable. The interpolation result has comparatively high accuracy, particularly when the source and target units are closer in scale levels. For instance, the interpolation between C.T.s and B.G.s generated preferred accuracy rates.

However, as we iterated many times, the validation is not feasible on most occasions because the data at the target unit do not exist. Therefore, we call the validation in Chap. 6 a partial validation. The recommendation is to limit the areal interpolation between two areal units that are not far apart in size and are relatively uniform. One future experiment shall focus on how administrative hierarchical level difference affects spatial interpolation accuracy. In other words, we shall design and conduct a routine test to estimate interpolation accuracies by interpolating between immediate adjacent levels, one and two levels apart, and so on.

Furthermore, we can expand this systematic accuracy assessment to various areal interpolation methods. We expect to better understand the areal interpolation accuracy by systematically assessing the combined hierarchical levels and interpolation methods. In addition, spatial interpolation or temporal filtering could be much compounded when missing data exists. How to handle missing data in areal interpolation or temporal filtering is beyond the scope of the current book. Interested readers can read the paper by Lan et al. (2022) for references.

There is another crucial areal interpolation experiment that should be conducted soon. The current experiments interpolate the overserved data from a source area that is often larger and has available data to a target area that is small and lacks data. However, as we discussed several times, the primary goal of areal interpolation is to disaggregate socioeconomic statistics enumerated over an administrative area to natural patches of environmental factors. The experiment in Chap. 7 interpolated a set of selected variables from the counties to the townships but did not interpolate them to land use and land cover (LULC) patches. Directly interpolating these variables to the LULC patches would enable the development of a panel regression model over the LULC patches. The model results based on LULC patches should differ from those from townships. It should be noted that there should be good interest in examining the impacts of socioenvironmental drivers on grassland deterioration from LULC patches.

TSES also improves two statistical model paradigms for cross-scale regression analyses. The first one is the panel data regression analysis (PDRA). PDRA analyzes the interactions between environmental factors and socioeconomic drivers across regions as panels and over time as temporal slices (Gorus & Aslan, 2019). PDRA can analyze data with more sample variability and degrees of freedom

(Hsiao, 2014). Particularly, PDRA enables a simultaneous examination of disaggregated and aggregated data in one model (Hsiao et al., 2005). For example, you can simultaneously analyze the township-based and township-aggregated county data (assuming the counties consist of a set of townships). It is also true that PDRA supports examining the monthly-based data together with the aggregated yearly data. The capacity to handle the disaggregate and aggregate data simultaneously dramatically improves the applicability of PDRA and the efficiency of model coefficient estimates (Zhou et al., 2022).

From this perspective, PDRA can be called hierarchical PDRA. For example, Chap. 8 has developed a town–county two-level panel model, the first fine-scale panel regression model to study coupled human–nature systems in Mongolian grasslands. This model can disclose subtle interactions between grassland productivity, climate, LULC, and socioeconomic factors. In other words, the model can differentiate local and regional impacts of climate, LULC, and socioeconomic factors on grassland deterioration. As a result, this model clarifies several confusions about the effects of socioeconomic development, climate factors, and LULC changes on grasslands. The findings confirm that the interactions between ecosystem functions (grassland productivity) and climate, LULC, and socioeconomic factors vary at different geographic scales. We must examine the dynamics of total socioenvironmental systems with fine-resolution data and at multiple geographical scales.

The second one is the hierarchical linear mixed model (HLM), which we have developed to examine urban shrinking in Metropolitan Detroit. This HLM is the first empirical study of spatial processes and characteristics of urban shrinking. Detroit HLM, an urban shrinking model, has two distinct innovations compared to the current approaches. The logistic linear models of examining spatial inequality of urban shrinking treat hierarchical urban spatial structures as individual layers and analyze each layer to compare their similarities and differences. These models indirectly identify spatial connections between different spatial structures by analyzing the model outcomes to deduce how these models at different spatial units affect each other.

On the other hand, the spatial econometric models (SpEM) use spatial weight matrices to analyze spatial relationships. SpEM can use different orders of spatial adjacency matrices or spatial distance belts to model spatial structures. For example, SpEM spatial structure grows outwards from a basic spatial unit like a block group to the first group of block groups that touch the original block group, then the second group of block groups that touch the first group, and so on and so forth. However, the SpEM challenge is what this spatial structure indicates in terms of environmental, socioeconomic, or policy applications. The advantage of the Detroit HLM spatial structure has explicit implications in planning, policy, administration, or purposeful research. For example, the spatial structure of the Detroit case study is RNAZs, which are constructed from census tracts based on six dimensions of socioeconomic factors that affect urban shrinking. The term nested spatial structure is the unique characteristic of HLM deviating from other spatial structures. The spatial unit, RNAZs, is constructed from the networked urban system theory with big datasets and directly supports urban and regional planning practices.

However, the book does not compare the hierarchical panel data regression analysis (PDRA) and the hierarchical linear mixed model (HLM). It will be an exciting project to systematically compare and analyze differences between hierarchical PDRA and HLM to identify similarities and differences between the two modeling paradigms.

9.4 Combing Quantitative, Qualitative, and Data

Combing quantitative, qualitative, and data is one of the main innovations TSES advocates. TSES is the first comprehensive effort to integrate data processing, qualitative consideration, and quantitative analysis into a unified analytical framework. The TSES spatial (areal) interpolation generates spatial-unit-based new data, while the TSES temporal filtering creates temporal-rhythm-smoothed new data. We can analyze these newly unified data through three TSES empirical models (PDRA, cross-lagged SEM/panel data model, and HLM) to examine interactions between socioeconomic and environmental drivers and their spatial and temporal changes.

Qualitative data are conceptual or descriptive information or findings often compiled by interviews, observations, or questionnaire-based surveys. They are often questionnaires, narratives, in-depth descriptions of individuals, or interpretations of the relationships between natural and human environments (Elsawah et al., 2020; Smajgl & Barreteau, 2014). TSES has developed an SEM and panel-based causal construct model based on qualitative analysis of the relationships between natural environments and socioeconomic drivers in sustainable urban development. Moreover, mixed methods are commonly used as an alternative or complementary approach to qualitative analysis (Polhill et al., 2009; Midgley et al., 2013). TSES has adapted an HLM model to represent some qualitative elements in examining the urban shrinking in Metropolitan Detroit.

TSES does not systematically explore integrating qualitative analysis with quantitative models. The forward thoughts for enhancing qualitative research in TSES include the following. First, TSES includes survey data or new data from mass media, and these qualitative data can be transformed into categorical data, being integrated with TSES. Second, TSES integrates individual observations or fine-scale data, which can be easily coupled with three TSES analytical frameworks. Third, TSES can deploy more graphic forms to interpret model outcomes. However, three TSES modeling paradigms are empirical statistical models, limited to a full exploration of qualitative analysis. The TSES qualitative analysis is complementary but not integrative (Cheong et al., 2011).

9.5 Elevating the Adoption of TSES Models and Impacts on Policy

TSES aims to promote its adoption by researchers and practitioners and enhance its implications on policymaking. The four TSES case studies are the empirical investigation of time-series impacts of climate and LULC changes and socioeconomic transformations on grassland deterioration (Chap. 4), the assessment of urban sustainability from underlying causal structures and legacy effects in Chinese prefecture cities (Chap. 5), the exploration of spatial processes and disparities of urban shrinking in Metropolitan Detroit (Chap. 7), and modeling grassland deterioration considering spatial variations and time-lag effects at a finer geographical scale (Chap. 8). All the four case studies provide comprehensive analyses of the environmental and socioeconomic indications of the model outcomes and comparative evaluations of the findings in the context of current literature. These case studies also conclude with the policy implications of the results.

For example, Chap. 4 has developed a causal framework to examine grassland deterioration from four major components of a grassland ecosystem: global climate change, grassland deterioration, ecological/environmental consequences, and socioeconomic adaptations. The chapter comprises a comprehensive database reported in the current literature, including 33 variables in all 89 counties and covering 2001–2017. The chapter conducts an extensive and full-blown experiment with detailed methodological discussions, result validations, and finding discussions. The findings confirm that the EMD-Lagged Model generated similarly low total root mean squared error (RMSE) values as the SVR-A-Lagged Model did. However, the spatial patterns of prediction errors of the EMD-Lagged Model were less conforming with the linear relationships between grassland productivity and grassland types than the SVR-A-Lagged Model.

Furthermore, the EMD-Lagged Model was the sole model among the 18 models that revealed some causal relationships between grassland deterioration and almost all socioeconomic and LULC variables (except for 3 out of 28). In other words, this model significantly augmented the causalities between grassland deterioration and socioeconomic and LULC variables. Therefore, the spatial errors of this model were less conforming with the regional patterns of climate changes but revealing the impacts of socioeconomic and policy factors on grassland productivity.

The urban sustainability case study has illustrated six complementary approaches for a comprehensive measurement of sustainable urban system (SUS) level: (1) developing a conceptual framework based on cross-lagged panel model for urban sustainability (CLPM-US) to measure SUS level; (2) examining SUS spatial variations and temporal changes simultaneously; (3) using CLPM-US to reveal how legacy effects of a sustainable urban system and its underlying causal constructs affect the present conditions and future directions of urban sustainability; (4) using CLPM-US to compute summative SUS scores over times to measure SUS levels and their time-series changes; (5) supplementing the summative scores with their underlying causal structures to assess unique contributions of urban causal constructs to sustainability scores; and (6) validating and comparing the urban sustainability rankings of CLPM-US with the rankings from the multi-objective optimization problems algorithm.

Several findings are exciting and informative. First, Chinese prefecture cities' SUS levels heavily depend on natural resource consumption. This dependence itself reveals that urban growth is still in an early stage in China. Second, the relationships between SUS levels and clean production states are generally nonsignificant or negative. This finding indicates that the priority of clean production in urban development is still lower than the priority of economic development. Third, the role played by the central state investment in urban sustainability fluctuates and is not stable. The changing nature of state funding concerning urban sustainability further confirms that sustainable urban growth is not a high priority in China during the study period.

The case study of Metropolitan Detroit urban shrinking through the hierarchical linear mixed model has revealed several important findings. The term mixed indicates the addition of fixed effect and random effect. Since the causal relationships exist between inequality, socioeconomic driving factors, and urban shrinking, these relationships can be modeled as fixed effects. On the other hand, these relationships vary across hierarchical spatial structures, which can be examined through random slope models as random effects. In other words, we can investigate the causal relationships between urban shrinking and its driving forces through the fixed model (the random intersect model) and the spatial processes and interactions of urban shrinking by using the random model (the random slope model). The random models reveal that urban shrinking in Metropolitan Detroit expands into three directions: the primary direction toward the west, the city of Livonia; the secondary one toward the northwest, the city of Southfield; and the third toward the southwest, the city of Dearborn. In addition, urban shrinking extended toward the northeast but soon stopped in the cities of Clinton and Sterling Heights. Urban shrinking in Metropolitan Detroit exhibit obvious spatial processes and dynamics. Therefore, the HLM spatial structure has explicit meanings in planning, policy, administration, or purposeful research.

The book also adopts many graphics to interpret and explain the findings of the four models. The visual presentations of the findings help readers understand the research questions, the methods applied to solve them, and the results in the context of decision-making. Moreover, the book provides step-by-step implementation strategies, all associated data files, and Python and Stata codes. The book intends to reduce the learning curves and make it easier for readers to adopt methodologies, data, and processing.

However, the policy implication is far from implementing the decision-making process. The writing and development of the book are based on scholarly research but not on participatory considerations or inputs. The validation and the discussion of the limitations are conducted by chapters or by each empirical model. There is a lack of systematic or comprehensive validation protocols.

References

Cheong, S.-M., Brown, D. G., Kok, K., & Lopez-Carr, D. (2011). Mixed methods in land change research: Towards integration. *Transactions of the Institute of British Geographers, 37*(1), 8–12. https://doi.org/10.1111/j.1475-5661.2011.00482.x

de Smith, M. J., Goodchild, M. F., & Longley, P. A. (2021). *Geospatial analysis: A comprehensive guide to principles, technologies and software tools*. Troubador Publishing Ltd.

Elsawah, S., Filatova, T., Jakeman, A. J., Kettner, A. J., Zellner, M. L., Athanasiadis, I. N., Hamilton, S. H., Axtell, R. L., Brown, D. G., Gilligan, J. M., Janssen, M. A., Robinson, D. T., Rozenberg, J., Ullah, I. I. T., & Lade, S. J. (2020). Eight grand challenges in socio-environmental systems modeling. *Socio-Environmental Systems Modelling, 2*, 16226. https://doi.org/10.18174/sesmo.2020a16226

Gorus, M. S., & Aslan, M. (2019). Impacts of economic indicators on environmental degradation: Evidence from MENA countries. *Renewable and Sustainable Energy Reviews, 103*, 259–268. https://doi.org/10.1016/j.rser.2018.12.042

Hsiao, C. (2014). *Analysis of panel data* (3rd ed., p. 562). Cambridge University Press.

Hsiao, C., Shen, Y., & Fujiki, H. (2005). Aggregate vs disaggregate data analysis—A paradox in the estimation of money demand function of Japan under the low interest rate policy. *Journal of Applied Econometrics, 20*, 579–601. https://doi.org/10.1002/jae.806

Lan, H., Stewart, K., Sha, Z., Xie, Y., & Chang, S. (2022). Data gap filling using cloud-based distributed Markov chain cellular automata framework for land use and land cover change analysis: Inner Mongolia as a case study. *Remote Sensing, 2022*(14), 445. https://doi.org/10.3390/rs14030445

Midgley, G., Cavana, R. Y., Brocklesby, J., Foote, J. L., Wood, D. R. R., & Ahuriri-Driscoll, A. (2013). Towards a new framework for evaluating systemic problem structuring methods. *European Journal of Operational Research, 229*(1), 143–154.

Polhill, J. G., Sutherland, L.-A., & Gotts, N. M. (2009). Using qualitative evidence to enhance an agent-based modelling system for studying land use change. *Journal of Artificial Societies and Social Simulation, 13*(2), 10.

Smajgl, A., & Barreteau, O. (2014). Empiricism and agent-based modelling. In A. Smajgl & O. Barreteau (Eds.), *Empirical agent-based modelling – Challenges and solutions* (pp. 1–26). Springer. https://doi.org/10.1007/978-1-4614-6134-0_1

Xie, Y., Yu, M., Tian, G., & Xing, X. (2005). Socioeconomic driving forces of arable land conversion: A case study of Wuxian City, China. *Global Environmental Change, 15*, 238–252.

Zhou, C., Xie, Y., Zhang, A., Liu, C., & Yang, J. (2022). Spatiotemporal analysis of interactions between seasonal water, climate, land use, policy, and socioeconomic changes: Hulun-Buir steppe as a case study. *Water Research, 209*, 117937.

Index

Y. Xie, *Total Socioenvironmental Systems*,
https://doi.org/10.1007/978-3-031-39594-9